Causality from the Point of View of Statistics

José A. Ferreira

Causality from the Point of View of Statistics

WIPF & STOCK · Eugene, Oregon

CAUSALITY FROM THE POINT OF VIEW OF STATISTICS

Wipf & Stock
An Imprint of Wipf and Stock Publishers
199 W. 8th Ave., Suite 3
Eugene, OR 97401

www.wipfandstock.com

PAPERBACK ISBN: 978-1-6667-7708-6
HARDCOVER ISBN: 978-1-6667-7709-3
EBOOK ISBN: 978-1-6667-7710-9

In memory of my parents,
Elsa Maria and José,
and maternal grandparents,
Lucinda and António.

Contents

Preface

Il est plus aisé de dire des choses nouvelles
que de concilier celles qui ont été dites.

Vauvenargues

Statistics has always been concerned with 'causality'. With the exception of the more technological applications that require nothing else than the summary of data or the making of predictions, applications of statistics to scientific problems are bound up with the explanation of certain phenomena at some level of detail and hence with ideas of cause and effect. And this is true in a less obvious sense: On the one hand, every application of statistics is founded on probability models, and every model, however simple it may be, tells a story of how empirical data are supposed to come into being; in particular, limit theorems and theorems characterizing distributions and random processes involve formalized notions of causality, with certain variables causing certain effects or resulting from certain chains of events (see, for example, the books by Billingsley [11], Brémaud [14], Feller [29], Karlin and Taylor [51] and Ross [86] on probability or 'applied probability', those of Galambos and Kotz [38] and Leadbetter, Lindgren and Rootzen [52] on characterizations, and that of Brockwell and Davis [15] on time series models). On the other hand, the methods used to estimate and assess such models (through diagnostic or goodness-of-fit techniques, through forecasting exercises) usually take their aspects of cause and effect into account, and, to the extent that the models are correct and incorporate the factors pertaining to the phenomena being studied, conclusions drawn from their statistical analysis express relations of cause and effect (see for instance the books by Basawa and Prakasa Rao [7], Billinsgley [10], Heyde [43] and Lehmann [54, 55] on statistical methods, and those by [12], Coles [17], Cox and Snell [19], Daniel and Wood [21], Guttorp [40], Rao [77] and Tanur *et al.* [103] on applications).

Admittedly, the best possible contribution of statistics to a scientific problem is often the establishment of associations and the provision of estimates or predictions rather than the revealing, in one stroke, of which factors cause what. But that is only natural because theories are normally established in stages and at any given stage the knowledge built into probability models and statistical analyses is necessarily guided and restricted by the knowledge available in the field where they are being applied (cf. pp. 163–4 of Lehmann [56]). It is also true that the use of statistics is typically far from being the best possible (as cogently argued by Diaconis [25] and made patent by Freedman [33, 36, 37] and by many other statisticians), and in the midst of all visible efforts one can count relatively few significant applications of statistics. But that is perhaps what one should expect from science in general—and especially from modern science and from a discipline like statistics whose applications are determined by other disciplines.

Nevertheless, it is undeniable that even 'classical' statistical methods (covered by Lehmann's [53, 54, 55] texts) incorporate elements of causality and can lead to successful applications. In the analysis of experiments (in biology, chemistry, medicine, etc.) and of industrial processes, the idea that 'treatments' may or may not have certain effects at different dose levels and under certain background conditions epitomizes what one normally understands by 'causality'. In genetics, many detailed and accurate explanations would be unthinkable without a combination of empirical observation, 'probability modelling', experimentation and statistics (e.g. Feller [29], Sturtevant [102]). In observational studies, too, and despite the regular company of the much-abused linear regression model, causality can make a respectable appearance: apart from studies of industrial processes where the conditions underlying the data cannot be fully controlled (exemplified in Daniel and Wood [21]), and from studies where one empirical quantity can be described accurately by an equation involving other quantities known to determine it (illustrated by the 'trees data' on pp. 126–134 of Jørgensen [49]), one must mention Gauss's regression analysis of planetary orbits, eloquently summarized by Diaconis's [25], and the many important practical applications described in Rao [77] and Tanur *et al.* [103].

In view of these observations it is surprising to learn that in the last three decades or so a number of research groups have been developing methods designed to establish causality in applications of statistics while entertaining the idea that the discipline of statistics, or at least some wide 'classical' streak of it, has always been divorced from problems of causality—and that, in part, because the traditional language and tools of statistics (which of course include the basic notions of sets and functions) do not allow one to formulate notions of causality.[1] **(1)** Had this been an opinion held by a group of isolated eccentrics busying themselves with irrelevant questions, it would have been of little or no consequence to the practice of statistics; but it has hindered the work of the researchers involved, led to the creation of more or less different schools of 'statistical causality' which to this day cannot agree with (nor understand) each other completely, has alienated most mainstream statisticians from the work being carried out by those schools, and confused many potential users of their work, and for that reason it is unfortunate and requires some explanation. **(2)**

It would be a task for the historian of ideas in statistics to try to elucidate the social and technical factors behind this state of things, and we shall not embark on that. Yet, two of the more technical factors are so obvious and so fully acknowledged in this work that they must be spelled out. One is a general discomfort on the part of the schools indicated above with the notion of random variable, patent, in different forms and varying degrees, in nearly all their works. The disregard for standard notation and the shunning of the mathematical definition of random variable—which is not only natural but also elementary in the discrete case and, ironically, just what is needed to express causal aspects of statistics—and of its relationship with empirical data in most accounts of statistical causality, are all symptoms and consequences of that discomfort, and they have been the reason for much aversion on the part of mainstream statisticians for all things 'causal'.

[1] See notes on pp. xv–xvii.

Now if f is a real-valued function of four variables—say defined on $\{0,1\}\times\mathbb{R}^3$—and T, B_1, B_2 and ε are random variables defined on a sample space Ω, then the functions R, R_1, R_2 and R_3 defined for $\omega \in \Omega$ by

$$R(\omega) = f(T(\omega), B_1(\omega), B_2(\omega), \varepsilon(\omega)), \quad R_1(\omega) = f(t, B_1(\omega), B_2(\omega), \varepsilon(\omega)),$$

$$R_2(\omega) = f(T(\omega), b_1, B_2(\omega), \varepsilon(\omega)), \quad R_3(\omega) = f(t, b_1, b_2, \varepsilon(\omega)),$$

for fixed numbers t, b_1 and b_2, are all random variables on the same space, and they may be taken to represent a response measured with random error ε on an individual following respectively exposure to a randomly determined treatment T under randomly determined background conditions (B_1, B_2), exposure to a fixed treatment t under random background, exposure to a random treatment under partly fixed and partly random background, and exposure to a fixed treatment under a fixed background. Moreover, the possibility that $f(t, b_1, b_2, \varepsilon(\omega))$ may vary with t for some numbers b_1 and b_2 and for ω in a set of positive probability corresponds to the empirical notion that a treatment may have an effect—a *causal effect*—on a response; and to investigate the existence or magnitude of such an effect from 'observations' on $(R, T, B_1, \ldots)$ is to investigate whether or to what extent variations in R arise from variations in T proper rather than from variations in (B_1, B_2, ϵ) via some additional (known or unknown) relationship, such as $T = \tau(B_1, B_2, \epsilon, \ldots)$—a relationship that *confounds* the workings of T on R. Once one realizes this and, in a given real-life problem, sets out to relate the theoretical objects with empirical data and the phenomena underlying those data, deciding which random variables should be regarded as observable and which as unobservable, which arguments should be fixed at what values, to what extent f is known, and so on, one is ready to formulate and try to study questions of statistical causality. Any insecurity or bewilderment over R being caused by (T, B_1, B_2) rather than the converse (the model is quite explicit about that, even if it does not fit the real-life problem), over such things as $f(T, B_1, B_2, \varepsilon)$, $f(1-T, B_1, B_2, \varepsilon)$, $f(t, B_1, B_2, \varepsilon)$ and $f(1-t, B_1, B_2, \varepsilon)$ being all defined on the same space, and hence pertaining to the same individual, while only one of the first two and at most one of the other two may be regarded as observable in a real sample, will also disappear as a result. **(3)**

The other factor contributing to the image of statistics as a discipline that is unwilling and unequipped to deal with problems of causality is the quality of the overwhelming majority of applications of statistics. Although different schools of statistical causality will have different opinions about actual statistical practice, they all seem to agree that the most popular approach to studying the effect of a treatment based on observational data, namely that of fitting a regression model to the response, treatment and explanatory or confounding variables and then testing whether the coefficient of treatment is zero, is not 'causal'. **(4)** Some, moreover, practically equate our discipline with this approach; according to them, statistics has really been busy with computing correlations and finding associations rather than with trying to establish whether a certain factor indeed causes a certain response—which should call for substantial changes in the discipline, maybe nothing short of a revolution. **(5)** The reasons behind this verdict are not always clear, and they appear to differ somewhat in different schools. They cannot be completely in tune with the careful arguments presented in Diaconis [25] and Freedman [33, 36, 37], since those arguments also apply to alternatives proposed by some schools of statistical causality (e.g. pp. 243–278 of Freedman [37] and Subsect. 2.2, p. 121, of Ferreira [31]). And they do not seem to take into account that when a model is approximately correct, as in many applications to genetics, in the outstanding example of Gauss, and in the more ordinary examples mentioned above, 'traditional' statistics is thoroughly engaged with causality. **(6)**

Whatever explanations one may find for it, the present situation of statistical causality within the mainstream of statistics is regrettable. During the last few years I have met several statisticians who were genuinely interested in learning about the works of Pearl, Robins, Rubin and others, but felt frustrated by the difficulties in understanding them. Even some of those who did work in line with the systems of Pearl and Robins have confessed to not understanding them well. Others are positively sceptical about the worth of anything related to statistical causality. Those who are not statisticians but use or would like to use statistics react both with credulity and puzzlement to the promises, prophetic announcements and polemics from the various camps of statistical causality; and in the midst of it all statisticians must feel a bit lost and guilty, not quite sure whether they are missing out on something and wondering whether they could indeed achieve much more in their statistical analyses. This is a pity because the various schools of statistical causality did consider some interesting questions and obtained results that are worth knowing about and studying, which may sometimes be applied in real-life problems, and which certainly help us to think about aspects of causality in applications of statistics. **(7)**

My purpose in writing this treatise has been to explain (so that I myself could understand) the main ideas developed by the various schools of statistical causality in a unified way and as clearly as possible. That the different approaches to causality can be put together and made to fit in with classical statistical theory is not very surprising, but it may be surprising to see how their unification rests on the careful definition and interpretation of sets of random variables, as opposed to distributions. A probability model meant to represent causal aspects of a real-life problem must be based on sets of random variables defined as functions of 'primitive' random variables in a sequential order that mimics purported relations of cause and effect, together with 'structural' assumptions about the functions and distributional assumptions about the random variables. Distributional assumptions alone are not sufficient to specify such a model because they only contain information about *correlation*, i.e. about the dependence between random variables. Backed up by definitions of random variables, they embody and may provide information on *causation*, but at a price: those definitions must rest on some external basis, on postulates pertaining to the real-life problem and which cannot be verified prior to the application of the model (though they may be at a later stage, following other research efforts)—and this is only natural since in applications the observable data are the empirical analogues of putative random variables observed at some point ω rather than the random variables proper (the functions defined on the sample space as opposed to their values at a given element in it), which are designed to represent the physical mechanisms generating the data. In principle, then, many frequently used models, such as the linear regression and analysis of variance models, are 'causal', and they have been used legitimately so in much experimental research during the last 200 years; and queueing models, epidemic models (e.g. Isham [47]), including D. Bernoulli's model and its variants (e.g. Dietz and Heesterbeek [26]), and many other models designed for non-experimental studies, can also be 'causal' at some level of detail. But all these models cease to be 'causal' whenever their application is not supported by clearly formulated postulates or if the postulates on which they are based are not justified in earnest, which is what happens in many applications of statistics to observational studies in medicine, economics and sociology, of which Yule's study on the causes of poverty is an early example (Sect. 1.4 of Freedman [36]). **(8)**

The approach to 'causality' proposed here is based on the explicit—and yet completely flexible—definition of various sets of random variables pertaining to a real-life problem and on the calculation of expectations of functions of those variables; so it may be legitimately placed under the heading of what (at least since the 1960's) is often called "applied probability" or "probability modelling" and identified with such books as Karlin and Taylor [51] and Ross [85]. The novelties in it arise from the particularities of the problems considered by the various schools of statistical causality. If we exclude problems of statistical inference (which may be reduced to more or less familiar problems of estimation and testing), these problems are essentially of two kinds: (i) specification of a causal effect pertaining to a given model and determination of conditions under which that effect can be quantified and estimated from observable data; (ii) calculation of probabilities pertaining to *altered versions* of a given model as a means of predicting the effects of interventions—of alterations to a 'system'.

Plan of the book; suggestions for the reader[2]

In view of their importance for the study of causality in statistics, we have collected some results and observations about random variables in Chap. 1. Our main purpose here is to lay the foundation for the models of statistical causality introduced later (in which the random draw ω is to make a few but important appearances) by recalling that the very definition of random sequence incorporates ideas akin to those normally used to describe causality; for example, the idea of independent random variables corresponds to that of real-life events or phenomena that do not influence, affect or cause each other, and the idea of a random sequence defined recursively corresponds to that of a 'chain of events' or sequence of interrelated events or phenomena. Another purpose is to illuminate the connection between the mathematical ideas and the empirical objects to which they are meant to apply, and to show why many real-life problems require (though many others may dispense with) the definition of explicit functional relations between variables. Finally, a third purpose is to provide the reader who is less familiar with random variables and their construction with a picture that makes them look as 'real' or 'concrete' as the number π, the exponential function or any other concept involving an infinite

[2] The interdependence of the chapters is summarized on p. xxiii.

number of operations. Although the material in Chap. 1 is well known and can be found in Billingsley's [11] beautiful book and in many other texts on probability, our presentation is a little original and, despite being intuitive rather than rigorous, somewhat more comprehensive than is usually the case. Chap. 2 treats the simplest aspects of statistical causality, those needed to understand Rubin's approach to the subject and the corresponding views of Pearl and Robins. Here it is shown how the problems of type (i) and type (ii) mentioned above are related in a certain sense. Chap. 3 treats more general aspects of causality needed to understand the problems and methods studied by Pearl, Robins and others. Chap. 4 collects several examples, adapted from the literature, illustrating calculations typically required in problems of causality. Finally, Chap. 5, which can be read after Chap. 2 and Sects. 4.6 and 4.7, describes some basic methods of estimating and testing for the effect of a treatment based on stratification or matching.

The material presupposes a knowledge of statistics at the level of Mood, Graybill and Boes [59] and Rice [79], supplemented perhaps by the review of probability in Chap. 1 of Brémaud [14] and by parts of Chaps. 1–6 of Ferguson [30]. Readers armed with this knowledge should be familiar with the notation used, except perhaps for a few special symbols listed on p. xxi. Although the level of probability is elementary, the presentation is quite abstract at times; this seems necessary to me because questions of cause and effect in statistical models can be surprisingly slippery and (at least for those of us who are not particularly talented) seem to require a strict formalism in order to be tackled systematically. Readers may need some patience before they get used to this formalism, but I believe that after studying Chap. 2 with care they will be rewarded with the right tools for thinking about causal aspects of statistics and for discussing them efficiently with others. Because my intention is to explain fundamental ideas as concisely as possible, and because those ideas presuppose some ability to frame specific empirical problems in abstract terms, no *realistic* examples based on real or simulated data or situations are provided in the main text. Nevertheless, the exercises (especially those of Chaps. 3, 4 and 6), which are meant to be a continuation of the main text and are straightforward or contain detailed instructions, include a variety of such examples, some of which illustrate the difficulties of real-life problems even in favourable circumstances and the limits and imperfections of the methods. In showing that a model can be written in two ways or that two models are *equivalent*—which is to say that homologous sets of variables in them are defined in the same order (so as to represent the same relations of cause and effect) and have the same distribution—we often use the device of turning one standard uniform random variable into two or more standard uniform random variables by sharing the digits in the decimal expansion of the former with the latter; although this is efficient and seems perfectly natural to us, readers who find it strange may try, as yet another exercise, to get to the same results by other means. I have adopted the more old-fashioned and less impressive 'descriptive' style of exposition rather than that of 'theorem-proof', but I think that the results contained in the book (identities, Pearl's criterion, properties of tests and estimates) and the conditions under which they are valid are stated accurately (or with the appropriate degree of latitude) and clearly.

Postscript

I have benefited greatly from studying the works of David Freedman. Freedman did much to improve statistical practice, and he was and remains a generous and disinterested source of enlightenment to applied and theoretical statisticians alike. He was also one of the first to make a serious effort to understand, explain and improve the statistical causality being produced by others. I am not sure that Freedman would have liked to read this little book, but I hope he would not have disapproved of it.

De Bilt,
April 2023

José A. Ferreira

Notes

1 This idea is expressed, in full or fragmented form and with varying degrees of nuance, in many papers and books, but the preface of Pearl, Glymour and Jewel [73], the first two sections of Pearl [68], the discussion of Pearl [65] and the comment of Pearl [70] provide a good summary of it. That other researchers have partly accepted this point of view or brought up similar misconceptions in poorly conducted discussions and in quotations taken out of context seems clear from Lindley's [57] review of Pearl's work. It has been observed and adduced as sign of this disengagement on the part of statistics that 'causal', 'causality', 'causation' and similar terms hardly appear in statistics books, but one may equally note that they seldom appear in books on classical mechanics such as Goldstein's [39], where even 'cause' is used sparingly and usually in a rather informal way; it may be said that in applied mathematics causality is implied already by the formulation of a model—whether one needs to take or actually takes one's formulation seriously when applying the model is another matter. Incidentally (see p. 81 of Brockwell and Davis [15]), in the theory of time series a special role is played by so-called 'causal processes', whose values are interpreted as being successively generated from past events.

2 It is well known that there are different schools of statistical causality, the most influential of which are those associated with J. Pearl and D. Rubin; indications of their positions, of the disagreements between them, of the alienation of mainstream statisticians from the newest research in statistical causality, and of the confusion of potential users of it, are sprinkled in Pearl's review [68] and in the literature cited in it. Some of the authors we cite speak of 'statisticians', 'serious statisticians', etc., and they often speak in the name of statisticians. We follow this custom here but try to give a reasonably precise definition of 'statistician': Our statistician is someone who has studied, at a university or independently, the equivalent of three to five years of essentially mathematical subjects and who has done some applied or theoretical work in statistics. In this sense, those who have studied non-mathematical subjects and then gained some sort of statistical status through statistical work or through postgraduate studies without having studied a substantial amount of undergraduate mathematics do not qualify as statisticians.

Our purpose is not to discriminate against some practitioners of statistics but to be clear about those we regard as 'mainstream', however heterogeneous a group they may form. Using this definition one can, for example, make the observation that nowadays the vast majority of statistical work is carried out by non-statisticians, or emit the opinion that in order to use statistics well one does not need to be a statistician.

3 In the preface to his famous introduction to probability [29], W. Feller warned against the tendency to "reduce probability problems to pure analysis as soon as possible" which comes with a "poorly defined notion of random variables". That this tendency had been quite widespread is suggested by J. Doob's account ([27], [101]) of how he used to surprise even first-rate statisticians by announcing that $\sin 2\pi\omega$ and $\cos 2\pi\omega$ are uncorrelated but dependent random variables on the unit interval. That the tendency persists today and may have even increased in the last years—perhaps because the majority of users of statistics are 'discontents' from other disciples, perhaps because undergraduate educational programmes have become shorter and less mathematical—is suggested by the present account. And yet, with so many fine books around, one ought to be aware that any worthwhile result in statistics—including very elementary ones—benefits from, or even relies on, the interplay between random variables and 'ordinary' functions such as distributions functions, densities and measures.

4 Implicit in this is the belief in what we may call the 'law of universal regression', according to which most random functions of several variables, no matter what the variables are and what phenomena they pertain to, could be replaced with little loss of accuracy by what is essentially a linear combination of those variables. Of course, it has been known for hundreds of years that a differentiable function of several variables can be approximated by a constant plus a linear combination of the variables in a neighbourhood of a point, and that the constants involved in that approximation generally depend on the neighbourhood. By the law of universal regression, most phenomena in the world could be studied by assuming that those constants are essentially independent of the neighbourhood. This might very well be true, but it has never been proved.

5 It must be admitted that if one takes into account the sheer number of publications written under the influence of the 'law of universal regression' just mentioned in Note 4 and forgets that the majority of users of statistics are not statisticians in the sense of the definition presented in the second half of Note 2, then it may be quite natural to equate our discipline with that law.

6 Whether the models involved in an analysis are approximately correct has always been an important question in statistics (as witnessed by Lehmann [56]), and one that underlies the possibility of causal interpretation; but in the majority of applications it is not even considered. The recent indictment of the p-value as a statistical tool is symptomatic: although the possibility of p-values being unreliable owing to their being calculated under the wrong models is evident to statisticians (e.g. Wasserstein and Lazar [107]), the claimants have hardly said a word about it (e.g. Nuzzo [64] and Trafimow and Marks [105]).

7 It must be said that the work of Rubin's school fits in much better with traditional statistical theory and therefore is better understood by mainstream statisticians, but in our view it too benefits from the clarification attempted in this book.

8 If the postulates have a rational justification and are open to criticism then such models are still 'causal', even if there are competing models based on alternative postulates, even if they turn out to be incomplete, insufficiently detailed, somewhat inaccurate or simply wrong. Like much else in science, the formulation, justification and evaluation of postulates and theories involve a certain amount of speculation and subjectivity—but no arbitrariness.

Acknowledgements

In the last years and at various times, several people at the RIVM (the National Institute for Public Health and the Environment, in the Netherlands) have thought it wise that we at the statistics department should be able to dedicate part of our working time to studying subjects new to us and to creating courses and holding meetings on those subjects; for this opportunity, which has helped me to learn something about 'statistical causality', I am grateful to Sophie Deleu, Dirk-Jan Griffioen, Inez Joung, Tom van Venrooij and Albert Wong. To Albert I am also indebted for encouragement and for many interesting and entertaining conversations about applied statistics.

In preparing this book I have benefited from discussions with Xiaoheng Huang, Remie Janssen, Lotta Meijering and Jeroen Rodenburg, and from the comments that Peter Engelfriet and Alina Nicolaie made on the text and on the exercises; Alina, especially, solved a number of problems in detail, and Peter's careful reading yielded plenty of corrections and valuable suggestions. All these colleagues have helped me in making important emendations, and to them go my sincere thanks.

The book is written in Latex; I should like to acknowledge my debt to the creators and distributors of this wonderful computer program, of its MiKTex implementation, of the TeXworks editor, and of the very many accompanying packages, such as pict2e (which I have used to draw the graphs), and to the legion of authors of Latex macros and palimpsests in circulation. I owe no lesser debt to the creators and distributors of the statistical computing system R, which I have used to make calculations and produce diagrams.

Finally, I am grateful to my wife, Sofia, for making life as pleasant as life on earth can be.

A list of special symbols

- := is the sign of assignment in the Pascal programming language: it is used sometimes to emphasize that an equality is being used to define a new object rather than to state an identity between two objects, so that $a := b$ means that a is defined as b, and it may be used in reverse, as in $a =: b$.
- $\equiv$ is used sometimes to emphasize that two objects are the same object written in somewhat different ways.
- # is used to denote the cardinality of sets: if A is a set, $\#A$ is the number of its elements.
- $\mathbf{1}_E$ is the indicator (or characteristic) function of the event or set E: $\mathbf{1}_E(\omega) = 1$ if $\omega \in E$, $\mathbf{1}_E(\omega) = 0$ otherwise.
- $\delta_{x,y}$, where x and y are two elements of an arbitrary set, denotes 'Kronecker's delta': $\delta_{x,y} = 1$ if $x = y$ and $\delta_{x,y} = 0$ otherwise.
- F^{-1} denotes the 'inverse' of a distribution function F, defined in Prob. 1.1.
- $\mathcal{L}$ is used to indicate the law or probability distribution of a random variable: $\mathcal{L}(Y)$ is the law of the random variable Y, namely the set of probabilities $\mathbf{P}(Y \in S)$ as S ranges over sets of real numbers; if X is a random variable and x a number then $\mathcal{L}(Y|X = x)$ is the set of probabilities $\mathbf{P}(Y \in S|X = x)$.
- $(x_i)_{i \in I}$ is an abbreviation of the vector or ordered set $(x_{i_1}, x_{i_2}, \ldots, x_{i_k})$ of numbers or functions indexed by the numbers $i_1 < i_2 < \cdots < i_k$ in $I = \{i_1, i_2, \ldots, i_k\}$.
- $\mathbb{N}$ is the set of natural numbers; $\{1, 2, 3, \ldots\}$); $\mathbb{N}_0 = \{0\} \cup \mathbb{N}$; $\mathbb{Q}$ is the set of rational numbers; $\mathbb{R}$ is the set of real numbers and $\mathbb{R}_+$ the set of non-negative real numbers.
- T is used as a superscript to denote the transpose of a matrix or vector.
- $\mathcal{T} \times \mathcal{X}$, the Cartesian product of two sets $\mathcal{T}$ and $\mathcal{X}$, is the set of ordered pairs (t, x) with $t \in \mathcal{T}$ and $x \in \mathcal{X}$; products $\mathcal{T} \times \mathcal{T} \cdots \times \mathcal{T}$ with the same set $\mathcal{T}$ repeated n times are defined similarly and abbreviated to $\mathcal{T}^n$.

Interdependence of the chapters

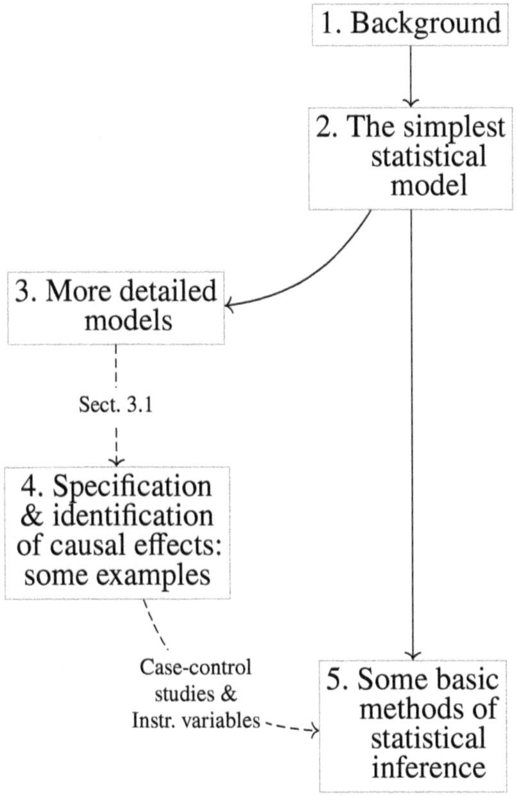

CHAPTER 1

Background

Although theories are rooted in the empirical world, there is no practice without some sort of theory. Thus, applications of statistics are useful only thanks to a probability model whose elements are made to correspond to empirical data in a convincing manner. In studying statistical aspects of causality, random variables and relations between random variables are essential elements of the probability model, and they must bear some correspondence to the causes and effects observed or assumed to exist in the empirical world. Alas, in the empirical world one seldom gets to see the analogues of random variables; one rather senses analogues of their probability distributions—and so it seems all the more important to know about the connections between random variables and distributions, about how random variables arise from distributions and distributions from random variables.

1.1 Random variables and distributions

Let $\Omega = [0, 1]$, $\mathcal{F}$ be the family of Borel subsets of Ω (which includes the intervals and countable unions and intersections of intervals), $\mathbf{P}$ Lebesgue measure on Ω, and consider the probability space $(\Omega, \mathcal{F}, \mathbf{P})$. In this space one can compute probabilities of events such as $\{\omega\}$, $E = \bigcup_n [a_n, b_n]$, $\Omega \cap \mathbb{Q} \equiv \{\omega \in \Omega : \omega \in \mathbb{Q}\}$, or

$$N_j = \left\{\omega \equiv \sum\nolimits_{m\geq 1} \frac{\omega_m}{10^m} \in \Omega : \lim_{n\longrightarrow\infty} \frac{1}{n} \sum\nolimits_{m=1}^{n} \delta_{\omega_m, j} = \frac{1}{10}\right\}$$

($j = 0, 1, ..., 9$). Indeed, $\mathbf{P}(\{\omega\}) = 0$, so that the probability of drawing any particular element of Ω is 0, $\mathbf{P}(E) = \sum_n \mathbf{P}([a_n, b_n])$ provided the subintervals of Ω in the union are disjoint, $\mathbf{P}(\Omega \cap \mathbb{Q}) = 0$, so that the event '$\omega$ is rational' has probability 0, and $\mathbf{P}(N_j) = 1$ for all j—Borel's *normal number* theorem. We interpret each $\omega \in \Omega$ as a possible 'random draw' from the unit interval and say that the probability that the draw falls between a and b is $\mathbf{P}([a, b]) = b - a$ ($a \leq b$), that it is impossible to draw any particular ω, that the draw is certain to be irrational, etc.

If $X : \Omega \to \mathbb{R}$ is a random variable—that is, X is such that for every real number x the set $\{\omega \in \Omega : X(\omega) \leq x\}$ belongs to $\mathcal{F}$, so that its probability can be calculated (as is the case with most functions one can conceive of in applications)—then the real number $X(\omega)$ is called the 'realization' of X associated with the random draw ω; and if A is an arbitrary Borel subset of $\mathbb{R}$, the event that a realization of X falls in A, defined as $X^{-1}(A) = \{\omega \in \Omega : X(\omega) \in A\}$ and often abbreviated as $\{X \in A\}$ or 'X is in A', has probability $\mathbf{P}(X \in A) \equiv \mathbf{P}(X^{-1}(A))$.

In particular, if U is the identity mapping on Ω then $\mathbf{P}(U \le u) = u$ $(u \in [0,1])$, $\mathbf{P}(U \text{ is irrational}) \equiv \mathbf{P}(U \in \Omega \backslash \mathbb{Q}) = 1$, and so on; U is said to be a *standard uniform* random variable, or a random variable with the *standard uniform distribution*, the distribution being the mapping $A \to \mathbf{P}(A)$. Moreover, if $D_n(\omega)$ denotes the n-th digit in the non-terminating decimal expansion of $U(\omega)$, so that

$$U(\omega) = \sum\nolimits_{n \ge 1} \frac{D_n(\omega)}{10^n} \equiv 0 \cdot D_1(\omega) D_2(\omega) D_3(\omega) \cdots, \tag{1.1}$$

then $D_1, D_2, \ldots$ are random variables on Ω and, for instance, the probability that the second and fourth digits equal 6 and 3, $\mathbf{P}(D_2 = 6, D_4 = 3)$, is

$$\sum\nolimits_{j,k=0}^{9} \mathbf{P}\left(\frac{j}{10} + \frac{6}{10^2} + \frac{k}{10^3} + \frac{3}{10^4} < U \le \frac{j}{10} + \frac{6}{10^2} + \frac{k}{10^3} + \frac{4}{10^4} \right) = \frac{1}{10^2},$$

which, because $\mathbf{P}(D_n = j) = 1/10$ for all j, shows that $\mathbf{P}(D_2 = 6, D_4 = 3) = \mathbf{P}(D_2 = 6)\mathbf{P}(D_4 = 3)$; and since the digits 3 and 6 play no special role in this calculation it follows that D_2 and D_4 are independent. More generally, $D_1, D_2, \ldots$ are independent random variables, each *uniform on* $\{0, 1, \ldots, 9\}$.

Thus a standard uniform random variable may be identified with a non-terminating decimal expansion consisting of independent random digits uniform on $\{0, 1, \ldots, 9\}$. But the digits of U can be arranged in a two-dimensional array by the 'diagonal method' in such a way that for each ω the correspondence

$$\begin{array}{lll}
U_1(\omega) & \leftrightarrow & 0 \cdot D_1(\omega)\ D_3(\omega)\ D_6(\omega)\ D_{10}(\omega) D_{15}(\omega) D_{21}(\omega) D_{28}(\omega) \cdots \\
U_2(\omega) & \leftrightarrow & 0 \cdot D_2(\omega)\ D_5(\omega)\ D_9(\omega)\ D_{14}(\omega) D_{20}(\omega) D_{27}(\omega) \quad \cdots \quad \cdots \\
U_3(\omega) & \leftrightarrow & 0 \cdot D_4(\omega)\ D_8(\omega)\ D_{13}(\omega) D_{19}(\omega) D_{26}(\omega) \quad \cdots \quad \cdots \quad \cdots \\
U_4(\omega) & \leftrightarrow & 0 \cdot D_7(\omega)\ D_{12}(\omega) D_{18}(\omega) D_{25}(\omega) \quad \cdots \quad \cdots \quad \cdots \quad \cdots \\
U_5(\omega) & \leftrightarrow & 0 \cdot D_{11}(\omega) D_{17}(\omega) D_{24}(\omega) \quad \cdots \quad \cdots \quad \cdots \quad \cdots \quad \cdots \\
U_6(\omega) & \leftrightarrow & 0 \cdot D_{16}(\omega) D_{23}(\omega) \quad \cdots \quad \cdots \quad \cdots \quad \cdots \quad \cdots \quad \cdots \\
U_7(\omega) & \leftrightarrow & 0 \cdot D_{22}(\omega) \quad \cdots \quad \cdots \quad \cdots \quad \cdots \quad \cdots \quad \cdots \quad \cdots \\
\multicolumn{3}{c}{\cdots\cdots\cdots\cdots\cdots\cdots\cdots\cdots\cdots\cdots\cdots\cdots\cdots\cdots\cdots}
\end{array}$$

determines a sequence $U_1(\omega), U_2(\omega), \ldots$ of numbers in Ω, and thereby (the random digits making up the decimal expansions being independent) a sequence $U_1, U_2, \ldots$ of standard uniform random variables, all defined on the same probability space $(\Omega, \mathcal{F}, \mathbf{P})$. This construction is concrete to the extent that numbers such as $\pi - 3$ and $e - 2$, say, are concrete and that to each of them there corresponds a countable sequence of irrational numbers obtained in the same manner as $U_1(\omega), U_2(\omega), \ldots$ are obtained from $U(\omega)$. Whether a particular $U(\omega) = \omega$ is a normal number or not is quite irrelevant (and probably unascertainable); what is important is that a draw from the probability space yields a countable number of independent draws with the same distribution, that sets of draws satisfying certain properties have probability 1 or 0, etc.

If F is an arbitrary right-continuous distribution function on $\mathbb{R}$ and its 'inverse' or 'quantile function' F^{-1} is defined by $F^{-1}(u) = \min\{x \in \mathbb{R} : F(x) \geq u\}$, $0 < u < 1$, then $\mathbf{P}(F^{-1}(U) \leq x) = F(x)$, i.e. the random variable $X = F^{-1}(U)$ has distribution function F (see Prob. 1.1). It follows that the correspondence between each $U(\omega) = \omega$ and $X_1(\omega), X_2(\omega), \ldots$, where $X_n(\omega) = F^{-1}(U_n(\omega))$ for each ω and n, determines a sequence $X_1, X_2, \ldots$ of independent random variables on the same probability space, each of which has distribution function F. More generally, a sequence $X_1, X_2, \ldots$ of independent random variables with distribution functions $F_1, F_2, \ldots$ is defined by the transformations $X_n = F_n^{-1}(U_n)$.

Still more generally, if F_1 is a distribution function and, for each $n \in \mathbb{N}$ and each $\mathbf{x} \in \mathbb{R}^n$, $F_{n+1,\mathbf{x}}$ is a distribution function in $\mathbb{R}$ then $X_1 := F_1^{-1}(U_1)$ and

$$X_{n+1} := F_{n+1,\mathbf{X}_n}^{-1}(U_{n+1}), \quad \mathbf{X}_n = (X_1, \ldots, X_{n-1}, X_n), \quad n \in \mathbb{N},$$

defines a sequence of *generally dependent* random variables on $(\Omega, \mathcal{F}, \mathbf{P})$ with joint distribution functions given by $G_1 = F_1$,

$$\begin{aligned} G_n(\mathbf{x}_n) := \; & \mathbf{P}(X_1 \leq x_1, X_2 \leq x_2, \ldots, X_n \leq x_n) \\ = \; & \int_{-\infty}^{x_{n-1}} \cdots \int_{-\infty}^{x_1} F_{n,\mathbf{x}'_{n-1}}(x_n) \mathrm{d}G_{n-1}(\mathbf{x}'_{n-1}), \end{aligned}$$

where $\mathbf{x}_n = (x_1, x_2, \ldots, x_n) \in \mathbb{R}^n$, $n = 2, 3, \ldots$ (see Probs. 1.2, 1.3). If for instance $F_{n+1,\mathbf{x}}$ does not depend on n and if it depends on $\mathbf{x} \in \mathbb{R}^n$ only through the last coordinate then the sequence $X_1, X_2, \ldots$ is a homogeneous Markov chain.

Evidently, the mapping of each draw $U(\omega) = \omega$ to the sequence $X_1(\omega), X_2(\omega), \ldots$ is not just a rearrangement of the digits of ω, as is the case with $U_1(\omega), U_2(\omega), \ldots$; but, again, the unfolding of a number such as $\pi - 3$ diagonally into a two-dimensional array followed by the transformation of the rows of the array by a real-valued function is hardly less concrete than $\pi - 3$ itself.

Finally, one can define an uncountable set $Y = (Y_t)_{t \geq 0}$ of random variables—a random function or 'continuous-time' stochastic process on $\mathbb{R}_+$—on $(\Omega, \mathcal{F}, \mathbf{P})$. Indeed, the procedure used above to get $U_1, U_2, \ldots$ from U can be applied to each of the U_n to yield an independent sequence of sequences $(V_{n,m})_{m \geq 1}$ of independent standard uniform random variables, and these in turn can be transformed into an independent sequence of sequences $(X_{n,m})_{m \geq 1}$ of independent random variables with any given distribution functions. But then, for example, two such sequences, say $(X_{1,m})_{m \geq 1}$ and $(X_{2,m})_{m \geq 1}$, the first of which may be assumed strictly positive, suffice to determine for each ω a right-continuous step function

$$Y_t(\omega) = \sum\nolimits_{m \geq 1} X_{2,m}(\omega) \mathbf{1}_{\{X_{1,1} + \cdots + X_{1,m} \leq t\}}(\omega), \tag{1.2}$$

whose jumps $X_{2,m}(\omega) - X_{2,m-1}(\omega)$ occur at the points $X_{1,1}(\omega) + \cdots + X_{1,m}(\omega)$; in particular, if the first sequence has the same exponential distribution and the second is degenerate at 1 then Y is a Poisson process (e.g. Sect. 23 of Billingsley [11]).

Typically, these processes arise from postulates suggested by empirical notions, such as regularity properties of $t \longrightarrow Y_t(\omega)$ for fixed ω, but first and foremost by 'structural' properties such as independence of increments in the case of the Poisson process (e.g. $Y_v - Y_u$ and $Y_t - Y_s$ are independent for $0 \leq s < t \leq u < v$, etc.), which may not be apparent from a definition such as (1.2). The following remark gives other examples of continuous-time processes defined as functions $\omega \longrightarrow \{Y_t(\omega) : t \geq 0\}$ on the unit interval, but will not be mentioned in the sequel.

Remark 1.1 For a second class of processes based on all the independent sequences $(X_{n,m})_{m\geq 1}$ obtained above, consider for each $n \in \mathbb{N}$ some real-valued functions $(f_{m,n})_{m\geq 1}$ on $\mathbb{R}_+$ such that $f_{m,n}(t) = 0$ for $t \notin]n-1, n]$ and set

$$Y_t^{(n)}(\omega) = \sum\nolimits_{m\geq 1} f_{m,n}(t) X_{n,m}(\omega)$$

for $t \in [n-1, n]$ and $\omega \in \Omega$ for which the series converges and $Y_t^{(n)}(\omega) = 0$ for other (t, ω). Under certain conditions on the $f_{m,n}$ and on the $X_{n,m}$ (see Ito and Nisio [48], pp. 41–3, 265–7 of Ash and Gardner [5], or Chap. 3 of Schilling and Partzsch [91]), for each n the function $t \longrightarrow Y_t^{(n)}(\omega)$ is continuous on $[n-1, n]$ for ω in a set of probability 1. Under such conditions, for ω in a set of probability 1 the function

$$Y_t(\omega) = \sum\nolimits_{n\geq 1} \left(Y_t^{(n)}(\omega) + \sum\nolimits_{j=1}^{n-1} Y_j^{(j)}(\omega) \right) \mathbf{1}_{[n-1,n[}(t)$$

is continuous on $\mathbb{R}_+$. For instance, if the $X_{n,m}$ are normally distributed with $\mathbf{E}X_{n,m} = 0$, $\mathrm{Var}(X_{n,m}) = 1$, and if $f_{n,m}(t) = \frac{\sqrt{2}}{\pi m}\,[1 - \cos(m\pi\{t - (n-1)\})]\,\mathbf{1}_{[n-1,n[}(t)$, then the $Y = (Y_t)_{t\geq 0}$ thus defined is a Wiener (or Brownian motion) process, which, like the Poisson process, has independent increments. □

1.2 Random variables and empirical data

In applications of statistics to real-life problems one identifies empirical data with the values $\mathbf{X}_1(\omega), \mathbf{X}_2(\omega), \ldots$ assumed by a sequence $(\mathbf{X}_m)_{m\geq 1}$ of random vectors (say) at some point ω which may or may not be empirically observable. Although different sequences defined in different probability spaces may be essentially equivalent to each other and any of them may serve equally well in a given application, *some* such sequence must be made to correspond (even if only implicitly) with the empirical data if anything nontrivial is to be inferred from these. For it is only thanks to a sort of approximate isomorphism between the empirical world and the ideal world by which we make empirical data correspond to theoretical objects, perform operations on the latter and deduce propositions from them, and then translate those propositions into empirical terms, that we can overstep the concrete but limited realm of data. Thus, for instance, a confidence interval for a parameter, or the very mention of the parameter, are usually meaningless without reference to some sequence of random variables.

The probability space on which the sequence is defined is usually irrelevant, and the particular ω is almost always irrelevant even when it exists and can be identified (as is the case, for example, in the drawing of samples from a finite set of objects). The distribution of the sequence, on the other hand, is always important, since any probabilistic statement one may wish to apply in the real world depends on whether its elements are independent, identically distributed, Markov, stationary, Gaussian, etc. The extent to which the definition of random variables—including possible functional relationships between them—matters depends on the problem. If the data are drawn randomly or almost randomly from some conceptually infinite population their definition is quite irrelevant. But if they are obtained successively in time and data obtained at different times are physically related—say humidity at a certain location after midnight and humidity and other atmospheric quantities before midnight—then it is essential that the elements of our sequence reflect those aspects by means of such relations as $\mathbf{X}_{m+1} = f_m(\mathbf{X}_m, \mathbf{X}_{m-1}, \ldots, \theta)$, say, where a lot or very little may be known or assumed about the functions f_m and the parameter θ. It is for this reason that time series and Markov chain models, and stochastic processes in general, are most often defined by means of relationships between random variables rather than purely by means of their joint distributions.

One may, of course, acknowledge that there is no way of checking whether a particular probability model provides a sufficiently accurate picture of a given real-life problem (e.g. pp. 323–334 of Freedman [37]), but often there are tests (including prediction exercises) which, together with plausibility arguments and a little faith, provide partial, provisional checks. For example, one can verify that many empirical (statistical) properties of a sequence of sexes of newborns in a given population, of a sequence of 'phenotypic determinations' in newborns of a given species, of the sequence of digits in the decimal expansion of π, etc., are almost perfectly consistent with the corresponding theoretical (probabilistic) properties of a sequence of independent and identically distributed random variables. Surely one must dismiss the idea that, since no model can be proven correct for a situation, all models are equally wrong, from which it is not a long way to concluding that all models are equally legitimate for a given real-life problem.

1.3 Notes

The ideas sketched here go back to the work of Hugo Steinhaus a century ago; for a history of their development leading to Kolmogorov's axiomatization of probability see Kahane [50] and Shafer and Vovk [95].

In his book [94], Glenn Shafer writes that

> Although causal ideas permeate the use of statistics in all branches of industry, commerce, government, and science, they have not been incorporated into the basic theory of the subject.

and that

> Because the older sciences—pure and practical—all demand a relatively high level of precision and security, they can successfully study only causal structures that are sufficiently uniform to permit relatively precise statistical estimation, and these structures can often be described adequately in the existing language of stochastic processes, without any explicit use of causal ideas.

Although we agree with most of this, it seems to us from what has been said in this chapter that probability spaces and random variables (hence stochastic processes) implicitly contain ideas of causality and that applied probability models, just as linear regression models (whether practitioners of statistics be aware of it or not), explicitly incorporate such ideas; after all, even the calculus, which is no more 'dynamic' than probability, has for centuries been used successfully to deal with problems of cause and effect.

In addition to "[...] bringing causal ideas into the foundations of probability", the purpose of Shafer's book was to create "[...] a framework within which causal conjectures can be more clearly quantified, debated, and confronted by statistical evidence". We believe that the ideas expounded in the present book constitute such a framework, and accordingly that they are compatible with Shafer's own theory, just as they are with the theories of Pearl, Robins and Rubin.

1.4 Problems

1.1 Let F be an arbitrary distribution function on the real line (non-decreasing, right-continuous and such that $F(-\infty) = 0$, $F(\infty) = 1$).

(a) Show that the 'inverse' or 'quantile function' of F, defined by

$$F^{-1}(u) := \min\{x : F(x) \geq u\} = \inf\{x : F(x) \geq u\}, \quad 0 < u < 1,$$

satisfies $F(x) \geq u \Leftrightarrow x \geq F^{-1}(u)$. Conclude that if U is a standard uniform random variable—i.e. such that $\mathbf{P}(U \leq u) = u$ for $0 \leq u \leq 1$—then $X := F^{-1}(U)$ is a random variable with distribution function F.

(b) Show that $F(F^{-1}(u)) \geq u$ $(0 < u < 1)$ and that $F(F^{-1}(u) - \varepsilon) < u$ for all $\varepsilon > 0$; conclude that if a random variable X has continuous distribution function F then $F(X)$ is standard uniform.

1.2 How can you define explicitly—or simulate data from—an arbitrary sequence of random variables $X_1, X_2, \ldots$? *Suggestion:* Using Prob. 1.1, define first X_1, then X_2 by means of its distribution function conditional on X_1, and so on. With some work one can prove that this method yields a sequence with the desired distributions in the discrete case, but the general case is difficult.

1.3 An important result about multivariate normal random vectors implies that if $\mathbf{X}_d = (X_1, X_2, \ldots, X_d)^{\mathrm{T}}$ $(d > 1)$ is normally distributed with mean column vector $\boldsymbol{\mu}_d$ and full rank covariance matrix $\boldsymbol{\Sigma}_{d,d}$, and if we write these as

$$\mu_d = \begin{pmatrix} \mu_{d-1} \\ \mu_d \end{pmatrix} \quad \text{and} \quad \begin{pmatrix} \Sigma_{d-1,d-1} & \Sigma_{d-1,1} \\ \Sigma_{1,d-1} & \sigma_d^2 \end{pmatrix},$$

where $\mathbf{E}(X_d) = \mu_d$ and $\text{Var}(X_d) = \sigma_d^2$, then the distribution of X_d conditional on $\mathbf{X}_{d-1}^{\text{T}} \equiv (X_1, \ldots, X_{d-1}) = \mathbf{x}_{d-1}^{\text{T}} \in \mathbb{R}^{d-1}$ is normal with

$$\mathbf{E}\left[X_d | \mathbf{X}_{d-1} = \mathbf{x}_{d-1}\right] = \mu_d + \Sigma_{1,d-1}\Sigma_{d-1,d-1}^{-1}(\mathbf{x}_{d-1} - \mu_{d-1})$$

and

$$\text{Var}\left[X_d | \mathbf{X}_{d-1} = \mathbf{x}_{d-1}\right] = \sigma_d^2 - \Sigma_{1,d-1}\Sigma_{d-1,d-1}^{-1}\Sigma_{d-1,1}$$

(see pp. 36–7 of Brockwell and Davis [15], for example, for a proof). Explain how one can use this result to define a sequence of normally distributed random variables, each finite subset of which has a nondegenerate multivariate normal distribution.

1.4 **(a)** Give an example of two dependent standard uniform random variables. *Suggestion:* Let $\tilde{U}$ and $\tilde{V}$ be independent standard uniforms, and put $U = \max\{\tilde{U}, \tilde{V}\}^2$ and $V = 1 - (1 - \min\{\tilde{U}, \tilde{V}\})^2$.

(b) Consider the random variable $R = \rho(X, U, V)$, where ρ is some real-valued function, U and V are standard uniforms, not necessarily independent, and X is another random variable independent of (U, V). Show that there is a function $\tilde{\rho}$ and a standard uniform random variable W defined on the same space as U, V and X such that $\tilde{R} = \tilde{\rho}(X, W)$ has the same distribution as R. *Suggestion:* Let $\tilde{U}$ and $\tilde{V}$ be independent standard uniforms yielding $(\check{U}, \check{V})$ as $\check{U} = \varphi_1(\tilde{U}, \tilde{V})$ and $\check{V} = \varphi_2(\tilde{U}, \tilde{V})$, say, with the same distribution as (U, V), and let $W = 0.\tilde{U}_1\tilde{V}_1\tilde{U}_2\tilde{V}_2\cdots$ where $\tilde{U}_1, \tilde{U}_2, \ldots$ and $\tilde{V}_1, \tilde{V}_2, \ldots$ are the digits in the decimal expansions of $\tilde{U}$ and $\tilde{V}$.

(c) With U, V, X and R as in **(b)**, is it possible to have $\rho(X(\omega), U(\omega), V(\omega)) = \rho'(X(\omega), W(\omega))$ for all ω, some function ρ' and a standard uniform random variable W, if necessary by imposing a condition on U, V? *Suggestion:* Let W' be obtained by alternating the digits in the decimal expansions of U and V, assume that its distribution function, F, is continuous and strictly increasing, put $W = F(W')$, and use Prob. 1.1 **(b)**.

CHAPTER 2

The simplest statistical model

In the simplest statistical model of cause and effect, a random 'treatment' influences a random 'response' and treatment and response are influenced by a 'situation' or set of 'background' variables. Because variations in the treatment occur together with variations in the situation, it may be difficult to disentangle the corresponding variations in response that are due to the treatment from those that are purely random or due to the situation, a problem known as 'confounding'. But if the background variables that confound the treatment effect are known then one may condition upon them, holding them fixed in some sense, and yet vary the treatment somewhat so as to determine its effect under each of a number of situations. This chapter treats the simplest aspects of causality in statistics, namely those covered by Rubin's theory and by the corresponding versions of Pearl and Robins; they are the most important for applications and include the interpretation of causal effect in real-life problems.

2.1 The basic model; unconfoundedness

The simplest statistical model of cause and effect is

$$R_n = \rho\big(V_n, \mathbf{X}_n, T_n\big), \quad T_n = \tau\big(U_n, \mathbf{X}_n\big), \quad n \in \mathbb{N}, \tag{2.1}$$

where $(\mathbf{X}_n)_{n\geq 1}$ is a sequence of random 'situations', or vectors of 'background variables', taking values in $\mathcal{X}$ (e.g. in $\mathbb{R}^d$), $(U_n)_{n\geq 1}$ and $(V_n)_{n\geq 1}$ are sequences of random variables such that U_n and V_n are both standard uniform conditionally on $\mathbf{X}_n$, all defined on our probability space $(\Omega, \mathcal{F}, \mathbf{P})$, $(T_n)_{n\geq 1}$ is a sequence of random 'treatments' taking values in $\mathcal{T}$ (e.g. in $\mathbb{N}_0$), $(R_n)_{n\geq 1}$ is the corresponding sequence of random 'responses' (e.g. taking values in $\mathbb{R}$), and ρ and τ are functions defined on $[0,1] \times \mathcal{X} \times \mathcal{T}$ and $[0,1] \times \mathcal{X}$, respectively.

There is no danger of confusing cause and effect in equations (2.1): R_n is a function of $(\mathbf{X}_n, T_n)$ and represents an event occurring after the determination of the latter; that is, the calculation of $R_n(\omega)$ for each $\omega \in \Omega$ is preceded by that of $T_n(\omega)$ and both are preceded by that of $U_n(\omega)$, $V_n(\omega)$ and $\mathbf{X}_n(\omega)$; and if there is an 'ultimate cause of things' it is the drawing of ω. We say that this is the simplest model because in some real-life problems one may wish that the treatments too affect some of the coordinates of the background variables, and hence that the second equation in (2.1), which we mostly think of as a scalar equation, be replaced by a system of equations involving T_n and $\mathbf{X}_n$; more complicated models in which part of $\mathbf{X}_n$ is a function of T_n and T_n a function of the remainder of $\mathbf{X}_n$ will be considered in Chaps. 3 and 4, where it will be seen that for studying the effect of a treatment the more general models can often be reduced to the present model.

Evidently, ρ may be constant in some or all of the coordinates of $\mathbf{X}_n$, and the same applies to τ; if τ is constant in its second argument then the treatments, even if randomly selected, are not subject to restrictions imposed by the situation. Note that the sequence $(\mathbf{X}_1, T_1), (\mathbf{X}_2, T_2), \ldots$ of pairs of situations and treatments may or may not be endowed with such properties as independence and stationarity, which are important in the problems of estimation and testing considered later.

Practically everything that follows applies with ρ and τ replaced by functions ρ_n and τ_n indexed by n; in fact, as the definitions stand, a coordinate of $\mathbf{X}_n$ may be reserved for choosing (randomly or not) a particular pair (ρ_n, τ_n) for (ρ, τ). The simplified version of (2.1) is used mainly for the convenience of assuming, in order to alleviate notation or motivate a definition or result, that the distribution of $(\mathbf{X}_n, T_n, R_n)$ is independent of n. The more general version is adopted explicitly only in the first part of Chap. 5, where it is needed to cover some real-life problems. Whatever version is meant, we shall refer to it as the *basic model*, to distinguish it from the more detailed models introduced in Chap. 3.

The functions ρ and τ are often regarded as unknown; the extent to which they are so (whether their expressions are unknown, whether some parameters of their partially known expressions are unknown) depends on the real-life problem to which the model is applied, and it is mostly irrelevant to our general discussion.

Example 2.1 When $\mathcal{X} = \mathbb{R}^d$ and $\mathcal{T} = \mathbb{R}$, the basic model includes a response like

$$R_n = \beta_0 T_n + \beta \cdot \mathbf{X}_n + \Phi^{-1}(V_n, 0, \sigma^2),$$

where T_n is as above, $\beta_0 \in \mathbb{R}$, $\beta \in \mathbb{R}^d$, $\sigma^2 > 0$, and Φ^{-1} is the inverse of the normal distribution function—i.e. the inverse of $\Phi(x, \mu, \sigma^2) = (\sigma\sqrt{2\pi})^{-1} \int_{-\infty}^{x} e^{-\frac{1}{2}(\frac{y-\mu}{\sigma})^2} dy$. This very particular model is not a linear regression model unless U_n and V_n are independent conditionally on $\mathbf{X}_n$—a condition that will be seen to play a crucial role in the study of the *treatment effect*, which will be defined soon. Of course, (2.1) is much more general: as mentioned in Chap. 1, each V_n can be used to generate a random function (cf. (1.2)); so ρ may, for example, be taken as a function acting on such a random function and having $\mathbf{X}_n$ and T_n as parameters. □

Example 2.2 The basic model also includes the binary response

$$R_n = \begin{cases} 1 & \text{if } V_n \leq \Pi(\beta_0 T_n + \beta \cdot \mathbf{X}_n), \\ 0 & \text{otherwise,} \end{cases}$$

with β_0 and β as above, Π a distribution function, and again T_n as in (2.1). This too is a very particular model; if U_n and V_n are independent conditionally on $\mathbf{X}_n$, then

$$\begin{aligned}\mathbf{P}(R_n = 1|\mathbf{X}_n = \mathbf{x}, T_n = t) &= \mathbf{P}(V_n \le \Pi(\beta_0 t + \beta \cdot \mathbf{x})|\mathbf{X}_n = \mathbf{x}, \tau(U_n, \mathbf{x}) = t) \\ &= \mathbf{P}(V_n \le \Pi(\beta_0 t + \beta \cdot \mathbf{t})|\mathbf{X}_n = \mathbf{x}) \\ &= \Pi(\beta_0 t + \beta \cdot \mathbf{x}),\end{aligned}$$

and when Π is the logistic distribution function, given by $\Pi(x) = \exp x/(1 + \exp x)$ for $x \in \mathbb{R}$, it may be recognized as the logistic regression model. □

Although we shall often think of R_n as a scalar, most of the ideas in this chapter apply to responses which are random vectors:

Example 2.3 When $\mathcal{X} = \mathbb{R}^d$ and $\mathcal{T} = \mathbb{R}$, the basic model includes the response

$$R_n = (\min\{S_n, C_n\}, \mathbf{1}_{\{S_n \le C_n\}}), \quad S_n := \sigma(V_n'', \mathbf{X}_n, T_n), \quad C_n := \chi(V_n', \mathbf{X}_n),$$

where V_n' and V_n'' are obtained from two infinite, disjoint sets of digits of the decimal expansion of V_n, and hence are independent and standard uniform conditionally on $\mathbf{X}_n$, and χ and σ are some functions. This model is used in 'survival analysis', where S_n represents the survival time of an individual, part of an equipment, etc., and C_n represents a 'censoring time'—a time beyond which survival may extend without its actual value being known, which happens when $\mathbf{1}_{\{S_n \le C_n\}} = 0$. Although the effect of the treatment is on survival, S_n is not always observable: what one observes is $\min\{S_n, C_n\}$ and $\mathbf{1}_{\{S_n \le C_n\}}$. The independence of U_n and V_n conditionally on $\mathbf{X}_n$ is needed in order to ensure that the censoring and survival times are independent conditionally on $(\mathbf{X}_n, T_n)$, which in turn is essential for applications. □

There is no loss of generality in assuming that U_n and V_n are standard uniform conditionally on $\mathbf{X}_n$, because, as explained in Sect. 2.5, whatever distribution they be given, we can always reconcile them with an *equivalent version* of the basic model, namely a model in which the background variables, the treatments and the responses are defined in the required order and the resulting sequence of random vectors has the same distribution as the given sequence $(\mathbf{X}_1, T_1, R_1), (\mathbf{X}_2, T_2, R_2), \ldots$ But at least in specific examples *some* distributions must be chosen, and taking a 'canonical' distribution (namely a *single* continuous distribution) such as the standard uniform facilitates the statement of definitions and results. There is even a good psychological reason for using a function such as Φ^{-1} in combination with V_n in the example just given: doing so reminds us and brings home to us that in a real-life problem mathematical ideas need to be mapped into empirical objects, and that the mapping requires justification.

We must make a number of other remarks:

Remark 2.1 The causal interpretation of (2.1) is not invalidated by the possibility of talking of the distribution of $(\mathbf{X}_n, T_n)$ conditionally on R_n and of constructing a vector $(\tilde{\mathbf{X}}_n, \tilde{T}_n)$ with the same distribution as $(\mathbf{X}_n, T_n)$ from a function of R_n—for example (see Probs. 1.1 and 1.2), in the case where $\mathcal{X}, \mathcal{T} \subset \mathbb{R}$, by

$$\tilde{T}_n = F_{R_n}^{-1}(\xi_n), \quad \tilde{\mathbf{X}}_n = G_{\tilde{T}_n, R_n}^{-1}(\xi_n'),$$

where

$$F_r(t) = \mathbf{P}(T_n \le t | R_n = r), \quad G_{t,r}(x) = \mathbf{P}(\mathbf{X}_n \le x | T_n = t, R_n = r),$$

and ξ_n, ξ'_n are independent standard uniforms independent of R_n. The $(\tilde{\mathbf{X}}_n, \tilde{T}_n, R_n)$ constructed in this way has the same distribution as $(\mathbf{X}_n, T_n, R_n)$, but these random vectors are quite different and in general represent different relations of cause and effect in a real-life problem (nevertheless, see Prob. 2.1); in addition, the *sequences* $(\mathbf{X}_1, T_1, R_1), (\mathbf{X}_2, T_2, R_2), \ldots$ and $(\tilde{\mathbf{X}}_1, \tilde{T}_1, R_1), (\tilde{\mathbf{X}}_2, \tilde{T}_2, R_2), \ldots$ may have different distributions unless the first is an independent sequence and $\xi_1, \xi'_1, \xi_2, \xi'_2, \ldots$ too are independent. Because observed data correspond to random variables evaluated at a sample point ω, in general they cannot tell us whether they originate by a process analogous to that used to define $(\mathbf{X}_n, T_n, R_n)$—namely by generating $\mathbf{X}_n$, then T_n based on $\mathbf{X}_n$, and finally R_n based on $\mathbf{X}_n$ and T_n—or by a process analogous to that used to define $(\tilde{\mathbf{X}}_n, \tilde{T}_n, R_n)$; typically, they only tell us whether the distribution of our random vectors is likely to be correct. Consequently, the basis and justification for the basic model are not to be found in observed data—the empirical analogs of $(\mathbf{X}_n(\omega), T_n(\omega), R_n(\omega))$ for some ω—but in knowledge of the field of application, part of which may, of course, have a statistical origin or nature. □

Remark 2.2 If the τ of (2.1) is constant in the first argument and not constant in the second then the treatment is fully determined by the situation, the response is mostly determined by the situation, and there is no treatment to talk of. When using the basic model as a model for observational data we normally want to exclude this situation and assume that τ is not constant in its first argument if it is not constant in the second; this implies that the distribution of T_n conditional on $\{\mathbf{X}_n = \mathbf{x}\}$ is not degenerate at least for some $\mathbf{x}$ and corresponds to Rubin's notion of 'probabilistic assignment' (p. 38 of Imbens and Rubin [46]). □

An interesting variant of the basic model is obtained by letting T_n be constant with probability 1, or letting it take a fixed number of values at fixed values of n, while allowing the $\mathbf{X}_n$ to be nondegenerate. If we regard our model as representing a self-contained 'world' or 'system' evolving from a random draw ω, we can think of this option as the external forcing of a fixed treatment, or of a fixed pattern of treatments, upon the system—which without that external *intervention* would have evolved randomly according to its own laws. Thus, we will see that when a real-life problem is represented by the basic model it is useful to consider the associated *intervention model*

$$R_n = \rho(V_n, \mathbf{X}_n, t_n), \quad n \in \mathbb{N}, \tag{2.2}$$

where $(\mathbf{X}_n)_{n\ge1}$, $(V_n)_{n\ge1}$ have the same distributions as in (2.1) and $(t_n)_{n\ge1}$ is a sequence of numbers in $\mathcal{T}$—which in particular may be set equal to a constant in order to represent the forcing of the same treatment in every situation.

Remark 2.3 When the t_n are all equal, the intervention model corresponds to an application of Pearl's '*do* operator'—"which simulates physical interventions by deleting certain functions [in this case τ] from the model, replacing them by a constant [...], while keeping the rest of the model unchanged"—to (2.1); see p. 107 of Pearl [68] or pp. 54-57 of Pearl, Glymour and Jewel [73]. Of course, in many non-experimental real-life situations the enforcement of a treatment upon a system does affect other aspects of that system (for instance, forcing an individual to stop smoking may cause him to increase his consumption of alcohol or foodstuffs, or to recalcitrate in unanticipated ways; cf. Sect. 8.2 of Cox and Wermuth [20]), but, as we shall see later on, an intervention model, even if purely hypothetical or unrealizable, may, in principle, serve a good purpose. Note that (2.2) really defines a *family of models*, one model for each sequence $(t_n)_{n\geq 1}$, and that (2.2) could also be seen as a special case of (2.1) where τ depends only on the index n of the pair $(U_n, \mathbf{X}_n)$. □

The models now defined bring with them a distinction between *observable* and *unobservable* random variables: in (2.1), the situations, treatments and responses are observable because they represent empirical data connected with a cause and effect relationship; the U_n and V_n are unobservable because they represent all that which cannot be observed empirically and which one hopes constitutes 'purely random' factors. Of course, unobservable random variables are as mathematical and as 'real' as observable random variables; the main reason for distinguishing them is that only the observable ones can be used in statistical procedures—such as versions of the matching and stratification procedures treated in Chap. 5, or procedures to estimate the function $(\mathbf{x}, t) \to \mathcal{L}[R_n|\mathbf{X}_n = \mathbf{x}, T_n = t]$, which plays an important role in the sequel.[1]

To say that treatment has an effect on the response, or that there exists a *treatment effect*, in connection with (2.1) or (2.2) is essentially to say that the response function ρ is not constant in its last argument. We shall say that treatment has an effect on the response if, for some n and some $\mathbf{x}$ such that $\mathbf{P}(\mathbf{X}_n \in N) > 0$ for every neighbourhood N of $\mathbf{x}$, the function $(v, t) \to \rho(v, \mathbf{x}, t)$ varies with t for some $v \in [0, 1]$. Except in special cases which are irrelevant for applications, if treatment has an effect then

$$\mathcal{L}\big(\rho(V_n, \mathbf{X}_n, t)\big|\mathbf{X}_n = \mathbf{x}\big) = \mathcal{L}\big(\rho(V_n, \mathbf{x}, t)\big|\mathbf{X}_n = \mathbf{x}\big) = \mathcal{L}\big(\rho(V_n, \mathbf{x}, t)\big)$$

and

$$\mathcal{L}(R_n|\mathbf{X}_n = \mathbf{x}, T_n = t),$$

both of which have a greater chance of being related to empirical data than $\rho(v, \mathbf{x}, t)$ does, are not constant in t for the n and $\mathbf{x}$ in question, so our definition is more practical than it may appear; see Prob. 2.2.

Remark 2.4 Because U_n and V_n are standard uniform conditionally on $\mathbf{X}_n$, each is standard uniform unconditionally as well, so the conditioning on $\{\mathbf{X}_n = \mathbf{x}\}$ can be removed from $\mathcal{L}\big(\rho(V_n, \mathbf{x}, t)\big|\mathbf{X}_n = \mathbf{x}\big)$ and from other expressions involving the law of $\rho(V_n, \mathbf{x}, t)$, as done in the last identity. □

In real-life problems the $(\mathbf{X}_n, T_n, R_n)$ are meant to represent observations on 'individuals'—or 'units', 'patients', etc.—and serve as models for the responses of

[1] See the list of special symbols on p. xxi for the meaning of $\mathcal{L}$.

those individuals to treatments under certain conditions. The hypothesis that the treatment has no effect on the response then means that the response of the n-th individual is fully determined by the situation and by incidental factors represented by V_n. On the other hand, it is enough that a single individual's response be affected by the treatment in a 'realizable' situation for the treatment to have an effect.

It follows from all this that a study of the treatment effect on the response amounts to a study of the random variables $\rho(V_n, \mathbf{X}_n, t)$, $t \in \mathcal{T}$, the so-called *potential outcomes*, and of related random variables such as $\rho(V_n, \mathbf{X}_n, t) - \rho(V_n, \mathbf{X}_n, t')$ or $\rho(V_n, \mathbf{X}_n, t)/\rho(V_n, \mathbf{X}_n, t')$, $t \neq t'$. But these are unobservable (only $\rho(V_n, \mathbf{X}_n, T_n)$ is observable), so it is not obvious that their laws can be estimated from observed data. In fact, we will now see that *the possibility of estimating* $\mathcal{L}(\rho(V_n, \mathbf{X}_n, t) | \mathbf{X}_n = \mathbf{x})$ *from data*—hence of studying the effect of treatment on response—corresponds to *the possibility of removing confounding* in the basic model; for the $\mathbf{X}_n$ are confounders of treatment because unless one is somehow able to 'fix' them it is generally impossible to know whether differences in response are due to differences in the situations or to differences in the treatment.

Ideally, the study of the treatment effect would involve the comparison between the values of $\rho(V_n, \mathbf{X}_n, t)$ for various $t \in \mathcal{T}$—the potential outcomes of the n-th individual. But at each draw ω we only get to see $\rho(V_n(\omega), \mathbf{X}_n(\omega), t)$ for a single t, namely $t = T_n(\omega)$; we do not see any of its *counterfactuals*, i.e. the set of $\rho(V_n(\omega), \mathbf{X}_n(\omega), t)$ for $t \neq T_n(\omega)$—if we did, then the differences

$$\rho(V_n(\omega), \mathbf{X}_n(\omega), T_n(\omega)) - \rho(V_n(\omega), \mathbf{X}_n(\omega), t)$$

for $t \neq T_n(\omega)$ would reveal the effect of the treatment. Even in the intervention model (2.2) the most that one is given to see and is allowed to compare (e.g. when the $\mathbf{X}_n$ are discrete) at a single draw ω are pairs of responses $\rho(V_m(\omega), \mathbf{X}_m(\omega), t_m)$ and $\rho(V_n(\omega), \mathbf{X}_n(\omega), t_n)$ such that $\mathbf{X}_m(\omega) = \mathbf{X}_n(\omega) = \mathbf{x}$ for some $\mathbf{x} \in \mathcal{X}$ and $t_m \neq t_n$, which, because they generally differ in the first argument of ρ, will, on their own, seldom tell us whether differences between them are caused by the treatment or by V_m and V_n (whose realizations are unobservable). It seems evident, then, that in order to study the effect of the treatment based on a single realization $(\mathbf{X}_1(\omega), T_1(\omega), R_1(\omega))$, $(\mathbf{X}_2(\omega), T_2(\omega), R_2(\omega)), \ldots$ of the basic model one has to make do with the comparison of empirical conditional averages such as

$$\frac{\sum_{n=1}^N \mathbf{1}_{\{\mathbf{X}_n=\mathbf{x}, T_n=t\}}(\omega)\rho(V_n(\omega), \mathbf{x}, t)}{\sum_{n=1}^N \mathbf{1}_{\{\mathbf{X}_n=\mathbf{x}, T_n=t\}}(\omega)} \quad \text{and} \quad \frac{\sum_{n=1}^N \mathbf{1}_{\{\mathbf{X}_n=\mathbf{x}, T_n=t'\}}(\omega)\rho(V_n(\omega), \mathbf{x}, t')}{\sum_{n=1}^N \mathbf{1}_{\{\mathbf{X}_n=\mathbf{x}, T_n=t'\}}(\omega)}$$

for $t \neq t'$ and various $\mathbf{x}$—and hope that the $V_n(\omega)$ for which $T_n(\omega) = t$ are not essentially different, as far as their role in ρ is concerned, from those for which $T_n(\omega) = t'$. But there is nothing in the conditions introduced so far that prevents us from having, for instance, $V_n(\omega) < 1/2$ for n such that $\mathbf{1}_{\{\mathbf{X}_n=\mathbf{x}, T_n=t\}}(\omega) = 1$ and $V_n(\omega) \geq 1/2$ for n such that $\mathbf{1}_{\{\mathbf{X}_n=\mathbf{x}, T_n=t'\}}(\omega) = 1$, an occurrence that would *confound* the workings of t in $\rho(V_n(\omega), \mathbf{x}, t)$ and could produce a difference between the two averages even if ρ were constant in its last argument. It is therefore necessary to require, for the purpose of studying the treatment effect in the basic model, that V_n be independent of T_n conditionally on $\mathbf{X}_n$, or, what is equivalent, that U_n *and* V_n *be independent conditionally on* $\mathbf{X}_n$. This condition, which we call *unconfoundedness*, should rule out confounding because each V_n contributing to one average has the same distribution as each V_n contributing to the other, so that differences between averages that seem too extreme can be correctly attributed to the different values of the treatment rather than to randomness.

Remark 2.5 If U_n and V_n are independent given $\mathbf{X}_n$ then the set $\big(\rho(V_n, \mathbf{X}_n, t)\big)_{t\in\mathcal{T}}$ of potential outcomes is independent of $T_n = \tau(U_n, \mathbf{X}_n)$ given $\mathbf{X}_n$. In particular, if $\mathcal{T} = \{0, 1\}$ this means that $(R_1^{(n)}, R_0^{(n)})$, where $R_t^{(n)} \equiv R_t^{(n)}(\mathbf{X}_n) := \rho(V_n, \mathbf{X}_n, t)$, is independent of T_n given $\mathbf{X}_n$, which is Rubin's condition of unconfoundedness (e.g. Rubin [88] or p. 38 of Imbens and Rubin [46]). Conversely, if Rubin's condition holds then we can define

$$\tilde{R}_t^{(n)} \equiv \rho(\tilde{V}_n, \mathbf{X}_n, t) := G_{t,\mathbf{X}_n}^{-1}(\tilde{V}_n), \quad \tilde{T}_n \equiv \tau(\tilde{U}_n, \mathbf{X}_n) := F_{\mathbf{X}_n}^{-1}(\tilde{U}_n),$$

where $G_{t,\mathbf{x}}$ is the distribution function of $R_t^{(n)}$ given $\{\mathbf{X}_n = \mathbf{x}, T_n = t\}$, $F_{\mathbf{x}}$ that of T_n given $\{\mathbf{X}_n = \mathbf{x}\}$, and $\tilde{U}_n$ and $\tilde{V}_n$ are independent standard uniform random variables independent of $\mathbf{X}_n$, to get 'probabilistic copies'

$$(\mathbf{X}_n, \tilde{T}_n, \tilde{R}_0^{(n)}, \tilde{R}_1^{(n)})$$

of the $(\mathbf{X}_n, T_n, R_0^{(n)}, R_1^{(n)})$. Moreover, if the given $(T_1, R_0^{(1)}, R_1^{(1)}), (T_2, R_0^{(2)}, R_1^{(2)}), \ldots$ are independent conditionally on $\mathbf{X}_1, \mathbf{X}_2, \ldots$, we may arrange for the new sequence to have the same distribution and to follow an equivalent version of the given basic model with unconfoundedness. This shows that (2.1) in conjunction with the conditional independence of U_n and V_n is essentially equivalent to the so-called Neyman-Rubin model—in fact, it is a slight generalization of it through the more general nature of $\mathcal{T}$, $\mathcal{X}$, etc. □

When and to what extent it is possible to study the effect of the treatment by comparing empirical averages also depends on the properties of those averages, which in turn depend on properties of the sequence $(\mathbf{X}_1, T_1, R_1), (\mathbf{X}_2, T_2, R_2), \ldots$ other than unconfoundedness. Thus, although we shall not consider the question of estimating and testing for a treatment effect until Chap. 5, we must note that when the distribution of $(\mathbf{X}_n, T_n, R_n)$ is independent of n, and hence is the same as that of a generic vector $(\mathbf{X}, T, R)$ defined on the same probability space, empirical conditional averages of the form

$$\frac{\sum_{n=1}^{N} \mathbf{1}_{\{\mathbf{X}_n=\mathbf{x}, T_n=t\}} f(R_n)}{\sum_{n=1}^{N} \mathbf{1}_{\{\mathbf{X}_n=\mathbf{x}, T_n=t\}}}, \tag{2.3}$$

where f is some real-valued function, converge under certain conditions and in a certain sense to

$$\mathbf{E}[f(R)|\mathbf{X} = \mathbf{x}, T = t] = \mathbf{E}[f(R_n)|\mathbf{X}_n = \mathbf{x}, T_n = t] \tag{2.4}$$

as $N \to \infty$ (e.g. with probability 1 if the $(\mathbf{X}_n, T_n, R_n)$ are independent).[2] In the sequel we shall often refer to the possibility of estimating expectations such as these *from observed data*, by which we imply that the right-hand side of (2.4) is independent of n and can be estimated consistently by (2.3) with samples—not necessarily random samples—$(\mathbf{X}_1, T_1, R_1), \ldots, (\mathbf{X}_N, T_N, R_N)$.

In order to confirm that under unconfoundedness the comparison of empirical conditional averages—those of (2.3) for varying t—provides a way of studying the effect of the treatment, consider (2.4) with $f = \mathbf{1}_A$ and varying A, i.e. the law

$$\mathcal{L}(R_n|\mathbf{X}_n = \mathbf{x}, T_n = t) = \mathcal{L}(\rho(V_n, \mathbf{x}, t)|\mathbf{X}_n = \mathbf{x}, T_n = t).$$

As we have indicated, when based on a single realization of $(\mathbf{X}_n, T_n, R_n)_{n\geq 1}$ the study of the treatment effect amounts to the comparison of the conditional laws

$$\mathcal{L}(\rho(V_n, \mathbf{x}, t)|\mathbf{X}_n = \mathbf{x}, T_n = t) \quad \text{and} \quad \mathcal{L}(\rho(V_n, \mathbf{x}, t')|\mathbf{X}_n = \mathbf{x}, T_n = t') \tag{2.5}$$

for $t \neq t'$ and varying $\mathbf{x}$. By the conditional independence of U_n and V_n,

$$\begin{aligned} \mathcal{L}(R_n|\mathbf{X}_n = \mathbf{x}, T_n = t) &\equiv \mathcal{L}(\rho(V_n, \mathbf{X}_n, T_n)|\mathbf{X}_n = \mathbf{x}, T_n = t) \\ &= \mathcal{L}(\rho(V_n, \mathbf{x}, t)|\mathbf{X}_n = \mathbf{x}, \tau(U_n, \mathbf{x}) = t) \\ &= \mathcal{L}(\rho(V_n, \mathbf{x}, t)|\mathbf{X}_n = \mathbf{x}) \\ &= \mathcal{L}(\rho(V_n, \mathbf{x}, t)), \end{aligned} \tag{2.6}$$

so any difference between the two laws of (2.5) is purely a result of the difference between taking t and t' in the last argument of the response function, and it follows that the function $t \to \mathcal{L}(\rho(V_n, \mathbf{X}_n, t)|\mathbf{X}_n = \mathbf{x})$—which then provides a complete description of the causal effect of the treatment on the response under a given situation—can be estimated from observed data in the guise of $\mathcal{L}(R_n|\mathbf{X}_n = \mathbf{x}, T_n = t)$.

Let us emphasize that without unconfoundedness the first law in (2.6) normally differs from the last one; one may very well be able to estimate the former from data, but the estimate will not pertain to the latter, which represents the causal effect of the treatment. Thus it is significant that the model of Example 2.1 is a linear regression model only when unconfoundedness holds, and that the least-squares estimator of β_0 is inconsistent if unconfoundedness fails, as illustrated by Prob. 2.9.

Thanks to (2.6), one can estimate from data such things as

$$\mathbf{P}(\rho(V_n, \mathbf{X}_n, t) \leq r) = \sum_{\mathbf{x}} \mathbf{P}(\rho(V_n, \mathbf{x}, t) \leq r|\mathbf{X}_n = \mathbf{x})\mathbf{P}(\mathbf{X}_n = \mathbf{x}), \tag{2.7}$$

the distribution function of potential outcomes when ρ is scalar, and

$$\sum_{\mathbf{x}} \{\mathbf{E}(\rho(V_n, \mathbf{x}, t)|\mathbf{X}_n = \mathbf{x}) - \mathbf{E}(\rho(V_n, \mathbf{x}, t')|\mathbf{X}_n = \mathbf{x})\} \, \mathbf{P}(\mathbf{X}_n = \mathbf{x}), \tag{2.8}$$

---

[2] This is generally true if the $\mathbf{X}_n$ and T_n are discrete, which we assume for simplicity in much of the book; if the $\mathbf{X}_n$ are not discrete, the indicators $\mathbf{1}_{\{\mathbf{X}_n=\mathbf{x}, T_n=t\}}$ in (2.3) are replaced by $\mathbf{1}_{\{\mathbf{X}_n \in B_N(\mathbf{x}), T_n=t\}}$, where $B_N(\mathbf{x})$ is a neighbourhood shrinking to $\mathbf{x}$ as $N \to \infty$.

the overall mean difference between the responses to treatments t and t', which is $\mathbf{E}\big[\rho(V_n, \mathbf{X}_n, t) - \rho(V_n, \mathbf{X}_n, t')\big]$—the expected difference between the potential outcomes to t and t' of *the same* arbitrary individual, and a general measure of treatment effect. Observe that when the T_n are discrete the possibility of estimating the last law in (2.6) for each t presupposes $\mathbf{P}(T_n = t|\mathbf{X}_n = \mathbf{x}) > 0$, and the possibility of estimating overall mean differences presupposes $\mathbf{P}(T_n = t|\mathbf{X}_n = \mathbf{x}) < 1$ for all t and some $\mathbf{x}$; cf. Remark 2.2.

Remark 2.6 The law of $\rho(V_n, \mathbf{X}_n, t)$ may be of interest conditionally as well as unconditionally on $\mathbf{X}_n$: the conditional law provides complete information about how the response varies in direction and magnitude by varying the treatment and the situation, but the unconditional law, which is involved in (2.7)–(2.8), can be useful for making general statements. Evidently, if $\mathcal{L}\big(\rho(V_n, \mathbf{X}_n, t)\big)$, the 'unconditional law', is not constant in t for some n then treatment has an effect, but the converse is not true: see the end of Example 2.4 later on. □

Remark 2.7 The probability in (2.7) corresponds to Pearl's 'adjustment formula' (e.g. Sect. 3.2 of Pearl, Glymour and Jewel [73]); when treatment is binary and $t = 1$, $t' = 0$, the integrated difference (2.8) between conditional expectations corresponds to Rubin's 'super population average treatment effect' (e.g. Sect. 6.7 of Imbens and Rubin [46]). For simplicity, when presenting expressions such as these we will often assume that the $\mathbf{X}_n$ and T_n are discrete, but it will be easy to guess from them in each case what the appropriate expression is when the $\mathbf{X}_n$ take values in $\mathbb{R}^d$. □

We have seen that under the basic model with unconfoundedness 'correcting for' $\mathbf{X}_n$ makes it possible to study the effect of T_n on R_n. Ignoring $\mathbf{X}_n$, on the other hand, or some of its components, may bring confounding back and remove that possibility. For if $\mathbf{X}_n = (\mathbf{X}'_n, \mathbf{X}''_n)$, say, ignoring $\mathbf{X}'_n$ means that this vector becomes invisible in (2.1), which then is equivalent to

$$\begin{aligned} R_n &= \rho(V_n, \varphi(W_n, \mathbf{X}''_n), \mathbf{X}''_n, T_n) \equiv \tilde{\rho}(\tilde{V}_n, \mathbf{X}''_n, T_n), \\ T_n &= \tau(U_n, \varphi(W_n, \mathbf{X}''_n), \mathbf{X}''_n) \equiv \tilde{\tau}(\tilde{U}_n, \mathbf{X}''_n), \end{aligned} \tag{2.9}$$

where φ is some function yielding $\mathbf{X}'_n = \varphi(W_n, \mathbf{X}''_n)$ unbeknownst to us, U_n, V_n and W_n are independent standard uniforms, the standard uniform $\tilde{U}_n$ is obtained by alternating the digits in the decimal expansions of U_n and W_n, and the standard uniform $\tilde{V}_n$ by alternating the digits of V_n and W_n; and since this is a special case of the basic model *without* unconfoundedness, owing to the dependence between $\tilde{U}_n$ and $\tilde{V}_n$, the effect of the treatment on the response is *not* necessarily characterized by

$$\mathcal{L}(R_n|\mathbf{X}''_n = \mathbf{x}'', T_n = t) = \mathcal{L}\big(\tilde{\rho}(\tilde{V}_n, \mathbf{x}'', t)\big|\mathbf{X}''_n = \mathbf{x}'', \tilde{\tau}(\tilde{U}_n, \mathbf{x}'') = t\big),$$

because the presence of t in the conditioning event induces an effect of the treatment also through the first argument of the response function.

Thus, correcting for anything less than $\mathbf{X}_n$ in the basic model may lead to bias in the study of the treatment effect, and it is in that sense that $\mathbf{X}_n$ is a set of confounders.

2.2 Examples

Two examples will help us summarize the ideas considered so far. In each case, we examine a particular version of the basic model with unconfoundedness and show that failing to condition on the background variables creates confounding and gives a distorted view of the effect of the treatment on the response. Confounding is illustrated further in Probs. 2.9–2.10.

Example 2.4 **Simpson's paradox**
Patients may or may not take a new medicine in order to recover from an illness, and the indicator R_n of recovery of the n-th patient satisfies

$$R_n = B^{-1}(V_n, p(T_n, \mathbf{X}_n)), \quad T_n = B^{-1}(U_n, q(\mathbf{X}_n)), \tag{2.10}$$

where B is the Bernoulli distribution function (with 'success rate' indicated in the second argument), B^{-1} its 'inverse', p and q are functions with values in $[0, 1]$, T_n is the indicator of treatment ($T_n = 1$ if the patient takes the medicine), $\mathbf{X}_n$ is the patient's sex (say $\mathbf{X}_n = 1$ if the patient is a man), and U_n, V_n are independent standard uniforms independent of $\mathbf{X}_n$.[3] It follows from these definitions that model (2.1) holds with unconfoundedness and that

$$\mathbf{P}(R_n = 1\,|T_n = t, \mathbf{X}_n = x) = p(t, x), \quad \mathbf{P}(T_n = 1\,|\mathbf{X}_n = x) = q(x) \tag{2.11}$$

($t, x \in \{0, 1\}$). Suppose that

$$p(1, x) > p(0, x) \tag{2.12}$$

for $x = 0, 1$, i.e. that the medicine makes recovery more likely for both men and women; is it possible that

$$\mathbf{P}(R_n = 1\,|T_n = 1) \le \mathbf{P}(R_n = 1\,|T_n = 0), \tag{2.13}$$

i.e. that a patient randomly drawn from the population of patients governed by (2.10) is more likely to have recovered having *not* taken the medicine? It is possible—and that is the 'paradox'.

Indeed,

$$\mathbf{P}(R_n = 1\,|T_n = 1) = \sum_{x=0,1} \frac{\mathbf{P}(R_n=1, T_n=1, \mathbf{X}_n=x)}{\mathbf{P}(T_n=1)} = \sum_{x=0,1} \frac{p(1,x)q(x)\mathbf{P}(\mathbf{X}_n=x)}{\mathbf{P}(T_n=1)},$$

$$\mathbf{P}(R_n = 1\,|T_n = 0) = \sum_{x=0,1} \frac{p(0,x)\{1-q(x)\}\mathbf{P}(\mathbf{X}_n=x)}{\mathbf{P}(T_n=0)},$$

and we can show that $p(t, x)$ and $q(x)$ may be chosen so as to yield (2.13).

---

[3] Alternatively, $B^{-1}(w, \theta) = \mathbf{1}_{[0,p]}(w)$ ($0 \le p \le 1$). Evidently, in a real-life problem one does not need to assume or commit to a particular definition of the random variables which serve as a model for the empirical data (though *some* model must be assumed, even if only implicitly, since except in trivial situations empirical data alone are meaningless for purposes of inference). Any definition which is consistent with the order in which the empirical events are supposed to occur and with the postulated class of distributions may be admitted and will furnish the desired 'causal skeleton', as illustrated by the interpretation of confounding to be given shortly in the present example.

Suppose that

$$p(1,1) > p(0,1) > p(1,0) > p(0,0), \tag{2.14}$$

so that, although the medicine favours recovery in both sexes, recovery is more likely in *untreated* men than in *treated* women. The middle inequality implies $p(0,1) - p(1,0) > 0$, and it is obvious that

$$p(1,1) - p(0,1) > -\{p(1,0) - p(0,0)\}.$$

The second of these last two inequalities is equivalent to

$$p(1,1) - p(0,0) > p(0,1) - p(1,0),$$

which by the first can be written as

$$\theta := \frac{p(1,1) - p(0,0)}{p(0,1) - p(1,0)} > 1. \tag{2.15}$$

Suppose too that $q(0) = \beta$ and $q(1) = 1 - \beta$ for some $\beta \in]0,1[$ such that

$$\frac{\beta}{1-\beta} \geq \theta; \tag{2.16}$$

since $\theta > 1$, this implies $\beta > 1/2$, so *women are more likely to take the medicine than men are*.

Finally, suppose $\mathbf{P}(T_n = 1) = 1/2$: this would be the case if, for example, a *target population* were formed by recruiting the same number of treated and untreated individuals from a *population of interest* and if individuals were then randomly sampled from that target population. (More generally, one may take $q(0) = \beta_0$ and $q(1) = \beta_1$ for certain $\beta_0 > \beta_1$ and allow certain values of $\mathbf{P}(T_n = 1) \neq 1/2$, but this makes the calculations more complicated.)

Since by the choice of q

$$\begin{aligned}\mathbf{P}(T_n = 1) &= \sum\nolimits_{x=0,1} \mathbf{P}(T_n = 1\,|\mathbf{X}_n = x)\,\mathbf{P}(\mathbf{X}_n = x) = \sum\nolimits_{x=0,1} q(x)\mathbf{P}(\mathbf{X}_n = x)\\ &= \beta\,\mathbf{P}(\mathbf{X}_n = 0) + (1-\beta)\,\{1 - \mathbf{P}(\mathbf{X}_n = 0)\},\end{aligned}$$

the last assumption entails (by $\beta > 1/2$) $\mathbf{P}(\mathbf{X}_n = 0) = 1/2$. But then $\mathbf{P}(\mathbf{X}_n = x) = \mathbf{P}(\mathbf{T}_n = t)$ for all t, x and

$$\begin{gathered}\mathbf{P}(R_n = 1\,|T_n = 1) - \mathbf{P}(R_n = 1\,|T_n = 0) =\\ \sum\nolimits_{x=0,1} p(1,x)q(x) - p(0,x)\{1 - q(x)\} =\\ [p(1,1) - p(0,0)](1-\beta) - [p(0,1) - p(1,0)]\beta,\end{gathered}$$

which yields (2.13) because by (2.15) and (2.16)

$$[p(1,1) - p(0,0)] \leq [p(0,1) - p(1,0)]\frac{\beta}{1-\beta}.$$

What is happening? From (2.10) it is clear that the treatment has a positive effect on the response, because by (2.12) $p(t,x)$ increases with t for fixed x and $B^{-1}(v,p)$ increases with p for fixed v, so that $\rho(v,x,t) := B^{-1}(v, p(t,x))$ increases with t for fixed (v,x); and treatment has an effect despite the T_n in $\rho(V_n, \mathbf{X}_n, T_n)$ not being 'manipulable' but rather a random function of $(\mathbf{X}_n, U_n)$. Nevertheless, if we consider treated patients, i.e. condition on $\{T_n = 1\}$, the equation $T_n = B^{-1}(U_n; q(\mathbf{X}_n))$ constrains the patient to be a woman—forces $\mathbf{X}_n$ to be 0—more frequently than not, because $q(0) > q(1)$. Consequently, more frequently than not we have

$$p(T_n, \mathbf{X}_n) = p(1,0) \quad \text{and} \quad R_n = B^{-1}(V_n; p(1,0)),$$

so more frequently than not the recovery rate of a treated patient is the *third smallest* in the ranking determined by (2.14). It follows that when $q(0) \gg q(1)$, the chance of the treated patient being a man is small and the recovery rate will be *little more than* $p(1,0)$. Conversely, if we consider untreated patients then more frequently than not $\mathbf{X}_n = 1$,

$$p(T_n, \mathbf{X}_n) = p(0,1) \quad \text{and} \quad R_n = B^{-1}(V_n; p(0,1)),$$

more frequently than not the recovery rate of the untreated is the *third largest* in the ranking determined by (2.14), and when $q(0) \gg q(1)$ the probability of the untreated being male is high and the recovery rate will be *little less than* $p(0,1)$.

This confounding of the treatment effect by sex, which manages to counterfeit a *negative effect*, can be expressed more concisely: Although $p(t,x) \uparrow$ in t for fixed x, we also have $p(t,x) \uparrow$ in x for fixed t and $q(x) \downarrow$ in x, so if we look for a 'large' T_n in the hope of finding a 'large' $R_n = B^{-1}(V_n; p(T_n, \mathbf{X}_n))$ the constraint $T_n = B^{-1}(U_n; q(\mathbf{X}_n))$ gives us a 'small' $\mathbf{X}_n$, which in turn yields a 'small' R_n.

Confounding arises because if we ignore sex then $\mathbf{X}_n$ becomes invisible in (2.10), which then is equivalent to a special case of (2.9), namely

$$R_n = \tilde{\rho}(\tilde{V}_n, T_n), \quad T_n = \tilde{\tau}(\tilde{U}_n), \tag{2.17}$$

where $\tilde{U}_n$ and $\tilde{V}_n$ are dependent, so that the effect of the treatment on the response is *not* characterized by $\mathcal{L}(R_n | T_n = t)$, as basing one's conclusion on (2.13) implies. On the other hand, if we know that (2.10) holds for some known or unknown p and q—i.e. if we know that (2.1) holds with unconfoundedness—then we know that we can 'correct for' sex in order to estimate the causal effect of treatment on the response, since this is entirely embodied by the first equation in (2.11).

Incidentally, with $\mathbf{P}(\mathbf{X}_n = x) = 1/2$ and under (2.12) the overall treatment effect,

$$\begin{aligned}\mathbf{P}\left(B^{-1}(V_n, p(t, \mathbf{X}_n)) = 1\right) &= \sum_x \mathbf{P}(R_n = 1 \,|T_n = t, \mathbf{X}_n = x)\, \mathbf{P}(\mathbf{X}_n = x) \\ &= \frac{p(t,0) + p(t,1)}{2},\end{aligned}$$

is an increasing function of t; evidently (cf. Remark 2.6), this need not be the case if the effect of the treatment is negative for one sex and positive for the other.

Simpson's paradox can be sensed in real-life examples, though the explanation these afford is necessarily less clear-cut because there are usually more confounders than those represented in the data; for an example see Rice [79], pp. 422-3. □

Example 2.5 **A continuous version of Simpson's paradox**
The logarithm of the body mass index of an individual randomly drawn from a certain population, R_n, is related to a standardized measure of the intensity of his physical training, T_n, and to the logarithm of his age, $\mathbf{X}_n$, by

$$\begin{aligned} R_n &= \rho(V_n, \mathbf{X}_n, T_n) := \alpha \mathbf{X}_n - \beta T_n + \Phi^{-1}\left(V_n, 0, \sigma_1^2\right), \\ T_n &= \tau(U_n, \mathbf{X}_n) := \mu + \tfrac{\sigma_2}{\sigma_3}\left(\mathbf{X}_n - \gamma\mu\right) + \Phi^{-1}\left(U_n, 0, \sigma_2^2\right), \\ \mathbf{X}_n &= \Phi^{-1}\left(\xi_n, \gamma\mu, \sigma_3^2\right), \end{aligned} \tag{2.18}$$

where $\alpha, \beta, \gamma, \mu, \sigma_1, \sigma_2, \sigma_3 > 0$, $\Phi(\cdot, m, s^2)$ is the normal distribution function with mean m and variance s^2, $\Phi^{-1}(\cdot, m, s^2)$ its inverse, and ξ_n, U_n and V_n are independent standard uniform random variables.

These specifications—which constitute another example of the basic model with unconfoundedness—imply that $\mathbf{X}_n$, T_n and R_n are jointly normally distributed with

$$\begin{gathered} \mathbf{E}\,\mathbf{X}_n = \gamma\mu, \quad \mathbf{E}\,T_n = \mu, \quad \mathbf{E}\,R_n = (\alpha\gamma - \beta)\mu, \\ \operatorname{Var}(\mathbf{X}_n) = \sigma_3^2, \quad \operatorname{Var}(T_n) = 2\sigma_2^2, \\ \operatorname{Cov}(\mathbf{X}_n, T_n) = \sigma_2\sigma_3, \quad \operatorname{Corr}(\mathbf{X}_n, T_n) = \tfrac{\sqrt{2}}{2}, \\ \operatorname{Var}(R_n) = (\alpha\sigma_3 - \beta\sigma_2)^2 + \sigma_1^2 + \beta^2\sigma_2^2, \\ \mathcal{L}(R_n | \mathbf{X}_n = x) = \mathcal{N}\left(\mu\beta\left\{\gamma\tfrac{\sigma_2}{\sigma_3} - 1\right\} + \left\{\alpha - \beta\tfrac{\sigma_2}{\sigma_3}\right\}x, \sigma_1^2 + \beta^2\sigma_2^2\right), \end{gathered} \tag{2.19}$$

etc., and that the law of R_n conditional on $(\mathbf{X}_n, T_n) = (x, t)$ is

$$\mathcal{L}(R_n | T_n = t, \mathbf{X}_n = x) = \mathcal{N}(\alpha x - \beta t, \sigma_1^2), \tag{2.20}$$

i.e. normal with mean $\alpha x - \beta t$ and variance σ_1^2 (Prob. 2.4).[4]

It is clear from the first equation in (2.18) that (2.20) characterizes completely the joint effect of age and physical training on body mass, and that at any given age body mass tends to decrease with physical training. Thus, physical training—which may be thought of as potentially manipulable in reality, even though according to (2.18), the present model of reality, it is not—tends to reduce body mass at all ages. Nevertheless, because by the second equation in (2.18) physical training tends to increase with age in the population in question, it can be seen that, if α is sufficiently bigger than β, body mass will appear to increase rather than decrease with physical training when regarded as a function of this variable alone—as when data on (T_n, R_n) are used to 'regress' body mass on training without 'correcting for' age.

[4] By $\mathcal{N}(m, s^2)$ we mean the normal law of mean m and variance s^2.

Similarly to what happens in the previous example, confounding arises by ignoring age because doing so amounts to hiding $\mathbf{X}_n$ in (2.18), which then is equivalent to a model such as (2.17), for which unconfoundedness fails; unconfoundedness holds only if age is accounted for.

In order to exhibit the 'paradox' explicitly we need to find $\mathbf{P}(R_n > r|T_n = t)$ and show that for some choices of α and β this is an increasing function of t for each r. To this end we first note (see Prob. 2.5) that $(\mathbf{X}_n, T_n, R_n)$ retains its distribution if the last two equations in (2.18) are replaced by[5]

$$\mathbf{X}_n = \gamma\mu + \frac{\sigma_3}{2\sigma_2}(T_n - \mu) + \Phi^{-1}\left(U_n, 0, \frac{\sigma_3^2}{2}\right) \quad \text{and} \quad T_n = \Phi^{-1}\left(\xi_n, \mu, 2\sigma_2^2\right). \tag{2.21}$$

Then

$$\mathcal{L}(R_n|T_n = t) = \mathcal{L}\left(\alpha\mathbf{X}_n - \beta t + \Phi^{-1}\left(V_n, 0, \sigma_1^2\right)|T_n = t\right) =$$

$$\mathcal{L}\left(\alpha\left\{\gamma\mu + \tfrac{\sigma_3}{2\sigma_2}(t-\mu) + \Phi^{-1}\left(U_n, 0, \tfrac{\sigma_3^2}{2}\right)\right\} - \beta t + \Phi^{-1}(V_n, 0, \sigma_1^2)\right) =$$

$$\mathcal{N}\left(\mu\,\alpha\left\{\gamma - \tfrac{\sigma_3}{2\sigma_2}\right\} + \left\{\alpha\tfrac{\sigma_3}{2\sigma_2} - \beta\right\}t, \sigma_1^2 + \tfrac{\alpha^2\sigma_3^2}{2}\right),$$

so the paradox arises if and only if $\frac{\alpha}{\beta} > 2\frac{\sigma_2}{\sigma_3}$.

Of course, (2.18) is rather unrealistic as a model for body mass as a function of physical training and age; for instance, diet influences and is influenced by body mass and physical training, so even if its functional expressions were correct the V_n of the first equation would depend on the U_n of the second through diet. □

2.3 Questions of interpretation: Lord's paradox

Let us look again at the intervention model and reconsider its relation with the basic model. It is evident that unconfoundedness plays no role in the intervention model proper, since $\mathcal{L}(\rho(V_n, \mathbf{X}_n, t)|\mathbf{X}_n = \mathbf{x}) = \mathcal{L}(\rho(V_n, \mathbf{x}, t))$ can be estimated—hence the treatment effect investigated—from data following (2.2). In an observational study the data are assumed to follow (2.1), not (2.2); yet, under unconfoundedness, the estimates of $\mathcal{L}(R_n|\mathbf{X}_n = \mathbf{x}, T_n = t)$ obtained from observational data are also estimates of $\mathcal{L}(\rho(V_n, \mathbf{x}, t))$, these two laws being one and the same by (2.6). Thus, data following the 'observational model' under unconfoundedness provide us with estimates of the intervention model, hence with the means of predicting the *results of interventions*, i.e. predicting the consequences of forcing a particular treatment t upon an individual under a given situation, as opposed to letting the individual choose the treatment T_n.

[5] It is important to realize that we are not abandoning (2.18) as a model in favour of (2.21); the latter representation is just one 'photographic negative' of the *distribution* of $(\mathbf{X}_n, T_n, R_n)$ and is used merely as a convenient method of finding a conditional distribution (incidentally, a conditional distribution has to do with known information rather than with an antecedent).

In a given real-life problem one must have a clear idea about the reality that each of the two models is supposed to represent. While the basic model corresponds to the passive observation of situations, of the determination of the treatments and of the responses that follow (as if an observer locked out of 'the system' were able to witness some of the consequences of the draw ω 'causing everything' but unable to interfere with it), the intervention model corresponds to the forcing of fixed treatments upon situations (experimental or observational) and the observation of the ensuing results (as if the observer were able to prevent ω from dictating the treatments, and possibly the situations, but not from dictating the responses). Accordingly, a statistical analysis founded on the first model normally refers to *what happened*—to the influence that the treatment had in the past—whereas an analysis founded on the second normally refers to *what may happen*—to the possible influence that the treatment may have in the future. Yet, in a real-life problem in which both points of view are possible—say a study of a population whose individuals are free to choose their treatment at one stage but at a second stage may be assigned a particular treatment—the models have a legitimate parallel existence: even if it is clear that in the past the treatment has had a certain effect in a certain direction, it may or may not follow that a reversal of treatment at the time would have been possible *in reality*, just as it may or may not follow that a prescription of treatment in the future will *in reality* have the desired effect.

Our next example illustrates these points.

Example 2.6 **Lord's paradox**

As part of a programme to encourage his employees to live more healthily, at the beginning of the year the director of a company inaugurates a new canteen offering a rich variety of healthy dishes (salads, vegetable soups, grilled fish, etc.). Although the old canteen, which already had a substantial selection of healthy dishes on offer, undergoes some refurbishment and keeps its old faithful selection of unhealthy dishes (fat smoked sausages, fried meat and fish, chips, etc.), it is hoped that, helped by New Year resolutions, the gastronomical, social and architectural attractions of the new canteen will gradually win over most of the employees and eventually render the old one superfluous.

For logistic reasons, after a trial-period of two weeks employees are encouraged to opt for one of the two canteens during the rest of the year.

In order to investigate the possible beneficial effects of the opening of the new canteen, measurements of body mass index are taken on a large random sample of employees at the beginning of the year. At the end of the year body mass is measured again on the same employees, who are also asked which of the two canteens had their preference during the year.

The company's statistician called in to analyze the data happens to know that the logarithm of body mass index of the n-th employee at the end of the year, R_n, the logarithm of his body mass at the beginning of the year, $\mathbf{X}_n$, and the canteen preferred by him, T_n (equal to 1 if it is the new one, equal to 2 otherwise), obey the following model rather closely:

$$R_n = \mu_{T_n} + \varrho(\mathbf{X}_n - \mu_{T_n}) + \Phi^{-1}\left(V_n, 0, (1-\varrho^2)\sigma^2\right),$$

$$T_n = 1 + B^{-1}\left(U_n, q(\mathbf{X}_n)\right), \tag{2.22}$$

$$\mathbf{X}_n = B^{-1}(\xi_{1,n}, p)\Phi^{-1}(\xi_{2,n}, \mu_1, \sigma^2) + \left[1 - B^{-1}(\xi_{1,n}, p)\right] \Phi^{-1}(\xi_{2,n}, \mu_2, \sigma^2),$$

where

$$q(x) = \frac{(1-p)\phi(x, \mu_2, \sigma^2)}{p\,\phi(x, \mu_1, \sigma^2) + (1-p)\phi(x, \mu_2, \sigma^2)},$$

$\sigma > 0$, $p, \varrho \in]0, 1[$, μ_1 and μ_2 are parameters, $\mu_1 < \mu_2$, $\phi(\cdot, m, s^2)$ is the normal density function of mean m and variance s^2, $\xi_{1,n}$, $\xi_{2,n}$, U_n and V_n are independent standard uniform random variables, and B and Φ are as in examples 2.4 and 2.5 respectively. From (2.22) one can conclude that $\mathbf{X}_n$ and T_n are positively correlated: employees with a large body mass index at the beginning of the year are more likely to opt for the old canteen. But the model provides more, namely information about the relations of cause and effect between the variables.

After examining the data, the statistician realizes that the distribution of body mass does not seem to have changed among those who frequented the old canteen nor among those who frequented the new canteen, and, moreover, that the distribution of the *gain* in the logarithms of body mass, $R_n - \mathbf{X}_n$, seems to be the same in the two groups and to have mean 0. These observations correspond to the identities

$$\mathcal{L}(R_n | T_n = t) = \mathcal{N}(\mu_t, \sigma^2) = \mathcal{L}(\mathbf{X}_n | T_n = t), \quad \mathcal{L}(R_n) = \mathcal{L}(\mathbf{X}_n), \tag{2.23}$$

$$\mathbf{E}R_n = p\mu_1 + (1-p)\mu_2 = \mathbf{E}\mathbf{X}_n,$$

$$\mathrm{Var}R_n = \sigma^2 + p(1-p)(\mu_1 - \mu_2)^2 = \mathrm{Var}\mathbf{X}_n,$$

and

$$\mathcal{L}(R_n - \mathbf{X}_n | T_n = t) = \mathcal{L}(R_n - \mathbf{X}_n) = \mathcal{N}\left(0, 2(1-\varrho)\sigma^2\right), \tag{2.24}$$

which follow from (2.22) (see Prob. 2.6), and they suggest that the opening of the new canteen has had no effect in any of the two groups. On hearing the statistician present this conclusion, the director is sceptical: after all, he certainly lost some weight thanks to the healthy diet of the new canteen, and one of his assistants who had about the same height and weight as he at the beginning of the year and kept to the old canteen even grew slightly fatter. So he decides to consult a second statistician. After discussing things with the first statistician and accepting that model (2.22) provides a correct description of reality, the second statistician agrees that it is almost certainly true that the distribution of body mass in the whole population of employees has not changed throughout the year, neither among healthy eaters nor among unhealthy eaters, but observes that

$$\begin{aligned}\mathcal{L}(R_n | T_n = t, \mathbf{X}_n = x) &= \mathcal{L}\left(\mu_t + \varrho(x - \mu_t) + \Phi^{-1}\left(V_n, 0, (1-\varrho^2)\sigma^2\right)\right) \\ &= \mathcal{N}\left(\varrho x + (1-\varrho)\mu_t, (1-\varrho^2)\sigma^2\right),\end{aligned} \tag{2.25}$$

and that (based on the parameter estimates) it is almost certain that $\mu_2 > \mu_1$. This, he adds, is a strong indication that if a randomly sampled employee has a logarithm of body mass of x and opts for the healthy diet then after a year he will tend to have a logarithm of body mass of about $x - (1 - \varrho)(x - \mu_1)$, whereas if he keeps to the old canteen his body mass will be around $x - (1 - \varrho)(x - \mu_2)$; since the difference between these,

$$\mathbf{E}(R_n | T_n = 1, \mathbf{X}_n = x) - \mathbf{E}(R_n | T_n = 2, \mathbf{X}_n = x) = (1 - \varrho)(\mu_1 - \mu_2), \tag{2.26}$$

is almost certainly negative, there is, concludes the second statistician, a strong indication that adopting the healthy diet tends to lower one's body mass.

The director is relieved, but after a moment's thought directs a puzzled look at the first statistician. Well, says the first statistician, it is evident that no one has kept exactly his initial body mass, that some healthy eaters lost weight and some unhealthy eaters put on weight, but it is also true that some healthy eaters put on weight and some unhealthy eaters lost weight. Clearly, he goes on, it is almost certain that the choice of canteen has an effect on body mass, because *we know* that if employee n started with a logarithm of body mass x and opted for canteen t then his logarithm of body mass at the end of the year is given by

$$\rho(V_n, x, t) := (1 - \varrho)\mu_t + \varrho x + \Phi^{-1}\left(V_n, 0, (1 - \varrho^2)\sigma^2\right),$$

and that this random variable, which depends on t and x only in the first and second terms, almost certainly increases with t for fixed x; so, everything else (namely the V_n) being statistically equal of two employees with the same body mass, the one opting for the healthy diet will, a year later, tend to have a lower body mass than the one keeping to the unhealthy diet—which is what (2.26) indicates. It follows that if we had been able to foist $t = 1$ into the 'response function' of every employee without affecting its other aspects (embodied by V_n) we should have noticed a decrease in the body mass of the population. That this is not the case is explained by each individual having been left to choose the canteen at will; actually, he says pointing to (2.23), the equality of the distributions of the initial and final body mass (both the same mixture of normal distributions) and the correlation between body mass and canteen can be explained by most people having kept to the diet they followed before the opening of the new canteen, even if they changed restaurant—after all, the old canteen already served healthy dishes in the past, and it is evident from the knowledge contained in model (2.22) that the healthy eaters today are, for the most part, the healthy eaters of the past.

The second statistician and the director agree with this explanation and the latter realizes that a general reduction of body mass among his employees requires more drastic measures. Later, reflecting on the problem again, the statisticians conclude that what they have witnessed was an empirical equivalent of many realizations of the first two steps of a stationary sequence: although each two-step realization of such a sequence appears as two unique and to some extent erratic numerical observations, statistically the two sets of observations follow the same law. □

2.4 Unconfoundedness in real-life problems

As far as validity is concerned, there is a very substantial difference between using the basic model with unconfoundedness in a real-life observational study and using it or the intervention model in a real-life experimental study (questions of validity of the intervention model in an observational study have just been discussed and had been mentioned earlier in Remark 2.3). In fact, unconfoundedness makes it possible to treat observational data as if they were experimental (accordingly, the models and methods studied in this book serve both types of data), so as an assumption it must be quite strong. Although methods of data analysis, sketched in Sect. 5.5, can sometimes detect violations of unconfoundedness, one can seldom be quite certain about the validity of inferences on $\mathcal{L}(\rho(V_n, \mathbf{x}, t))$ based on estimates of $\mathcal{L}(R_n|\mathbf{X}_n = \mathbf{x}, T_n = t))$, so if unconfoundedness is assumed in a real-life, non-experimental problem then it has to be justified for the most part by means of evidence derived from the relevant fields of knowledge rather than from the data to which the basic model is being applied. That this is quite difficult to do, however, may be judged by simple, 'homemade' experiments in which unconfoundedness depends on one's ability of controlling the background conditions at different applications of the treatment. In the typically more difficult observational studies carried out in medicine, biology, sociology and economics, it has to be more difficult to argue that the factors affecting the determination of the treatment (the U_n and $\mathbf{X}_n$) and those affecting the response (the V_n and $\mathbf{X}_n$) have only the factors specified by the situation (the $\mathbf{X}_n$) in common. It is not surprising, therefore, that according to Freedman, who wrote so many lucid reviews of such studies (see Freedman [36, 37], the number of successful attempts at establishing causality in observational studies is "at best quite limited", and that the much-needed justifications for the assumptions of causal models are rare and generally unconvincing.[6]

Nevertheless, some wisdom has been gained about ways of planning studies, of selecting subpopulations, of selecting potentially confounding variables, of forming groups of units (strata or 'matched sets') within which treatments can be compared with relatively little bias, and such wisdom can help to improve a study and to assess its credibility; see Rosenbaum's book [83], in particular Sect. 3.13 and Chap. 8, for an overview.

According to some authors (e.g. pp. 708–9 of Pearl [66], pp. 129–130 of Pearl [68], or p. 5 of Pearl [69]), justifying unconfoundedness requires the laying bare of the relations of cause and effect between the confounders and the treatment and response variables, namely by proposing a detailed model involving all the variables thought to be relevant to the real-life problem in hand and founded principally on subject-matter (as opposed to statistical) knowledge—a model generalizing the basic model and whose correctness can be checked, in part, with data. It is to such models and to the questions they pose that we shall turn in the next chapter.

---

[6] Incidentally, the assumption of unconfoundedness as formulated here ought to make clear what needs to be checked or argued for in a given application; as suggested by Pearl (e.g. p. 129 of Pearl [68], p. 126 of Pearl, Glymour and Jewel [73]), Rubin's formulation of unconfoundedness, explained in Remark 2.5, appears to be less helpful in this respect.

2.5 Further remarks on the basic model

We have stated earlier that even if in (2.1) the U_n and V_n were to have distributions other than the standard uniform conditionally on $\mathbf{X}_n$ then we could always find an equivalent version of the corresponding model which would satisfy the conditions of the basic model. To explain what we mean by this, let us write

$$F(u;\mathbf{x}) = \mathbf{P}(U_n \le u|\mathbf{X}_n = \mathbf{x}), \quad G(v;\mathbf{x},u) = \mathbf{P}(V_n \le v|U_n = u, \mathbf{X}_n = \mathbf{x}),$$

and let ξ_n and ξ'_n be independent standard uniforms independent of $\mathbf{X}_n$. Then (Probs. 1.1 and 1.2) $\tilde{U}_n := F^{-1}(\xi_n;\mathbf{X}_n)$ and $\tilde{V}_{n,\tilde{U}_n} := G^{-1}(\xi'_n;\mathbf{X}_n,\tilde{U}_n)$ have the same joint distribution as U_n and V_n, and hence $\mathbf{X}_n$,

$$\tilde{R}_n = \rho\big(\tilde{V}_{n,\tilde{U}_n}, \mathbf{X}_n, \tilde{T}_n\big) \quad \text{and} \quad \tilde{T}_n = \tau(\tilde{U}_n, \mathbf{X}_n)$$

have the same joint distribution as the $\mathbf{X}_n$, T_n and R_n of (2.1) and their definition follows the required order. But $\tilde{R}_n$ and $\tilde{T}_n$ can also be written as

$$\tilde{R}_n = \tilde{\rho}\big(\xi''_n, \mathbf{X}_n, \tilde{T}_n\big) \quad \text{and} \quad \tilde{T}_n = \tilde{\tau}(\xi_n, \mathbf{X}_n)$$

where ξ''_n is the standard uniform random variable obtained by alternating the digits in the decimal expansions of ξ_n and ξ'_n,

$$\tilde{\tau}(w,\mathbf{x}) = \tau\left(F^{-1}(w;\mathbf{x}),\mathbf{x}\right) \quad \text{and} \quad \tilde{\rho}(w'',\mathbf{x},t) = \rho\left(G^{-1}(w';\mathbf{x},F^{-1}(w;\mathbf{x})),\mathbf{x},t\right),$$

with $w'' \in]0,1[$ made up by alternating the digits in the decimal expansions of w and w'; alternatively, if U_n and V_n are independent conditionally on $\mathbf{X}_n$, so that $G(v;\mathbf{x},u)$ is constant in u, then w plays no role in the definition of $\tilde{\rho}$, w'' is simply w', ξ''_n is ξ'_n, and therefore ξ''_n and ξ_n are independent uniforms.

If $(U_1,V_1),(U_2,V_2),\ldots$ are independent and independent of $\mathbf{X}_1,\mathbf{X}_2,\ldots$, then by taking $(\xi_1,\xi'_1),(\xi_2,\xi'_2),\ldots$ independent and independent of $\mathbf{X}_1,\mathbf{X}_2,\ldots$ the sequence $(\mathbf{X}_1,\tilde{T}_1,\tilde{R}_1),(\mathbf{X}_2,\tilde{T}_2,\tilde{R}_2),\ldots$ thus obtained (which need not be independent unless $\mathbf{X}_1,\mathbf{X}_2,\ldots$ are independent) follows the basic model and its distribution is the same as that of the given sequence. This is true even if the distribution of (U_n,V_n) varies with n, since when we speak of the basic model we include the more general form of it referred to on p. 10, in which the response and treatment functions are indexed as the variables they help to generate.

If on the contrary $(U_1,V_1),(U_2,V_2),\ldots$ are dependent conditionally on $\mathbf{X}_1,\mathbf{X}_2,\ldots$ then $(\mathbf{X}_n,\tilde{T}_n,\tilde{R}_n)$ and $(\mathbf{X}_n,T_n,R_n)$ have the same distribution for each n but the two *sequences* $(\mathbf{X}_1,\tilde{T}_1,\tilde{R}_1),(\mathbf{X}_2,\tilde{T}_2,\tilde{R}_2),\ldots$ and $(\mathbf{X}_1,T_1,R_1),(\mathbf{X}_2,T_2,R_2),\ldots$ may have different distributions. In this case there is another construction which leads to the desired new sequence but whose description is quite complicated and will not be attempted here. Yet let us mention that it involves the definition of a sequence $(\tilde{U}_1,\tilde{V}_{1,\tilde{U}_1}),(\tilde{U}_2,\tilde{V}_{2,\tilde{U}_2}),\ldots$ by recursion, with $\tilde{U}_{n+1}$ involving $\mathbf{X}_1,\ldots,\mathbf{X}_{n+1}$ and a number of independent uniforms used to define $(\tilde{U}_i,\tilde{V}_{i,\tilde{U}_i})$ for $i \le n$, with $\tilde{V}_{n+1,\tilde{U}_{n+1}}$

involving those same variables and still another independent uniform used in $\tilde{U}_{n+1}$, and that when $\tilde{U}_{n+1}$ and $\tilde{V}_{n+1,\tilde{U}_{n+1}}$ are inserted in τ and ρ they can be written as functions of standard uniforms obtained by alternating the digits from the uniforms used earlier and combined with new functions τ_{n+1} and ρ_{n+1} to yield $\tilde{T}_{n+1}$ and $\tilde{R}_{n+1}$.

Admittedly, these considerations are rather theoretical; yet they explain why the basic model can be seen as a *blueprint* of infinitely many models of cause and effect involving 'situations', 'treatments' and 'responses': if accompanied by appropriate changes in ρ and τ, changes in the distributions of the random variables may yield vectors with the same distribution as the $(\mathbf{X}_n, T_n, R_n)$ but which differ very much from them as functions on the sample space Ω. This is perfectly natural and does not invalidate the use of the model to represent relations of cause and effect; for any blueprint we may adopt (and one must be adopted in any given real-life problem) will share the essential features of causality with the others, namely the occurrence of a situation, followed by the occurrence of the treatment as a result of that situation and of extraneous factors, and the response generated as a result of the situation, of the treatment, and of further extraneous factors.

Finally, we note that the possibility of U_n and V_n being dependent conditionally on $\mathbf{X}_n$ is made more explicit by the above representations of $\tilde{T}_n$ and $\tilde{R}_n$, where $\tilde{U}_n$ and $\tilde{V}_{n,\tilde{U}_n}$ are dependent if and only if $G^{-1}(\,\cdot\,; \mathbf{x}, u)$ is not constant in u.

2.6 Mathematical models of causality

Notions of causality certainly precede notions of probability, so it is natural to try to situate our statistical models of causality among mathematical models which for ages have been used to treat problems involving relations of cause and effect.[7]

When one says that something affects, is the cause of, or has an effect on something else, then one implies that it does so after a certain moment and in a certain situation which may help bring about the effect and which up to that moment may itself affect and be affected by the cause. This idea can be expressed by the equations

$$r = \rho(\mathbf{x}, t), \quad \tau(\mathbf{x}, t) = 0, \tag{2.27}$$

where $\mathbf{x}$ describes the situation prior to the moment when the effect comes into force, t stands for the cause—called *treatment* here on account of problems of cause and effect typically encountered in applications of statistics—r stands for the joint result of, or the *response* to, $\mathbf{x}$ and t, and τ and ρ are some functions defined on a set $\mathcal{A} \subset \mathcal{X} \times \mathcal{T}$ of 'admissible' pairs of situations in a set $\mathcal{X}$ and treatments in a set $\mathcal{T}$.

In general, τ and $\mathbf{x}$ are both vector-valued, so that the second equation may define t implicitly as a function of $\mathbf{x}$ and the coordinates of the latter as functions of its other coordinates and of t. Thus the second equation is a *constraint* representing

[7] This section is meant to serve as a sort of 'foundational justification' for the models studied in the book, but it may be skipped or read later because its material is hardly mentioned elsewhere.

the interaction between the admissible t and $\mathbf{x}$ prior to the moment when the effect comes into force.

For example, the second equation may consist of a system of linear constraints in numerical variables, such as those arising in optimization problems in operations research and economics; and then $\mathcal{A}$ may be a closed bounded set and $\rho(\mathbf{x}, t)$ may represent a cost, say, incurred by opting for a treatment t under a situation $\mathbf{x}$.

If $\mathbf{x}$ is vector-valued and depends on a variable, then its coordinates may include a real-valued function and some of its derivatives, which may themselves involve the variable t; thus the second equation in (2.27) may consist of a system of differential equations, such as those arising in problems of physics, in which case $\mathcal{X}$ will be some set of functions.

Whatever the version of the model and the problem which it is meant to represent, the causality in it amounts to the way the value of $\rho(\mathbf{x}, t)$ varies with varying t for each $\mathbf{x}$; and, without the probabilities and random draws, confounding amounts to a limited freedom of varying t owing to the constraints on the pairs $(\mathbf{x}, t)$ and to the limits imposed by the set $\mathcal{A}$ on those pairs. Depending on how much can be assumed about ρ, τ, $\mathcal{X}$, $\mathcal{T}$ and $\mathcal{A}$ and on how much is observable of the data t, $\mathbf{x}$, $\tau(\mathbf{x}, t)$ and $\rho(\mathbf{x}, t)$ in the particular real-life problem, one may determine and quantify more or less of the effect of the treatment on the response. Thus, being able to observe two pairs $(\mathbf{x}, t_1), (\mathbf{x}, t_2) \in \mathcal{A}$ and comparing the corresponding responses $\rho(\mathbf{x}, t_1)$ and $\rho(\mathbf{x}, t_2)$ enables us to study the treatment effect at least in one situation; and being able to assume that $\rho(\mathbf{x}_1, t) = \rho(\mathbf{x}_2, t)$ for some $\mathbf{x}_1 \neq \mathbf{x}_2$ and t in a certain range gives us the possibility of studying the treatment effect by comparing $\rho(\mathbf{x}_1, t_1)$ and $\rho(\mathbf{x}_2, t_2)$ for certain $\mathbf{x}_1$, $\mathbf{x}_2$ and $t_1 \neq t_2$. These comparisons are akin to those done in the procedures of matching and stratification, already alluded to and explained in Chap. 5.

Given a model like (2.27), we may always consider that once the treatment has come into force the situation $\mathbf{x}$ changes into a new situation $\mathbf{x}'$, partly as a result of the treatment and of the response, which we represent by $\mathbf{x}' = \xi(\mathbf{x}, t, r)$ for some function ξ, and the treatment and the response change into $t' = \tau'(\mathbf{x}', \mathbf{x}, t, r)$ and $r' = \rho'(\mathbf{x}', t', \mathbf{x}, t, r)$ by means of some functions τ' and ρ'. But substitution gives

$$r' = \rho'\left[(\xi(\mathbf{x}, t, r), \tau' (\xi(\mathbf{x}, t, r), \mathbf{x}, t, r), \mathbf{x}, t, \rho(\mathbf{x}, t)\right] =: \tilde{\rho}(\mathbf{x}, t), \qquad (2.28)$$

which is of the same form as the first equation in (2.27) except that the response is now thought of as being obtained at a later stage and need not even be of the same nature as the first one—and, evidently, the effect of t may be amplified through ρ', τ' and ξ, but it may also vanish. This observation illustrates the idea, emphasized for example in Cox and Wermuth [20], that a causal formulation of events based on equations such as (2.27) "does not imply a notion of ultimate causation" because it "may itself have a further explanation at some deeper level"—thus the equations $r' = \tilde{\rho}(\mathbf{x}, t)$, $\tau(\mathbf{x}, t) = 0$ have a deeper explanation in terms of (2.27) together with the equations defining $\mathbf{x}'$ and t', but they themselves can serve to investigate the effect of the original treatment.

Because so many real-life problems have to do with some form of randomness, probabilistic models of causality deserve a place among the older, deterministic ones. But how should such probabilistic models look like? The natural thing is perhaps to make the response and constraints random, writing

$$R(\omega) = \rho(\omega, \mathbf{X}(\omega), T(\omega)), \quad \tau(\omega, \mathbf{X}(\omega), T(\omega)) = 0, \tag{2.29}$$

for ω in a sample space Ω, by which the situation $\mathbf{X}$, treatment T and the response R, and maybe even the set $\mathcal{A}$ of admissible points, become random variables. In a real-life problem one would regard these random elements as unknowable, at least partly, and only the values assumed by them, such as $r = R(\omega)$, $t = T(\omega)$, $\mathbf{x} = \mathbf{X}(\omega)$ for certain ω, as observable.

In this model the effect that a particular value t of the treatment has on the response in a particular situation $\mathbf{x}$ becomes random, and one can only speak of averages of it with respect to a probability measure $\mathbf{P}$, such as

$$\int_\Omega f(\rho(\omega, \mathbf{x}, t))\mathbf{P}(\mathrm{d}\omega) \tag{2.30}$$

for some functions f. The statistical aspect of the model appears with the idea that one can make a large number of draws ω from Ω and average the corresponding values of $f(\rho(\omega, \mathbf{x}, t))$ to get an estimate of the function of $(\mathbf{x}, t)$ in (2.30). Although this function is not random, we cannot revert to a deterministic model because $\rho(\omega, \mathbf{x}, t)$ is not necessarily observable: what is observable is $\rho(\omega, \mathbf{X}(\omega), T(\omega))$, and we are not able to pick elements of a fixed $\mathcal{A}$ at will in order to use them in ρ but rather are left to observe those $(\mathbf{X}(\omega), T(\omega))$ that happen to be drawn.

We can, however, *wait* for draws such that $(\mathbf{X}(\omega), T(\omega)) = (\mathbf{x}, t)$ for each $(\mathbf{x}, t)$, and average the corresponding values of $f(\rho(\omega, \mathbf{x}, t))$ to get an estimate of

$$\frac{\int_{\{\omega' \in \Omega : (\mathbf{X}(\omega'), T(\omega')) = (\mathbf{x}, t)\}} f(\rho(\omega, \mathbf{x}, t))\mathbf{P}(\mathrm{d}\omega)}{\mathbf{P}((\mathbf{X}, T) = (\mathbf{x}, t))}, \tag{2.31}$$

where for simplicity we assume that $(\mathbf{X}, T)$ is discrete; for this should be the next best thing, and on the set over which it is being integrated here the response is $\rho(\omega, \mathbf{x}, t) = \rho(\omega, \mathbf{X}(\omega), T(\omega))$ and corresponds to observed data $(\mathbf{X}(\omega), T(\omega), R(\omega))$. But the question is whether the function in (2.31) is the same as that in (2.30). It may be if the set $\{\omega' \in \Omega : (\mathbf{X}(\omega'), T(\omega')) = (\mathbf{x}, t)\}$ is 'neutral' regarding the role of t and $\mathbf{x}$ in ρ, as in the trivial case of a nonrandom response function; but in principle the ω in that set may implicitly involve $(\mathbf{x}, t)$, and if that is so then the values of $\rho(\omega, \mathbf{x}, t)$ being averaged in (2.31) with fixed $\mathbf{x}$ depend on t also through the first argument of ρ, while those in (2.30) depend on t only through the last argument of ρ. This phenomenon may be seen as a probabilistic analogue of the confounding in model (2.27), to the extent that the possibility of trying different values of t in (2.31) is limited by the pairs $(\mathbf{x}, t)$ that turn up and by the knowledge of whether the values of t that turn up affect $\rho(\omega, \mathbf{x}, t)$ through the last argument of ρ alone or rather through the first argument as well in the guise of the concomitant ω.

The requirement that allows one to 'correct for confounding' in this model is that the randomness be apportioned separately—or, more precisely, independently—to the constraints and to the response. In order to explain this, let us take, without loss of generality, $\Omega = [0, 1]$ and $\mathbf{P}$ such that $\mathbf{P}([a, b]) = b - a$ for all $[a, b] \subset \Omega$. Consider the points $\omega_0 = 0.d_1 d_3 \cdots$, $\omega_1 = 0.d_2 d_4 \cdots$, obtained from the digits in the decimal expansion of each draw $\omega = 0.d_1 d_2 \cdots \in \Omega$. As mentioned in Chap. 1, ω_0 and ω_1 can themselves be regarded as independent draws from Ω, so that $\mathbf{P}(\{\omega \in \Omega : \omega_0 \in [a_0, b_0], \omega_1 \in [a_1, b_1]\}) = (b_0 - a_0)(b_1 - a_1)$, etc. The same applies to any two points $\omega_{\mathbf{x},0}$ and $\omega_{\mathbf{x},1}$ obtained from two disjoint sequences of digits of ω possibly depending on $\mathbf{x}$. Consequently, if the T, $\mathbf{X}$ and ρ of (2.29) satisfy

$$\mathbf{X}(\omega) = \mathbf{x} \Rightarrow \mathbf{X}(\omega_{\mathbf{x},0}) = \mathbf{x} \wedge T(\omega) = T(\omega_{\mathbf{x},0}) \wedge \rho(\omega, \mathbf{x}, t) = \rho(\omega_{\mathbf{x},1}, \mathbf{x}, t)$$

for all $(\omega, \mathbf{x}, t)$, then the numerator of (2.31) is

$$\int_\Omega \mathbf{1}_{\{(\mathbf{X},T)=(\mathbf{x},t)\}}(\omega) f(\rho(\omega, \mathbf{x}, t))\mathbf{P}(\mathrm{d}\omega) =$$

$$\int_{\Omega^2} \mathbf{1}_{\{(\mathbf{X},T)=(\mathbf{x},t)\}}(\omega_{\mathbf{x},0}) f(\rho(\omega_{\mathbf{x},1}, \mathbf{x}, t))\mathbf{P}(\mathrm{d}\omega_{\mathbf{x},1})\mathbf{P}(\mathrm{d}\omega_{\mathbf{x},0}) =$$

$$\int_\Omega f(\rho(\omega_{\mathbf{x},1}, \mathbf{x}, t))\mathbf{P}(\mathrm{d}\omega_{\mathbf{x},1})\, \mathbf{P}((\mathbf{X}, T) = (\mathbf{x}, t)),$$

which divided by the last probability coincides with (2.30), as required.

In essence, this is the prototype of the probability model for studying the effect of a treatment on a response; together with the condition just introduced, which makes it possible to deal with confounding, it appears under various guises throughout the book.

2.7 Notes

Despite the second paragraph of this chapter, a concern about standard mathematical notation not being able to represent relationships of cause and effect appears to be quite common; see for instance the text leading to footnote 5 of Pearl [67] or the second paragraph on p. 291 of Cox and Wermuth [20].

Our formulation of unconfoundedness in connection with the basic model is based on Ferreira [31] and is similar to a condition proposed by Arjas and Parner [3, 4].

A short history of the ideas around potential outcomes can be found in Chap. 2 of Imbens and Rubin [46]. Potential outcomes and counterfactuals are perfectly well defined from a mathematical point of view, and they are utterly meaningful in a wide range of real-life situations, from the very prosaic or practical ("I should not have smoked"; see also p. 89 of Pearl, Glymour and Jewel [73]) to the more poetic or reflective (e.g. Robert Frost's poem "The road not taken"); nevertheless, it appears that some authors find potential outcomes problematic (e.g. Dawid [22, 23]).

It is important to note that the notion of causal effect, as embodied by the potential outcomes $\rho(V_n, \mathbf{X}_n, t)$ and their laws, can always be entertained in a real-life problem to which the basic model is applied: in this 'canonical' model the treatment has a *single*, definite place among the arguments of the response function and such things as the law of $\rho(V_n, \mathbf{X}_n, t) - \rho(V_n, \mathbf{X}_n, t')$ can be considered irrespective of whether or not they can be observed and estimated, of the concomitant value of T_n and of whether T_n provides information about V_n; see the more or less open Probs. 2.3 and 2.9 (b). Unconfoundedness, which allows the laws of $\rho(V_n, \mathbf{X}_n, t)$ and functionals of them to be estimated from data, may be plausible in an observational study, but it is usually difficult to argue that the most important confounders have been identified and measured and are represented in $\mathbf{X}_n$.

Simpson's paradox is due to E.H. Simpson; see pp. 240–1 of [100] and pp. 1–3 of Pearl, Glymour and Jewel [73]. The paradox may of course assume a less spectacular form than that exhibited in Example 2.4; for example, confounding may cause the effect to be missed rather than inverted, as on p. 241, Table 4, of Simpson [100]. A variant of Example 2.5 is sketched on pp. 3–4 of Pearl, Glymour and Jewel [73].

The observation that the results of interventions can, in principle, be predicted from observable data under unconfoundedness is due to Pearl (cf. pp. 55-7, 101-3 of [73]); it indicates the relation between determining a treatment effect by 'correcting for confounding' and determining the effect of an intervention.

The commonality between the models defined by (2.1) and (2.2), pointed out before Example 2.6, appears to have been overlooked by some authors. The basic model—which as we have noted can even be specialized to the intervention model—is consistent with the two points of view of 'what happened' and 'what may happen', and in a general discussion we need not exclude any of them. The two points of view may indeed be used in the same real-life problem: first in connection with an observational study and then in connection with an intervention guided by the conclusions of the observational study—though in the intervention the response function ρ and the distribution of the $\mathbf{X}_n$ required by reality may have changed (cf. second point of Remark 2.3). In the discussion around equation (6) of Cox and Wermuth [20] (which corresponds to our equation (2.7)), the law of $\rho(V_n, \mathbf{X}_n, t)$—and consequently (at least as far as realistic applications are concerned) the very concept of causal effect—is presented as if it were irremediably tied to interventions and to the conditions for their success in real-life problems. But if in a study this law is unbiasedly and accurately estimated from a sample, and if it appears very likely that it varies with t, then the treatment very probably had a causal effect on the response for the population underlying the sample (in the sense of $\rho(V_n, \mathbf{X}_n, t)$ being nonconstant in t for a substantial number of individuals). Whether one may have convinced the individuals of that population to opt for a particular value of the treatment instead of the treatments they chose at the time, and whether one may convince the individuals of that population to opt for a particular value of the treatment in the future, can be meaningful questions in a given real-life problem, but neither the definition of causal effect nor identity (2.6) are subordinated to them—indeed, the latter are theoretical notions, the former empirical.

Lord's paradox is due to F.M. Lord [58]; for other interpretations of it see Pearl [72] and Sect. 1.9 of Imbens and Rubin [46]. There has been some discussion about whether the analysis of the first statistician—based on estimating and interpreting the laws in (2.23) and (2.24)—is correct, whether it is the analysis of the second statistician—based on (2.25) and (2.26)—that is correct, whether both analyses are correct or both are incorrect; see Senn [93] and Eriksson and Haggstrom [28] and the literature cited there. Lord clearly assumes that the model used by the statisticians is the correct one, and, judging from his qualitative description, the paradox does not seem to rely on a particular specification of the assumed and true models (accordingly, our choice of distributions and functional relationships in (2.22) is not essential, though it is convenient); rather, it arises from the supposed conflict between the interpretations underlying the two different questions considered by the statisticians—whether the population as a whole *has changed* and whether the population as a whole *could have changed* (or *could change* in the future). Despite this, the paradox is sometimes confused with, and confused by, questions related to the correctness of the model assumed for the data and to the bias of a certain estimation method when the assumed model is incorrect. The consideration of a 'treatment' which is not readily conceived of as manipulable, such as 'sex', as in Lord's own version, or 'ability' can add to the confusion by making the causal interpretation of the second statistician look strange.

2.8 Problems

2.1 Define random variables $Z_1, Z_2, \ldots$ by

$$Z_n = \alpha + \beta Y_n + \varepsilon_n, \quad n = 1, 2, \ldots,$$

where Y_n is a random function of another random variable X_n, ε_n is a random function of Y_n and of X_n, and α and β are constants.

(a) Show that the (X_n, Y_n, Z_n) follow a model equivalent to a special version of the basic model.

(b) Show that

$$Y_n = \alpha^* + \beta^* Z_n + \varepsilon_n^*$$

for constants α^* and β^* and random variables ε_n^* and that the (X_n, Z_n, Y_n) too follow a model equivalent to a special version of the basic model, so that one may say that the Z_n cause the Y_n just as one may say that the Y_n cause the Z_n.

(c) Suppose unconfoundedness holds in the version of the basic model followed by the (X_n, Y_n, Z_n). Show that in general unconfoundedness does not hold in the version of the basic model followed by the (X_n, Z_n, Y_n).

(d) Suppose that the Z_n are defined instead by $Z_n = \varphi(Y_n, \varepsilon_n)$, where φ is some function but the other specifications are as before; do the (X_n, Z_n, Y_n) generally follow a model equivalent to a special version of the basic model?

2.2 Show that if $\mathcal{L}(\rho(V, \mathbf{x}, t))$ is constant in t for each $\mathbf{x}$ such that $\mathbf{P}(\mathbf{X} \in N) > 0$ for every neighbourhood N of $\mathbf{x}$, then a generic response R in the basic model *with unconfoundedness* may be replaced by an *equivalent response* $\tilde{R} := \rho(V, \mathbf{X}, T)$ defined on the same space as the generic triple $(\mathbf{X}, T, R)$ through a function $(v, \mathbf{x}, t) \to \tilde{\rho}(v, \mathbf{x}, t)$ such that $\mathcal{L}(\tilde{\rho}(V, \mathbf{x}, t)) = \mathcal{L}(\rho(V, \mathbf{x}, t))$ and $\tilde{\rho}(v, \mathbf{x}, t)$ is constant in t for each $(v, \mathbf{x})$, so that the generic triple $(\mathbf{X}, T, \tilde{R})$ has the same distribution as $(\mathbf{X}, T, R)$ and the random function $(\mathbf{x}, t) \to \tilde{\rho}(V, \mathbf{x}, t)$ is constant in t.

2.3 Suppose that unconfoundedness fails in the basic model; how would you define the effect of the treatment, which functions would you choose to quantify it, and could those functions be estimated from a random sample obeying the model?

2.4 Verify the identities in (2.19)–(2.20).

2.5 Show that substitution of the equations in (2.21) for the second and third equations in (2.18) does not change the distribution of $(\mathbf{X}_n, T_n, R_n)$. *Suggestion:* To get the first equation in (2.21), one writes $(\mathbf{X}_n - \gamma\mu) = c(T_n - \mu) + sZ_n$, where Z_n is standard normal and independent of T_n, multiplies both sides by $T_n - \mu$ and takes expectations to get $\mathrm{Cov}(T_n, \mathbf{X}_n) = c\,\mathrm{Var}(T_n)$, takes the variance on both sides to get $\mathrm{Var}(\mathbf{X}_n) = c^2\mathrm{Var}(T_n) + s^2$, solves these equations for c and s^2, and finally substitutes $\mathrm{Cov}(T_n, \mathbf{X}_n)$, $\mathrm{Var}(T_n)$ and $\mathrm{Var}(\mathbf{X}_n)$ in the solutions by their expressions in (2.19).

2.6 **(a)** Show that in Example 2.6 the densities of $\mathbf{X}_n$ and T_n satisfy

$$f_{\mathbf{X}_n}(x) = p\,\phi(x, \mu_1, \sigma^2) + (1 - p)\phi(x, \mu_2, \sigma^2),$$

$$f_{T_n|\mathbf{X}_n=x}(t) = \{1 - q(x)\}\delta_{t,1} + q(x)\delta_{t,2},$$

$$f_{T_n}(t) = p\delta_{t,1} + (1 - p)\delta_{t,2} \quad \text{and} \quad f_{\mathbf{X}_n|T_n=t}(x) = \phi(x, \mu_t, \sigma^2).$$

(b) Derive (2.23) and (2.24) from (2.22).

2.7 **(a)** Illustrate Simpson's paradox numerically by varying the parameters $q(0)$, $p(1, 1)$, etc. Allow for $q(0) = \beta_0$ to differ somewhat from $1 - q(1) = 1 - \beta_1$ and for $\mathbf{P}(\mathbf{X}_n = 0)$ to differ somewhat from $1/2$.

(b) Use a computer program to simulate data illustrating the continuous version of Simpson's paradox.

2.8 Use a computer program to simulate data illustrating Lord's paradox.

2.9 Let $\beta_1, \beta_2 \in \mathbb{R}$ be fixed and $(T_1, X_1, \varepsilon_1), \ldots, (T_N, X_N, \varepsilon_N)$ be independent random vectors, and define $R_1, \ldots, R_N$ by

$$R_i = \beta_1 X_i + \beta_2 T_i + \varepsilon_i, \quad i = 1, 2, \ldots, N. \tag{2.32}$$

(a) Suppose $\mathbf{E}(\varepsilon_i) = \beta_0$ for some fixed $\beta_0 \in \mathbb{R}$ and $\mathrm{Var}(\varepsilon_i) = \sigma^2$. When is the model thus defined a linear regression model?

(b) Call the R_i 'responses' and the T_i 'treatments'; can you define the causal effect of the treatment on the response? Since nothing is said about the relationship between ε_i, X_i and T_i you need to consider a couple of possibilities: For example, if $T_i = \tau(U_i, X_i)$ and $\varepsilon_i = \sigma\Phi^{-1}(V_i)$ with U_i and V_i standard uniform conditionally on X_i (not necessarily independent) then the treatment effect can be expressed very simply; but if instead $\varepsilon_i = \varphi(V_i, T_i)$ for some φ then the treatment effect is something else. Treatment effect can be defined whenever (2.32) can be expressed as a special case of the basic model.

In what follows, let (T, X, ε, R) have the same distribution as the $(T_i, X_i, \varepsilon_i, R_i)$ and assume that $\mathbf{E}(\varepsilon) = 0$, $|\mathrm{Corr}(T, X)| \neq 1$ and $\mathrm{Var}(X), \mathrm{Var}(T), \mathrm{Var}(\varepsilon) < \infty$.

(c) Write $\mathbf{T} = (T_1, \ldots, T_N)^{\mathrm{T}}$, $\mathbf{X} = (X_1, \ldots, X_N)^{\mathrm{T}}$, $\mathbf{T}^{\mathrm{T}}\mathbf{X} = \mathbf{X}^{\mathrm{T}}\mathbf{T} \equiv \mathbf{X} \cdot \mathbf{T} = \sum_{i=1}^{N} X_i T_i$, $\boldsymbol{\varepsilon} = (\varepsilon_1, \ldots, \varepsilon_N)^{\mathrm{T}}$, etc. Then (2.32) can be written more compactly as $\mathbf{R} = [\mathbf{X}\,\mathbf{T}]\boldsymbol{\beta} + \boldsymbol{\varepsilon}$, where $\boldsymbol{\beta} := (\beta_1, \beta_2)^{\mathrm{T}}$. Show that the least-squares estimator of $\boldsymbol{\beta}$ based on the (R_i, T_i, X_i) (and on the assumption that these data follow a linear regression model) is

$$\hat{\boldsymbol{\beta}} = \begin{bmatrix} \hat{\beta}_1 \\ \hat{\beta}_2 \end{bmatrix} = \frac{1}{(\mathbf{T}^{\mathrm{T}}\mathbf{T})(\mathbf{X}^{\mathrm{T}}\mathbf{X}) - (\mathbf{X}^{\mathrm{T}}\mathbf{T})^2} \begin{bmatrix} \mathbf{T}^{\mathrm{T}}\mathbf{T} & -\mathbf{X}^{\mathrm{T}}\mathbf{T} \\ -\mathbf{X}^{\mathrm{T}}\mathbf{T} & \mathbf{X}^{\mathrm{T}}\mathbf{X} \end{bmatrix} \begin{bmatrix} \mathbf{X}^{\mathrm{T}} \\ \mathbf{T}^{\mathrm{T}} \end{bmatrix} \mathbf{R}.$$

Show that as $N \to \infty$

$$\frac{1}{N}\mathbf{T}^{\mathrm{T}}\mathbf{T} \to \mathbf{E}(T^2), \quad \frac{1}{N}\mathbf{X}^{\mathrm{T}}\mathbf{T} \to \mathbf{E}(XT), \quad \frac{1}{N}\mathbf{X}^{\mathrm{T}}\mathbf{X} \to \mathbf{E}(X^2),$$

$$\begin{aligned} \frac{1}{N}\mathbf{X}^{\mathrm{T}}\mathbf{R} &= \frac{1}{N}\mathbf{X}^{\mathrm{T}}[\mathbf{X}\,\mathbf{T}]\boldsymbol{\beta} + \frac{1}{N}\mathbf{X}^{\mathrm{T}}\boldsymbol{\varepsilon} = \frac{1}{N}\mathbf{X}^{\mathrm{T}}\mathbf{X}\beta_1 + \frac{1}{N}\mathbf{X}^{\mathrm{T}}\mathbf{T}\beta_2 + \frac{1}{N}\mathbf{X}^{\mathrm{T}}\boldsymbol{\varepsilon} \\ &\to \mathbf{E}(X^2)\beta_1 + \mathbf{E}(XT)\beta_2 + \mathbf{E}(X\varepsilon), \\ \frac{1}{N}\mathbf{T}^{\mathrm{T}}\mathbf{R} &= \frac{1}{N}\mathbf{T}^{\mathrm{T}}[\mathbf{X}\,\mathbf{T}]\boldsymbol{\beta} + \frac{1}{N}\mathbf{T}^{\mathrm{T}}\boldsymbol{\varepsilon} = \frac{1}{N}\mathbf{T}^{\mathrm{T}}\mathbf{X}\beta_1 + \frac{1}{N}\mathbf{T}^{\mathrm{T}}\mathbf{T}\beta_2 + \frac{1}{N}\mathbf{T}^{\mathrm{T}}\boldsymbol{\varepsilon} \\ &\to \mathbf{E}(XT)\beta_1 + \mathbf{E}(T^2)\beta_2 + \mathbf{E}(T\varepsilon) \end{aligned}$$

(e.g. with probability 1), and hence that

$$\begin{aligned} \hat{\boldsymbol{\beta}} &\to \frac{1}{\mathbf{E}(T^2)\mathbf{E}(X^2) - \mathbf{E}(XT)^2} \begin{bmatrix} \mathbf{E}(T^2) & -\mathbf{E}(XT) \\ -\mathbf{E}(XT) & \mathbf{E}(X^2) \end{bmatrix} \begin{bmatrix} \mathbf{E}(X^2)\beta_1 + \mathbf{E}(XT)\beta_2 + \mathbf{E}(X\varepsilon) \\ \mathbf{E}(XT)\beta_1 + \mathbf{E}(T^2)\beta_2 + \mathbf{E}(T\varepsilon) \end{bmatrix} \\ &= \begin{bmatrix} \beta_1 \\ \beta_2 \end{bmatrix} + \frac{1}{\mathbf{E}(T^2)\mathbf{E}(X^2) - \mathbf{E}(XT)^2} \begin{bmatrix} \mathbf{E}(T^2) & -\mathbf{E}(XT) \\ -\mathbf{E}(XT) & \mathbf{E}(X^2) \end{bmatrix} \begin{bmatrix} \mathbf{E}(X\varepsilon) \\ \mathbf{E}(T\varepsilon) \end{bmatrix}. \end{aligned}$$

State conditions under which $\hat{\boldsymbol{\beta}}$ is consistent.

(d) Let U_i and V_i be standard uniform, (U_i, V_i) and X_i be independent, and set

$$\varepsilon_i = (\sigma + \theta X_i)\,\Phi^{-1}(V_i), \quad T_i = \mathbf{1}_{[0,X_i]}\left(\Phi^{-1}(U_i)\right) = \begin{cases} 1 & \text{if } U_i \le \Phi(X_i), \\ 0 & \text{otherwise,} \end{cases}$$

in (2.32), θ being a fixed number, Φ the standard normal distribution function, Φ^{-1} its inverse, and $\sigma > 0$. Verify that $\mathcal{L}(\varepsilon|X = x) = \mathcal{N}(0, (\sigma + \theta x)^2)$ and

$$\mathbf{P}(T = 1|X = x) = 1 - \mathbf{P}(T = 0|X = x) = \Phi(x).$$

With these additional specifications, is the model a special case of the basic model? Does it satisfy unconfoundedness? Is it a linear regression model with ε_i as error?

(e) With the specifications of **(d)** take X_i standard normal and $(\Phi^{-1}(U_i), \Phi^{-1}(V_i))$ bivariate normal with variances 1 and correlation coefficient ϱ. Show that[8]

$$\mathbf{E}(T\varepsilon) = -\frac{\varrho\,\sigma}{2\sqrt{\pi}}, \quad \mathbf{E}(X\varepsilon) = 0, \quad \mathbf{E}(TX) = \frac{1}{2\sqrt{\pi}}, \quad \mathbf{E}(T^2) = \mathbf{E}(T) = \frac{1}{2}, \quad \mathbf{E}(X^2) = 1,$$

and that the bias of $\hat{\boldsymbol{\beta}}$ satisfies

$$\text{bias}(\hat{\boldsymbol{\beta}}) := \mathbf{E}(\hat{\boldsymbol{\beta}} - \boldsymbol{\beta}) \to \begin{bmatrix} \frac{\varrho\,\sigma}{2\pi-1} \\ -\frac{2\sqrt{\pi}}{2\pi-1}\varrho\,\sigma \end{bmatrix}$$

When is $\hat{\beta}_1$ consistent? And $\hat{\beta}_2$?

(f) Instead of (2.32) take

$$R_i = \beta_1 X_i (T_i + 1) + \beta_2 T_i + \varepsilon_i, \quad i = 1, 2, \ldots, N, \tag{2.33}$$

with the specifications of **(d)** and **(e)** except that now

$$\varepsilon_i = \sigma\,\Phi^{-1}(V_i) + \theta X_i, \tag{2.34}$$

so that $\mathcal{L}(\varepsilon|X = x) = \mathcal{N}(\theta x, \sigma^2)$. Show that the least-squares estimator of $\boldsymbol{\beta}$ based on the (R_i, T_i, X_i) and on the assumption that these follow a linear regression model with the ε_i as errors is obtained by replacing $\mathbf{X}$ by $\mathbf{W} = [X_1(T_1+1) \cdots X_N(T_N+1)]^{\mathrm{T}}$ in the expression for $\hat{\boldsymbol{\beta}}$ in **(c)**. Derive the identities $\mathbf{E}(T) = \mathbf{E}(T^2) = 1/2$,

$$\mathbf{E}(T\varepsilon) = \frac{\theta - \varrho\,\sigma}{2\sqrt{\pi}}, \quad \mathbf{E}[X(T+1)T] = \frac{1}{\sqrt{\pi}}, \quad \mathbf{E}[X(T+1)\varepsilon] = \frac{3}{2}\theta, \quad \mathbf{E}[X^2(T+1)^2] = \frac{5}{2},$$

and show that as $N \to \infty$

$$\hat{\boldsymbol{\beta}} \to \begin{bmatrix} \beta_1 \\ \beta_2 \end{bmatrix} + \frac{1}{\frac{5}{4} - \frac{1}{\pi}} \begin{bmatrix} \frac{1}{2} & -\frac{1}{\sqrt{\pi}} \\ -\frac{1}{\sqrt{\pi}} & \frac{5}{2} \end{bmatrix} \begin{bmatrix} \frac{3}{2}\theta \\ \frac{\theta - \varrho\,\sigma}{2\sqrt{\pi}} \end{bmatrix} = \begin{bmatrix} \beta_1 \\ \beta_2 \end{bmatrix} + \begin{bmatrix} \frac{(3\pi-2)\theta+2\varrho\sigma)}{5\pi-4} \\ -\frac{(\theta+5\varrho\sigma)\sqrt{\pi}}{5\pi-4} \end{bmatrix}.$$

When are $\hat{\beta}_1$ and $\hat{\beta}_2$ consistent? When does the model satisfy *unconfoundedness*? When is it a linear regression model with the ε_i as errors?
Show that if $\varrho = 0$ then the (R_i, T_i, X_i) *do* follow a linear regression model; but then why are $\hat{\beta}_1$ and $\hat{\beta}_2$ not always consistent?
Can you define the causal effect of the treatment on the response? Does the definition of treatment effect rely on unconfoundedness?

(g) Discuss the application of regression models to real-life problems. Consider in particular problems in which one would like to avoid concluding that a treatment effect exists when it does not exist or is negligible.

---

[8] Some of these expectations and those in **(f)** require some skills in integration and their calculation may be omitted without detriment to the exercise.

2.10 Let $\mathbf{U}$ be a p-dimensional random vector following a mixture of normal distributions with mean vectors $\boldsymbol{\mu}_0$, $\boldsymbol{\mu}_1$, full rank covariance matrices Σ_0, Σ_1, and mixture parameter $p_0 = 1 - p_1 \in]0,1[$. Let $\mathbf{V}$ be a q-dimensional random vector defined from $\mathbf{U}$ by

$$\mathbf{V} = \boldsymbol{\xi} + \mathbf{B}(\mathbf{U} - \boldsymbol{\mu}),$$

where $\boldsymbol{\mu} = p_0\boldsymbol{\mu}_0 + p_1\boldsymbol{\mu}_1$, $\boldsymbol{\xi}$ is a q-dimensional normal random vector independent of $\mathbf{U}$, with mean vector $\boldsymbol{\nu}$ and covariance matrix $\mathbf{M}$, and $\mathbf{B}$ is a $q \times p$ matrix. Then $\mathbf{V}$ is normally distributed conditionally on $\mathbf{U}$, with $\mathbf{E}[\mathbf{V}|\mathbf{U}=\mathbf{u}] = \boldsymbol{\nu} + \mathbf{B}(\mathbf{u} - \boldsymbol{\mu})$ and $\mathrm{Var}[\mathbf{V}|\mathbf{U}=\mathbf{u}] = \mathbf{M}$. Let a binary variable T be defined from $\mathbf{U}$ by

$$T = \begin{cases} 1 & \text{if } U \leq \Pi_1(\mathbf{U}), \\ 0 & \text{otherwise,} \end{cases}$$

where U is a standard uniform random variable independent of $\mathbf{U}$, $\mathbf{V}$ and $\boldsymbol{\xi}$, and

$$\Pi_t(\mathbf{u}) = \frac{p_t|\Sigma_t|^{-1/2}e^{-\frac{1}{2}(\mathbf{u}-\boldsymbol{\mu}_t)^{\mathrm{T}}\Sigma_t^{-1}(\mathbf{u}-\boldsymbol{\mu}_t)}}{p_0|\Sigma_0|^{-1/2}e^{-\frac{1}{2}(\mathbf{u}-\boldsymbol{\mu}_0)^{\mathrm{T}}\Sigma_0^{-1}(\mathbf{u}-\boldsymbol{\mu}_0)} + p_1|\Sigma_1|^{-1/2}e^{-\frac{1}{2}(\mathbf{u}-\boldsymbol{\mu}_1)^{\mathrm{T}}\Sigma_1^{-1}(\mathbf{u}-\boldsymbol{\mu}_1)}}.$$

Then T depends on $\mathbf{U}$ but not on $\mathbf{V}$, and $\mathbf{P}(T = t|\mathbf{U} = \mathbf{u}) = \Pi_t(\mathbf{u})$ $(t = 0, 1)$. Finally, write $\mathbf{X} = (\mathbf{U}, \mathbf{V})$ and define R from $\mathbf{X}$ and T by

$$R = \mu + \sigma\Phi^{-1}(V) + (1 - T)A_0(\mathbf{X}) + TA_1(\mathbf{X}), \tag{2.35}$$

where μ and $\sigma > 0$ are parameters, A_0 and A_1 are some functions, and V is a standard uniform random variable independent of $(\boldsymbol{\xi}, \mathbf{X}, U)$; then the distribution of R conditionally on $\{T = t, \mathbf{X} = \mathbf{x}\}$ is normal with mean $\mu + A_0(\mathbf{x})(1-t) + A_1(\mathbf{x})t$ and variance σ^2:

$$\mathbf{E}[R|\mathbf{X} = \mathbf{x}, T = t] = \mu + A_0(\mathbf{x})(1-t) + A_1(\mathbf{x})t, \quad \mathrm{Var}[R|\mathbf{X} = \mathbf{x}, T = t] = \sigma^2.$$

In what follows we regard $\mathbf{X}$, T and R as being observable, all the other random variables as unobservable, and the parameters p_0, p_1, $\boldsymbol{\mu}_0$, $\boldsymbol{\mu}_1$, Σ_0, Σ_1, $\boldsymbol{\nu}$, μ and σ as unknown; and we consider a random sample $(\mathbf{X}_1, T_1, R_1), \ldots, (\mathbf{X}_N, T_N, R_N)$ of vectors defined in the same way as $(\mathbf{X}, T, R)$. As usual, T is interpreted as a treatment, R as a response, and $\mathbf{X}$ as a vector of background or confounding variables on an 'individual' or 'unit'.

(a) The model just defined is a special case of the basic model; give complete definitions of the functions ρ and τ corresponding to it. When is there no treatment effect? Does unconfoundedness hold? What is the law that characterizes the treatment effect on an individual with background variables equal to $\mathbf{x}$? What is the mean treatment effect on an individual with background variables equal to $\mathbf{x}$? And the *overall* mean treatment effect?

(b) Show that if $A_0(\mathbf{x}) = \mathbf{x} \cdot \boldsymbol{\beta}_0$ and $A_1(\mathbf{x}) = \mathbf{x} \cdot \boldsymbol{\beta}_1$ for some vectors $\boldsymbol{\beta}_0, \boldsymbol{\beta}_1 \in \mathbb{R}^{p+q}$ then the mean treatment effect and the overall mean treatment effect are

$$\mathrm{e}(\mathbf{x}) = \mathbf{x} \cdot (\boldsymbol{\beta}_1 - \boldsymbol{\beta}_0) \quad \text{and} \quad \epsilon = (\boldsymbol{\mu}, \boldsymbol{\nu}) \cdot (\boldsymbol{\beta}_1 - \boldsymbol{\beta}_0).$$

(c) Show that the model can be rewritten as a version of the basic model in which unconfoundedness holds with **X** replaced by a subset of it, write down the law that characterizes the treatment effect on an individual with given values of that subset and use it to obtain the overall mean treatment effect.

2.11 In the version of the basic model with unconfoundedness of Example 2.3, which law(s) would you use to quantify the causal effect of the treatment?

2.12 Consider the basic model with uncounfoundedness and

$$R_n = \rho(V_n, \mathbf{X}'_n, \mathbf{X}'''_n, T_n), \quad T_n = \tau(U_n, \mathbf{X}'_n, \mathbf{X}''_n),$$

where we write $\mathbf{X}_n = (\mathbf{X}'_n, \mathbf{X}''_n, \mathbf{X}''''_n)$.

(a) Show that U_n and V_n are independent, standard uniform random variables conditionally on $\mathbf{X}'_n$.

(b) Suppose that $\mathbf{X}''_n$ and $\mathbf{X}'''_n$ are independent conditionally on $\mathbf{X}'_n$. Show that the effect of the treatment on the response may be quantified by

$$\mathcal{L}(\tilde{\rho}_{\mathbf{X}'''_n}(V_n, \mathbf{x}', t)|\mathbf{X}'_n = \mathbf{x}'),$$

where $\tilde{\rho}_{\mathbf{x}'''}(v, \mathbf{x}', t) := \rho(v, \mathbf{x}', \mathbf{x}''', t)$; can one drop the event $\{\mathbf{X}'_n = \mathbf{x}'\}$ from this conditional law? From the text we know that $\mathcal{L}(R_n|\mathbf{X}_n = \mathbf{x}, T_n = t) = \mathcal{L}(\rho(V_n, \mathbf{x}, t))$ too characterizes the effect of the treatment; show that the two laws are related by

$$\mathbf{P}(\tilde{\rho}_{\mathbf{X}'''_n}(V, \mathbf{x}', t) \le r|\mathbf{X}'_n = \mathbf{x}') = \sum_{\mathbf{x}'',\mathbf{x}'''} \mathbf{P}(R_n \le r|\mathbf{X}_n = (\mathbf{x}', \mathbf{x}'', \mathbf{x}'''), T_n = t) \times \mathbf{P}(\mathbf{X}''_n = \mathbf{x}''|\mathbf{X}'_n = \mathbf{x}')\mathbf{P}(\mathbf{X}'''_n = \mathbf{x}'''|\mathbf{X}'_n = \mathbf{x}').$$

Consider the special case of $\mathbf{X}'_n = \xi'(U'_n)$, $\mathbf{X}''_n = \xi''(U''_n, \mathbf{X}'_n)$, $\mathbf{X}'''_n = \xi'''(U'''_n, \mathbf{X}'_n)$, with U'_n, U''_n and U'''_n independent standard uniforms independent of (U_n, V_n); rewrite $\mathcal{L}(\tilde{\rho}_{\mathbf{X}'''_n}(V_n, \mathbf{x}', t)|\mathbf{X}'_n = \mathbf{x}')$ so as to remove the conditioning event from it.

CHAPTER 3

More detailed models

More detailed models of cause and effect specify relations between background variables and allow for the treatment to influence other variables besides the response. Thus, the treatment can have an indirect influence on the response through other variables, which then represent the background only relative to the response. In fact, a variable which is influenced by the treatment and by background variables and in turn influences other variables may be seen as a second treatment, so the more detailed models are also able to represent a tandem of treatment-response relations. Such models bring up new questions about the definition and determination of causal effects and the possibility of estimating those effects from data. This chapter treats some of the most complicated aspects of causality from the point of view of statistics, considered for example in the works of Pearl and Robins but not in those of Rubin's school; yet there are connections between this and Chap. 2, because in a sense a detailed statistical model can be reduced to the basic one.

3.1 Levels of causality; more detailed models

We begin by introducing some terminology which qualifies the assumptions involved in a model and the degree to which those assumptions are thought to hold in a real-life problem. We speak of *zero-level causality* when a version of the basic model is assumed *without* unconfoundedness, so that conclusions based on the $(\mathbf{X}_n, T_n, R_n)$ may say something about the association between the treatment and the response but not necessarily about the causal effect of the former on the latter (since at least some confounding subsists). It seems that observational studies are mostly carried out at the zero-level, even if they do not explicitly invoke the basic model or do not discuss the extent of confounding, and even if they vary with respect to the statistical methods used. In fact, many studies assume (at least implicitly) that the data follow a regression model, a very particular version of the basic model with unconfoundedness (see Example 2.1); yet sometimes some of the actual data represented by $\mathbf{X}_n$ or T_n are preceded by, or at least are synchronous with, those represented by R_n, which not only invalidates unconfounding as it perverts the notion of causality.

In *first-level causality* one assumes the basic model with unconfoundedness—perhaps in a less explicit form, as Rubin's, discussed in Remark 2.5—and knows or believes that the model corresponds sufficiently well to the real-life problem. In particular, one has identified most of the factors influencing both treatment and response—the confounders in $\mathbf{X}_n$—and in principle can collect data on them in order to estimate or test for the causal effect of the treatment. Indeed, we have seen that under unconfoundedness the effect of the treatment on the response in the basic model, which is completely characterized by $\mathcal{L}(\rho(V_n, \mathbf{x}, t))$ for varying $(t, \mathbf{x}) \in \mathcal{T} \times \mathcal{X}$, can be estimated from observed data in the guise of $\mathcal{L}(R_n | \mathbf{X}_n = \mathbf{x}, T_n = t)$. Nevertheless, as suggested above, not many studies are carried out at the first-level, and few present a thorough argumentation to justify unconfoundedness based on a list of variables identified as confounders.

Identifying the variables that must be conditioned upon so as to make unconfoundedness tenable is recognized by many (e.g. Sect. 3.7 of Imbens and Rubin [46], pp. 76–7 of Rosenbaum [83]) as a difficult task, demanding substantial extra-statistical knowledge about the real-life problem in hand. As a general rule, Rosenbaum and Rubin recommend that "all the relevant covariates, that is, all the variables that may be associated with both outcomes and assignment to treatment", which correspond to the coordinates of $\mathbf{X}_n$ in our basic model, be considered possible confounders, and say that "there is little or no reason to avoid adjustment for a true covariate, a variable describing subjects before treatment" (see Imbens and Rubin [46] and Rosenbaum [83]). Pearl and others (e.g. [98], [69]), however, have pointed out that conditioning on some variables may create rather than remove confounding (as will be illustrated below after Example 3.2), so that in order to justify conditioning on a particular set of variables one generally needs to postulate (and defend, based on extra-statistical knowledge) a *detailed model* describing the relations between the various components of the situation, the treatment and the response—not just the model we have considered in Chap. 2 of the cause and effect relationship between treatment and response given a situation.

We speak of *second-level causality* whenever such a detailed model is assumed or can be justified in a real-life problem. Since, as noted in Sect. 2.6 (cf. the text around (2.27) and (2.28)), a given problem may admit various levels of causality, second-level causality may be based on several models of varying complexity (which of course should be consistent with each other), and, as we shall see below, it may be reduced to first-level causality.

As an example of the 'second-level model' we have in mind let us take

$$\begin{cases} X_1 = \varphi_1(U_1), X_2 = \varphi_2(U_2), \\ X_3 = \varphi_3(U_3, X_1, X_2), \\ X_4 = \varphi_4(U_4, X_1), \\ X_5 = \varphi_5(U_5, X_2), \\ T = \tau(U, X_3, X_4), \\ X_6 = \varphi_6(U_6, T), \\ R = \rho(V, X_3, X_5, X_6), \end{cases} \tag{3.1}$$

where $U, V, U_1, \ldots, U_6$ are independent standard uniform random variables, to be regarded as unobservable, purely random 'errors' and therefore usually referred to as *exogenous*, and $\tau, \rho, \varphi_1, \ldots, \varphi_6$ are real-valued functions.

The schematic representation of this set of equations in Fig. 3.1—a 'causal graph'—omits the essential information about the exogenous random variables but provides a more digestible summary of the essential aspects of it, namely of which variables 'influence' which. We shall use standard or self-explanatory terminology when referring to such graphs: the variables $X_1, X_2, \ldots, T, R$ correspond to *nodes* of the graph; the graph is *directed* because its edges are arrows; $X_1 \rightarrow X_4 \rightarrow T \rightarrow X_6 \rightarrow R$ is a *directed path* leading from X_1 to R, and we say that X_1 *leads to* R; $X_1 \rightarrow X_4 \rightarrow T \leftarrow X_3 \rightarrow R$ is a *path* (not a directed one); T is a *descendant* of X_1, and a *direct descendant* of X_4 (and of X_3), and X_1 and X_4 are *ancestors* of T (and of X_6 and R); $T \leftarrow X_4$ is a subpath of the path $T \leftarrow X_4 \leftarrow X_1$; and so on. Note that a causal graph is *acyclic*: no two nodes point arrows to each other, and, more generally, a directed path from a node cannot lead back to the node (a future event cannot influence a past event). Actually, whenever we talk of a path we really mean a *simple path*: in it, a node appears only once (so even among undirected paths we never consider closed ones—those that begin and end in the same node).

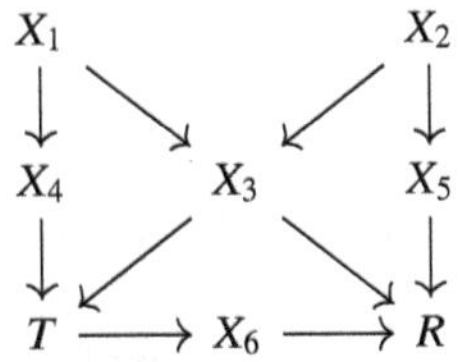

Fig. 3.1 Graph of the model at the second-level of causality defined by (3.1).

Remark 3.1 We do not regard the graph of Fig. 3.1 as a model: the model is the mapping that assigns a set or vector of numbers to each $\omega \in \Omega$, and the figure is a summary of this mapping which can serve as a tool to analyse it in some respects. Although in a general discussion one may consider Fig. 3.1 instead of (3.1), in a more specific example one may wish, for example, that τ be constant in the second argument, in which case the arrow from X_3 to T may be misleading. □

Model (3.1) represents second-level causality because it consists of more than just the two equations of (2.1); but it is not the only such model since the model without the equation $X_6 = \varphi_6(U_6, T)$ and with X_6 replaced by $\varphi_6(U_6, T)$ in the definition of R—a graph of which is obtained by deleting the node X_6 from Fig. 3.1 and linking T to R by a single arrow—is simpler but still of second-level causality. Evidently, if we discard X_6 and put $\mathbf{X} = (X_1, \ldots, X_5)$ then the model satisfies (2.1) with unconfoundedness and in essence reduces to first-level-causality.

More generally, if some of the functions and variables representing 'intermediate relations' are substituted into the functions they actuate, one may reduce the model to a first-level model:

Example 3.1 With $\mathbf{X} := (X_1, X_2, X_3)$ we have from (3.1)

$$R = \rho(V, X_3, X_5, X_6) = \rho\big(V, X_3, \varphi_5(U_5, X_2), \varphi_6(U_6, T)\big) \;=: \tilde{\rho}\big(\tilde{V}, \mathbf{X}, T\big),$$

$$T = \tau(U, X_3, X_4) = \tau\big(U, X_3, \varphi_4(U_4, X_1)\big) =: \tilde{\tau}\big(\tilde{U}, \mathbf{X}\big),$$

where and $\tilde{U}$ and $\tilde{V}$ are standard uniforms obtained from (U_4, U) and (U_5, U_6, V), respectively, independent and independent of $\mathbf{X}$, so the conditions of the basic model are satisfied with unconfoundedness and

$$\mathcal{L}\big(R\big|\mathbf{X} = \mathbf{x}, T = t\big) = \mathcal{L}\big(\tilde{\rho}\big(\tilde{V}, \mathbf{x}, t\big)\big|\mathbf{X} = \mathbf{x}, \tilde{\tau}\big(\tilde{U}, \mathbf{x}\big) = t\big) = \mathcal{L}\big(\tilde{\rho}\big(\tilde{V}, \mathbf{x}, t\big)\big). \tag{3.2}$$

Note that $\tilde{U}$ and $\tilde{V}$ are independent and independent of $\mathbf{X}$ because the event $\{\mathbf{X} = \mathbf{x}\}$ is equal to $\big\{\big(\varphi_1(U_1), \varphi_2(U_2), \varphi_3\big(U_3, \varphi_1(U_1), \varphi_2(U_2)\big)\big) = \mathbf{x}\big\}$ and hence involves (U_1, U_2, U_3), while $\tilde{U}$ involves (U_4, U) and $\tilde{V}$ involves (U_5, U_6, V). In particular, the conditional distribution of $\tilde{V}$ given $\{\mathbf{X} = \mathbf{x}\}$ is independent of $\mathbf{x}$, whence the last equality in (3.2) (cf. Remark 2.4). Note also (cf. Prob. 1.4) that $\tilde{V}$ may, for example, be defined by arranging the digits in the decimal expansions of V, U_5, U_6 alternately, so that $\tilde{\rho}$ involves first the 'unfolding' of a sequence in three sequences of digits.

As we shall see below, other reductions to a first-level model are possible. □

Despite the difficulty of setting up a convincing second-level model in a real-life problem, it seems that any earnest effort in that direction can only help in thinking about and defending the assumption of unconfoundedness in a corresponding first-level model—and in opening it to scrutiny and criticism. The study of such a second-level model can also render a statistical analysis more efficient if it is found that unconfoundedness holds by conditioning on a smaller or more easily available subset of variables, although the actual measure of causal effect depends somewhat on the subset chosen. Thus, the method used in Example 3.1 shows that both $\mathbf{X} = (X_1, X_3)$ and $\mathbf{X} = (X_2, X_3)$ guarantee unconfoundedness, but (3.2) could be easier to estimate with the latter and a somewhat different measure of causal effect could be estimated with the former:

Example 3.2 Take $\mathbf{X} = (X_2, X_3)$: then

$$\mathcal{L}(R|T = t, \mathbf{X} = \mathbf{x}) = \mathcal{L}\left(\rho(V, x_3, \varphi_5(U_5, x_2), \varphi_6(U_6, t)) \,\middle|\, \begin{matrix} \tau(U, x_3, \varphi_4(U_4, X_1)) = t, \\ \mathbf{X} = (x_2, x_3) \end{matrix}\right)$$

$$= \mathcal{L}\big(\rho(V, x_3, \varphi_5(U_5, x_2), \varphi_6(U_6, t))\big),$$

which yields the same measure of causal effect as conditioning on (X_1, X_2, X_3) does in Example 3.1. If we take $\mathbf{X} = (X_1, X_3)$ then

$$\mathcal{L}\big(R\big|T = t, \mathbf{X} = (x_1, x_3)\big) = \mathcal{L}\big(\rho(V, x_3, X_5, \varphi_6(U_6, t))\big|\varphi_3(U_3, x_1, X_2) = x_3\big),$$

which yields a different measure of causal effect, one that is obtained by integrating the first with respect to $\mathcal{L}(X_2|(X_1, X_3) = (x_1, x_3))$ (Prob. 3.6).

That different reductions to first-level causality may lead to different measures of causal effect will be further illustrated in Sect. 3.2. □

Another reason for trying to create a realistic second-level model is that the possibility mentioned above of the conditioning on certain variables creating rather than removing confounding, although believed to be mostly theoretical by some, cannot be ruled out. For instance, if we try to reduce model (3.1) to first-level causality by conditioning on $\mathbf{X} := X_3$ alone, writing, in notation analogous to that of Example 3.1,

$$R = \rho(V, X_3, X_5, X_6) = \rho\big(V, X_3, \varphi_5(U_5, X_2), \varphi_6(U_6, T)\big) =: \check{\rho}(\check{V}, \mathbf{X}, T),$$

$$T = \tau(U, X_3, X_4) = \tau\big(U, X_3, \varphi_4(U_4, X_1)\big) =: \check{\tau}(\check{U}, \mathbf{X}),$$

we find that $\check{U}$ and $\check{V}$ are no longer conditionally independent because $\check{U}$ involves (U_4, U, X_1), $\check{V}$ involves (U_5, U_6, V, X_2), and $\check{U}$ and $\check{V}$ are 'entangled' by X_1 and X_2 through the conditioning event $\{\mathbf{X} = \mathbf{x}\} = \{\varphi_3(U_3, X_1, X_2) = \mathbf{x}\}$, and therefore conclude that unconfoundedness fails if X_3 is the only variable to be corrected for in (3.1); but if ρ and τ are both constant in their second arguments (so that neither R nor T involve X_3) then (3.1) is automatically a first-level model whose unconfoundedness is destroyed by conditioning only on X_3.

Although a model at the second-level of causality ought to be based principally on extra-statistical knowledge, it has been pointed out by Pearl and others (e.g. Sect. 2.5 of Pearl, Glymour and Jewel [73]) that such a model implies the conditional independence of certain variables and hence can be checked, in part, by means of data on those variables. Thus, in model (3.1), X_3 and X_4 are independent conditionally on X_1 because

$$\begin{aligned}\mathcal{L}\big(X_4\big|X_1 = x_1, X_3 = x_3\big) &= \mathcal{L}\big(\varphi_4(U_4, x_1)\big|X_1 = x_1, \varphi_3(U_3, x_1, X_2) = x_3\big)\\ &= \mathcal{L}\big(\varphi_4(U_4, x_1)\big)\end{aligned}$$

does not depend on x_3, so if data are available on X_1, X_3 and X_4 a test of independence provides a test of the model; similarly, the independence of X_3 and X_5 conditionally on X_2, the independence of X_1 and X_6 conditionally on T, etc., are implications of the model that can be used to criticize it. On the other hand, as in other questions of goodness-of-fit, unless the sample size is very large the non-rejection of conditional independence does not imply the approximate correctness of the portion of the model being tested.

These considerations might tempt one to conclude that it is possible to infer the 'causal structure' underlying a certain phenomenon from data alone (i.e. without substantial knowledge about the phenomenon proper), even if only in an asymptotic sense. That this is not the case can be seen from the model $R = \rho(V, X, T)$, $T = \tau(U, X)$, $X = \varphi(W)$, say, with X, T and R taking values in $\mathbb{N}$, which is essentially the basic model and may be said to represent the simplest of all models containing the possibility of confounding. As pointed out in Remark 2.1, given a large random sample of vectors (X_i, T_i, R_i) with the same distribution as (X, T, R), it is impossible, without the knowledge of the order by which the variables are generated in reality, to find out whether the (X_i, T_i, R_i) are determined in the same fashion as (X, T, R) or rather by

$$X_i = H^{-1}(W_i; R_i, T_i), \quad T_i = G^{-1}(U_i; R_i), \quad R_i = F^{-1}(V_i),$$

in notation analogous to that of Remark 2.1. Even in special cases, as when ρ is constant in the third argument and hence there is no treatment effect, testing for the independence of T and R conditionally on X will not help us to discover the model behind the data because both sets of equations lead to the distribution of (X, T, R). The situation does not improve if one ignores that the confounder X plays a role in the generation of T and R and does not collect data on it; for then, as exemplified by Simpson's paradox, it is not even possible, in general, to find out whether the effect of T on R—in the event that the former is known to precede the latter (part of subject matter knowledge, not to be taken for granted in general)—is positive or negative.

In the following section we shall have more to say about the identification of subsets of variables which allow the reduction of a second-level model to a first-level model—sets of confounders such as $\{X_1, X_3\}$, $\{X_2, X_3\}$ or $\{X_1, X_2, X_3\}$ in the case of model (3.1). After that we shall consider other, more involved causal effects pertaining to a model at the second-level of causality.

3.2 Reduction to first-level causality; Pearl's criterion

Let us think of a generic model at the second-level of causality involving a number of random variables $X_1, X_2, \ldots, T, R$, such that T and R are functions of some of the X_j, none of the X_j is a function of (T, R) and R is a function of T, and consider the problem of finding a set or vector $\mathbf{X}$ of X_j in the model which allows the estimation of the effect of T on R by conditioning upon it—a set which provides the means of reducing the model to a first-level model, as illustrated in the preceding examples. There is at least one set with the required property, namely the union $\mathbf{X} = \mathbf{X}_1 \cup \mathbf{X}_2$ of the set $\mathbf{X}_1$ of X_j that affect T with the set $\mathbf{X}_2$ of X_j that affect R (and which by our definitions are not affected by T); for then $R = \rho(V, \mathbf{X}_2, T)$, $T = \tau(U, \mathbf{X}_1)$, $\mathbf{X}$ does not involve T, and so we have, with $\mathbf{x}$, $\mathbf{x}_1$ and $\mathbf{x}_2$ numerical vectors such that $\mathbf{x} = (\mathbf{x}_1, \mathbf{x}_2)$ and $\mathbf{X} = \mathbf{x} \Leftrightarrow \mathbf{X}_1 = \mathbf{x}_1 \wedge \mathbf{X}_2 = \mathbf{x}_2$,

$$\begin{aligned} \mathcal{L}(R|T = t, \mathbf{X} = \mathbf{x}) &= \mathcal{L}\big(\rho(V, \mathbf{x}_2, t)\big|\tau(U, \mathbf{x}_1) = t, \mathbf{X} = \mathbf{x}\big) \\ &= \mathcal{L}\big(\rho(V, \mathbf{x}_2, t)\big). \end{aligned} \tag{3.3}$$

by the independence of U, $V$and the exogenous variables involved in $\mathbf{X}$, and the last term here characterizes the causal effect of T on R completely and can be estimated from observed data in the guise of the first term. The point is that there may be a smaller or in some sense more convenient $\mathbf{X}$.

Remark 3.2 In the complements to this section we consider situations in which some of the X_j are functions of T. Note, however, that if some of the X_j are functions of T and are also involved in R then we can write them explicitly in terms of T and other variables in order to achieve a 'reduced' expression for R and so that the conditions just stated are satisfied. For instance, in model (3.1) the response can be written as $R = \rho\big(V, X_3, X_5, \varphi_6(U_6, T)\big) \equiv \tilde{\rho}\big(\tilde{V}, X_3, X_5, T\big)$ so to exclude X_6 from $\mathbf{X}_2$. □

In what follows, denote by $\mathbf{X}$ a (possibly empty) set with the required property, or a candidate for such a set. Let $\check{\mathbf{X}}$ be the set of variables not in $\mathbf{X}$ that affect the treatment as arguments of τ and $\hat{\mathbf{X}}$ the set of variables not in $\mathbf{X}$ that affect the response as arguments of ρ. Fig. 3.2 sketches a graph of the model with the sets $\mathbf{X}$, $\check{\mathbf{X}}$ and $\hat{\mathbf{X}}$, and with the double, dashed arrows indicating that some nodes $X_i \in \mathbf{X}$, $\check{X}_j \in \check{\mathbf{X}}$ and $\hat{X}_k \in \hat{\mathbf{X}}$ may be functionally related—so that we may have, for instance, $X_i = \varphi_i(U_i, \check{X}_j)$, $\check{X}_j = \check{\varphi}_j(\check{U}_j, \hat{X}_k)$, which in a detailed representation of the graph would correspond to the path $X_i \leftarrow \check{X}_j \leftarrow \hat{X}_k$. Since τ and ρ generally have common arguments, $\check{\mathbf{X}}$ and $\hat{\mathbf{X}}$ are typically not disjoint, but by definition none of their elements is to be conditioned upon and hence $\mathbf{X} \cap \hat{\mathbf{X}} = \mathbf{X} \cap \check{\mathbf{X}} = \emptyset$. Since not all variables need to be involved in τ and ρ, $\mathbf{X} \cup \check{\mathbf{X}} \cup \hat{\mathbf{X}}$ need not contain all the X_j in the model; thus Fig. 3.2 may be a somewhat incomplete representation of the model.

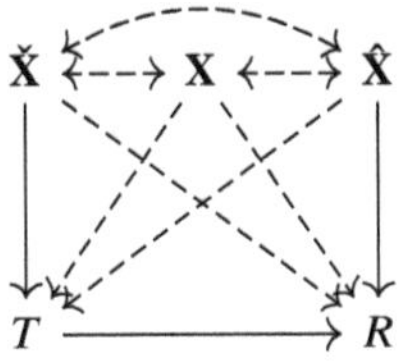

Fig. 3.2 Graphical representation of a second-level model to which Pearl's criterion applies.

To help fixing ideas, consider conditioning on $\mathbf{X} = \{X_2, X_3\}$ in model (3.1) without the equation defining X_6 and with $R = \rho\big(V, X_3, X_5, \varphi_6(U_6, T)\big)$. Since $T = \tau(U, X_3, X_4)$, we may take $\check{\mathbf{X}} = \{X_4\}$ and $\hat{\mathbf{X}} = \{X_5\}$. But because we can also write $T = \tau\big(U, X_3, \varphi_4(U_4, X_1)\big)$, we may take $\check{\mathbf{X}} = \{X_1\}$ and $\hat{\mathbf{X}} = \{X_5\}$ instead; and because we can also write $R = \rho\big(V, X_3, \varphi_5(U_5, X_2), \varphi(U_6, T)\big)$ we may take $\check{\mathbf{X}} = \{X_1\}$ and $\hat{\mathbf{X}} = \emptyset$ instead.

Conditioning on $\mathbf{X}$ will allow the estimation of the effect of T on R whenever

$$\begin{aligned} \mathcal{L}(R|T = t, \mathbf{X} = \mathbf{x}) &= \mathcal{L}\big(\rho(V, \hat{\mathbf{X}}, \mathbf{x}, t)\big|\tau(U, \check{\mathbf{X}}, \mathbf{x}) = t, \mathbf{X} = \mathbf{x}\big) \\ &= \mathcal{L}\big(\rho(V, \hat{\mathbf{X}}, \mathbf{x}, t)\big|\mathbf{X} = \mathbf{x}\big). \end{aligned} \tag{3.4}$$

But the second equality holds if and only if $\check{\mathbf{X}}$ and $\hat{\mathbf{X}}$ are independent conditionally on $\mathbf{X}$. Indeed, if this condition holds then $\tau(U, \check{\mathbf{X}}, \mathbf{x}) = t$ can be dropped from the second term in (3.4), because U and V are independent and they are not involved in $\mathbf{X}$, $\check{\mathbf{X}}$ or $\hat{\mathbf{X}}$; but if it fails then the treatment may, because of the constraint $\tau(U, \check{\mathbf{X}}, \mathbf{x}) = t$, exert an effect on the response through $\hat{\mathbf{X}}$—not only through the last argument of ρ.

The condition stated characterizes the desired sets $\mathbf{X}$, but it does not yet provide a direct, workable means of identifying them.[1]

As explained below, for the second equality in (3.4) to fail there must be at least two variables $\check{X} \in \check{\mathbf{X}}$ and $\hat{X} \in \hat{\mathbf{X}}$ which are dependent given $\mathbf{X}$. In the graph of a second-level model there is always a path (not necessarily a directed one) between nodes. Thus the dependence between $\check{X}$ and $\hat{X}$ implies that there is a path between them which links T and R (because $\check{X}$ points an arrow to T and $\hat{X}$ points one to R) and which represents a chain of random equations involving $\check{X}$ and $\hat{X}$ (hence also T and R) and other nodes.

As an example and in order to introduce some notation let us consider the path

$$T \leftarrow \check{X} \leftarrow N_1 \rightarrow N_2 \leftarrow N_3 \rightarrow N_4 \leftarrow N_5 \rightarrow \hat{X} \rightarrow R \tag{3.5}$$

with $N_1, N_2, \ldots$ standing for generic nodes (not necessarily belonging to $\mathbf{X}$, $\check{\mathbf{X}}$ or $\hat{\mathbf{X}}$). This path represents a chain of equations of the type

$$T = \tau(U, \check{X}, \ldots), \quad \check{X} = \check{\varphi}(\check{U}, N_1, \ldots), \quad N_2 = \psi_2(W_2, N_1, N_3, \ldots),$$

$$N_4 = \psi_4(W_4, N_3, N_5, \ldots), \quad \hat{X} = \hat{\varphi}(\hat{U}, N_5, \ldots), \quad R = \rho(V, \hat{X}, \ldots, T),$$

where $U, \check{U}, W_2, ..., V$ are independent standard uniforms, $\check{\varphi}, \hat{\varphi}, \psi_2, \ldots$ are certain functions, and the ellipsis after a list of arguments indicates the possible presence of other variables not in (3.5).

Pearl proposed what appears to be the most general criterion for deciding which variables to include in $\mathbf{X}$ (or not to include, when $\mathbf{X}$ is empty): *If* $\mathbf{X}$ *has the following property, then it secures the conditional independence of* $\check{\mathbf{X}}$ *and* $\hat{\mathbf{X}}$ *and hence the second identity in (3.4)*:

$\mathbf{X}$ is such that for each path P linking T to R with an arrow pointing at T and an arrow pointing at R one (hence only one) of these conditions holds:

(i) There is a node $X \in \mathbf{X}$ in P pointing an arrow to a node in P, so P contains a subpath of type $N' \rightarrow X \rightarrow N''$ or of type $N' \leftarrow X \rightarrow N''$;

(ii) No node of $\mathbf{X}$ in P points an arrow to a node in P, and P contains a subpath of the form $N' \rightarrow N \leftarrow N''$ where neither N nor any of its descendants is in $\mathbf{X}$.

Before proving the effectiveness of the criterion let us illustrate its application to the model of Fig. 3.1 without X_6: The path $T \leftarrow X_3 \rightarrow R$ does not permit **(ii)**, so **(i)** must hold and hence X_3 must be included in $\mathbf{X}$. But if **(i)** is to hold with the path

$$T \leftarrow X_4 \leftarrow X_1 \rightarrow X_3 \leftarrow X_2 \rightarrow X_5 \rightarrow R$$

then another node must be added to $\mathbf{X}$ (for otherwise X_3, the single node in it, would point no arrows in that path), and this may be X_1, X_2, X_4 or X_5. Since X_3 caters for the remaining two paths linking T to R, we conclude that if $\mathbf{X}$ consists of X_3 and at least one of X_1, X_2, X_4 or X_5 then it satisfies the criterion.

[1] Although $\mathbf{X} \cap \hat{\mathbf{X}} = \emptyset$, in general we are not allowed to drop the conditioning on $\{\mathbf{X} = \mathbf{x}\}$ from the last term of (3.4) because $\mathbf{X}$ and $\hat{\mathbf{X}}$ may be dependent.

Proof of Pearl's criterion:[2] Let $\mathbf{X}$ satisfy the criterion and suppose $\check{\mathbf{X}}$ and $\hat{\mathbf{X}}$ are non-empty and dependent given $\mathbf{X}$ (otherwise (3.4) holds and there is nothing to prove). We note first for later reference that if $\check{X} \in \check{\mathbf{X}}$ and $\hat{X} \in \hat{\mathbf{X}}$ then conditionally on $\{T = t, \mathbf{X} = \mathbf{x}\}$ we have

$$t = \tau(U, \check{\mathbf{X}}_1, \check{X}, \mathbf{x}), \quad R = \rho(V, \hat{\mathbf{X}}_1, \hat{X}, \mathbf{x}, t), \tag{3.6}$$

where $\check{\mathbf{X}}_1 = \check{\mathbf{X}} \setminus \{\check{X}\}$, $\hat{\mathbf{X}}_1 = \hat{\mathbf{X}} \setminus \{\hat{X}\}$, so

$$\mathcal{L}(R|T = t, \mathbf{X} = \mathbf{x}) = \mathcal{L}\big(\rho(V, \hat{\mathbf{X}}_1, \hat{X}, \mathbf{x}, t)\big|\tau(U, \check{\mathbf{X}}_1, \check{X}, \mathbf{x}) = t, \mathbf{X} = \mathbf{x}\big). \tag{3.7}$$

Conditionally on $\mathbf{X} = \mathbf{x} \equiv (x_1, x_2, \ldots)$ each pair of $\check{X} \in \check{\mathbf{X}}$ and $\hat{X} \in \hat{\mathbf{X}}$ satisfies

$$\check{X} = f(\check{W}, \check{W}_1, \check{W}_2, \ldots, \check{W}_p, x_{j_1} \ldots, x_{j_{p'}}) \tag{3.8}$$

and

$$\hat{X} = g(\hat{W}, \hat{W}_1, \hat{W}_2, \ldots, \hat{W}_q, x_{k_1} \ldots, x_{k_{q'}}) \tag{3.9}$$

for some functions f and g, some sets $\{\check{W}_1, \ldots, \check{W}_p\}$ and $\{\hat{W}_1, \ldots, \hat{W}_q\}$ of standard uniforms (none of which contains U or V), and certain coordinates $x_{j_1} \ldots, x_{j_{p'}}$, $x_{k_1} \ldots, x_{k_{q'}}$ of $\mathbf{x}$. This is because each node can be written as a composition of real-valued functions of uniform random variables. The conditional dependence of $\check{\mathbf{X}}$ and $\hat{\mathbf{X}}$ requires that a pair of constraints such as (3.8)–(3.9) hold for some $\check{X} \in \check{\mathbf{X}}$ and $\hat{X} \in \hat{\mathbf{X}}$ and that there be at least one constraint between some $\check{W}_j$ and some $\hat{W}_k$.

Suppose first that there exists a constraint determined by the single equation

$$\Psi_{j,k}(\check{W}_j, \hat{W}_k, \ldots) = 0, \tag{3.10}$$

where the ellipsis indicates a possible list of arguments consisting of uniforms and of x_i (this includes the possibility that $\check{W}_j = \hat{W}_k$). Then, writing $\check{N}_j$ for the unique node generated by $\check{W}_j$ and $\hat{N}_k$ for the unique node generated by $\hat{W}_k$, we have

$$\check{N}_j = \check{\psi}_j(\check{W}_j, \hat{W}_k, \ldots) \quad \text{or} \quad \hat{N}_k = \hat{\psi}_k(\hat{W}_k, \check{W}_j, \ldots), \tag{3.11}$$

or else

$$N_i = \psi_i(W_i, \check{W}_j, \hat{W}_k, \ldots) \tag{3.12}$$

for some node N_i. While (3.11) implies that $\check{N}_j$ is a function of $\hat{N}_k$ or $\hat{N}_k$ is a function of $\check{N}_j$, (3.12) implies that N_i is a function of both $\check{N}_j$ and $\hat{N}_k$.

In the first case it follows that there is a directed path leading from $\hat{N}_k$ to $\check{X}$ (via $\check{N}_j$) and a directed path leading from $\hat{N}_k$ to $\hat{X}$, or a directed path leading from $\check{N}_j$ to $\hat{X}$ (via $\hat{N}_k$) and a directed path leading from $\check{N}_j$ to $\check{X}$. Thus there is a path linking $\check{X}$ to $\hat{X}$ which includes no subpath of the type contemplated in condition **(ii)**; such a path must then satisfy condition **(i)** and hence contain a node $X_i \in \mathbf{X}$ leading to $\hat{X}$ or to $\check{X}$, or to both when X_i is an ancestor of both $\check{N}_j$ and $\hat{N}_k$.

---

[2] The ideas used in the proof are not needed in the sequel, but the remark that follows it is important for understanding the use of the criterion.

In the second case there is a path linking $\check{X}$ to $\hat{X}$ which consists of a directed subpath leading from $\check{N}_j$ to $\check{X}$, a directed subpath leading from $\hat{N}_k$ to $\hat{X}$, and an undirected subpath linking $\check{N}_j$ and $\hat{N}_k$. If any of these three subpaths contains a node $X_i \in \mathbf{X}$ different from N_i then it is essentially a path with the same properties as the path in the first case, the only difference being that it leads to one and only one of $\check{X}$ and $\hat{X}$. Otherwise, the third, undirected subpath must be $\check{N}_j \to N_i \leftarrow \hat{N}_k$; because this is a path contemplated in condition **(ii)**, neither N_i nor any of its descendants are in $\mathbf{X}$; but since $\check{W}_j$ and $\hat{W}_k$ are independent random variables unless the conditioning on $\mathbf{X}$ forces N_i or any of its descendants to take a numerical value, this possibility invalidates (3.12) as the constraint on $\check{W}_j$ and $\hat{W}_k$. Thus in the second case too the path linking $\check{X}$ to $\hat{X}$ satisfies **(i)**.

In the path in question let $X_i \in \mathbf{X}$ lead to $\check{X}$. Then the conditioning on $\mathbf{X}$ implies $X_i = x_i$ and $\check{X} = \tau_1(x_i, \ldots)$ for some function τ_1 whose arguments may include standard uniforms other than U, V and U_i (the uniform involved in the definition of X_i), and substitution in (3.8) shows that the first equation of (3.6) can be updated to

$$t = \tau(U, \check{\mathbf{X}}_1, \tau_1(x_i, \ldots), \mathbf{x}) \tag{3.13}$$

Similarly, if $X_i \in \mathbf{X}$ leads to $\hat{X}$ then substitution in (3.9) shows that the second equation of (3.6) can be updated to

$$R = \rho(V, \hat{\mathbf{X}}_1, \rho_1(x_i, \ldots), \mathbf{x}, t) \tag{3.14}$$

for some function ρ_1 sharing no random arguments with τ_1 (because the two paths from X_i to $\check{X}$ and from X_i to $\hat{X}$ go in opposite directions). Whatever the case (and both cases can occur under (3.10)), a number replaces a random variable in (3.6).

Now suppose that no two uniforms among those appearing in (3.8) and (3.9) can be found for which (3.10) holds. Then there is a chain of equations involving one $\check{W}_j$ from (3.8) and one $\hat{W}_k$ from (3.9), say

$$\Psi_{j,i_1}(\check{W}_j, W_{i_1}, \ldots) = 0, \quad \Psi_{i_1,i_2}(W_{i_1}, W_{i_2}, \ldots) = 0, \ldots,$$

$$\Psi_{i_{p-1},i_p}(W_{i_{p-1}}, W_{i_p}, \ldots) = 0, \quad \Psi_{i_p,k}(W_{i_p}, \hat{W}_k, \ldots) = 0,$$

rather than the single equation postulated in (3.11)–(3.12). Nevertheless, we can apply the previous argument with $\check{N}_j$ in place of $\check{X}$ and $\hat{N}_k$ in place of $\hat{X}$, with $\check{X}$ in place of T and $\hat{X}$ in place of R. In addition to equations (3.8) and (3.9) there will be a pair of analogous equations defining $\check{N}_j$ and $\hat{N}_k$ in terms of uniform random variables and coordinates of $\mathbf{x}$. If one of the analogues of (3.11)–(3.12) holds then there is a path linking $\check{X}$ to $\hat{X}$ in which $\check{N}_j$ leads to $\check{X}$, $\hat{N}_k$ leads to $\hat{X}$, or both, and hence as before (because $\check{X}$ leads to T and $\hat{X}$ leads to R) a path linking T to R in which $\check{N}_j$ leads to T and $\hat{N}_k$ leads to R; and then, as before, (3.13) or (3.14), or both, hold. If none of the analogues of (3.11)–(3.12) holds we can iterate the argument until one of them does, for the number of nodes is finite and eventually the analogues of $\check{N}_j$ and $\hat{N}_k$ must be the same or else be related by a single equation; when that happens the conclusion is again that (3.13) or (3.14), or both, hold.

What we have shown is that, whenever $\check{\mathbf{X}}$ and $\hat{\mathbf{X}}$ are dependent given $\mathbf{X}$, the first, the second or both equations in (3.6) may be reduced to (3.13), (3.14), or to these two equations combined. Since the reduction can be carried out with $\check{\mathbf{X}}_1$ and $\hat{\mathbf{X}}_1$ in place of $\check{\mathbf{X}}$ and $\hat{\mathbf{X}}$ and then repeated, we eventually conclude after so many steps that $\mathcal{L}(R|T = t, \mathbf{X} = \mathbf{x})$ has the form

$$\mathcal{L}\left(\rho\big(V, \hat{\mathbf{X}}_n, \rho_n(x_{i_1}, \ldots, x_{i_n}, \ldots), \mathbf{x}, t\big) \middle| \begin{array}{c}\tau\big(U, \check{\mathbf{X}}_m, \tau_m(x_{j_1}, \ldots, x_{j_m}, \ldots), \mathbf{x}\big) = t, \\ \mathbf{X} = \mathbf{x}\end{array}\right),$$

say, for some (possibly empty) sets $\check{\mathbf{X}}_m$ and $\hat{\mathbf{X}}_n$ no longer dependent given $\mathbf{X}$, and that from this last expression we can drop the constraint determined by τ to get[3]

$$\mathcal{L}(R|T = t, \mathbf{X} = \mathbf{x}) = \mathcal{L}\big(\rho(V, \hat{\mathbf{X}}_n, \rho_n(x_{i_1}, \ldots, x_{i_n}, \ldots), \mathbf{x}, t)\big|\mathbf{X} = \mathbf{x}\big).$$

Thus, after all, we conclude that $\check{\mathbf{X}}$ and $\hat{\mathbf{X}}$ are independent conditionally on $\mathbf{X}$, and therefore that (3.4) holds. □

Remark 3.3 The last term of (3.3) is the most basic characterization of the effect of treatment on response, since the only random element involved in it is the exogenous V. In general, as we have seen, the effect may be characterized by conditioning on a set $\mathbf{X}$ that is a proper subset of $\mathbf{X}_2$, so the last term in (3.4) may differ from the last term in (3.3), namely by being an integrated version of it.

For example, in model (3.1) without X_6 and with $R = \rho\big(V, X_3, X_5, \varphi_6(U_6, T)\big)$, we have $\mathbf{X}_1 = \{X_3, X_4\}$ and $\mathbf{X}_2 = \{X_3, X_5\}$, and if we take $\mathbf{X} = \mathbf{X}_1 \cup \mathbf{X}_2 = \{X_3, X_4, X_5\}$ in (3.3) then $\mathcal{L}(R|T = t, \mathbf{X} = (x_3, x_4, x_5)) = \mathcal{L}\big(\rho(V, x_3, x_5, \varphi_6(U_6, t))\big)$, which may be seen as the most basic characterization of the effect of treatment on response.

As seen earlier in Example 3.2, if in the same model we take $\mathbf{X} = \{X_1, X_3\}$ in (3.4), then $\check{\mathbf{X}} = \{X_4\}$, $\hat{\mathbf{X}} = \{X_5\}$ and the effect of the treatment on the response is characterized by

$$\mathcal{L}\big(R\big|T = t, \mathbf{X} = (x_1, x_3)\big) = \mathcal{L}\big(\rho(V, x_3, X_5, \varphi_6(U_6, t))\big|\varphi_3(U_3, x_1, X_2) = x_3\big),$$

while if we take $\mathbf{X} = \{X_2, X_3\}$ we obtain

$$\mathcal{L}\big(R\big|T = t, \mathbf{X} = (x_2, x_3)\big) = \mathcal{L}\big(\rho(V, x_3, \varphi_5(U_5, x_2), \varphi_6(U_6, t))\big);$$

if we take $\mathbf{X} = \{X_3, X_4\}$,

$$\mathcal{L}\big(R\big|T{=}t, \mathbf{X} = (x_3, x_4)\big) = \mathcal{L}\left(\rho(V, x_3, X_5, \varphi_6(U_6, t)) \middle| \begin{array}{c}\varphi_3(U_3, X_1, X_2) = x_3, \\ \varphi_4(U_4, X_1) = x_4\end{array}\right),$$

and if we take $\mathbf{X} = \{X_3, X_5\}$,

---

[3] If $\tau[U, \check{\mathbf{X}}_m, \tau_m(x_{j_1}, \ldots, x_{j_m}, \ldots), \mathbf{x}] = t$ involves some uniforms (which cannot be involved in $\rho[V, \hat{\mathbf{X}}_n, \rho_n(x_{i_1}, \ldots, x_{i_n}, \ldots), \mathbf{x}, t]$) also involved in $\mathbf{X} = \mathbf{x}$, then part of the latter constraint may have to be dropped together with the former, but after that it can be brought back into the conditioning event.

$$\mathcal{L}(R|T=t,\mathbf{X}=(x_3,x_5)) = \mathcal{L}(\rho(V,x_3,x_5,\varphi_6(U_6,t))).$$

Of these characterizations of treatment effect only the first and last coincide. If we integrate $\mathcal{L}(R|T=t,(X_3,X_4,X_5)=(x_3,x_4,x_5))$, the first measure of treatment effect, with respect to $\mathcal{L}(X_5|(X_3,X_4)=(x_3,x_4))$, we obtain $\mathcal{L}(R|T=t,(X_3,X_4)=(x_3,x_4))$, the penultimate measure. Similarly, the other measures of treatment effect can be obtained by integrating the first with respect to an appropriate distribution. This variety of effects reflects the different levels of detail contained in models at the second-level of causality.

The expressions of the treatment effect listed above are derived by a step-by-step substitution in ρ or τ of some of the $X_j \in \mathbf{X}$ by the values x_j they are conditioned to take (Prob. 3.7), a procedure formalized in the proof of Pearl's criterion. □

3.2.1 Complements

(A) With certain modifications the criterion can be applied to models in which some of the X_j are functions of T. As an example, consider the enlarged version of model (3.1) represented in Fig. 3.3 and to which the criterion does not apply: X_6, X_7, X_8 and X_9 are functions of T.

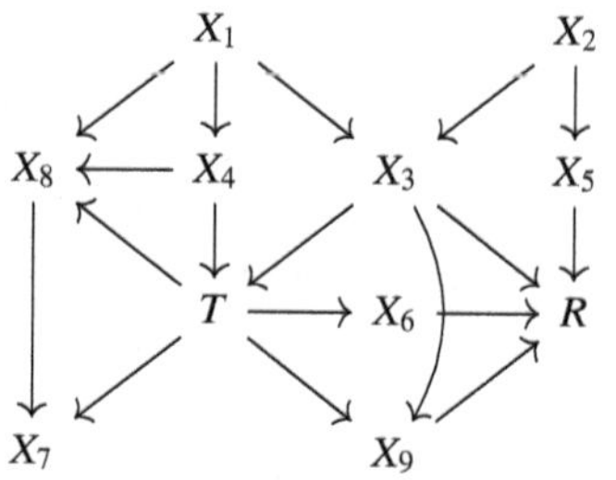

Fig. 3.3 Graph of a second-level model to which Pearl's criterion does not apply directly.

Suppose we include X_7 and X_8 in $\mathbf{X}$. Then in

$$\mathcal{L}(R|T=t,\mathbf{X}=\mathbf{x}) = \mathcal{L}(\rho(V,\hat{\mathbf{X}},\mathbf{x},t)|\tau(U,\check{\mathbf{X}},\mathbf{x})=t,\mathbf{X}=\mathbf{x}) \tag{3.15}$$

the conditioning event implies

$$T=\tau(U,X_3,X_4)=t, \quad X_7=\varphi_7(U_7,t,x_8)=x_7, \quad X_8=\varphi_8(U_8,X_1,X_4,t)=x_8.$$

The equality in the middle can be dropped from the right-hand side of (3.15) since it involves only an exogenous variable. Depending on which other variables are included in $\mathbf{X}$, the first equality may or may not be dropped; but if it is, then the third equality may subsist and, because it too depends on t, may affect the response, $\rho(V, X_3, X_5, \varphi_6(U_6, t), \varphi_9(U_9, X_3, t))$, as a function of t in the guise of X_3 or X_5.

Indeed, if we include X_3 and X_4 in $\mathbf{X}$ the conditioning event in (3.15) implies

$$\tau(U, x_3, x_4) = t, \quad \varphi_8(U_8, X_1, x_4, t) = x_8, \quad \varphi_3(U_3, X_1, X_2) = x_3, \quad \varphi_4(U_4, X_1) = x_4;$$

of these equalities, the first vanishes from the conditioning event in (3.15), but by the second and the third t may exert an effect on

$$\rho(V, x_3, \varphi_5(U_5, X_2), \varphi_6(U_6, t), \varphi_9(U_9, x_3, t))$$

also through X_2 (since X_1 and X_2 are entangled by the third equality and X_1 is provided information on t by the second). Thus, while the criterion shows that model (3.1)—and indeed the model represented by the graph of Fig. 3.3 *without* X_8—may be reduced to first level causality by conditioning on $\{X_3, X_4\}$, in the present model the addition of X_8 to $\{X_3, X_4\}$ prevents such a reduction.

To see how the criterion can deal with this situation, note that if part of $\mathbf{X}$, say $\mathbf{X}' \equiv \mathbf{X}'(T)$, is a function of T, and the rest, say $\mathbf{X}''$, is not, then, with $\mathbf{x}$, $\mathbf{x}'$ and $\mathbf{x}''$ numerical vectors such that $\mathbf{x} = (\mathbf{x}', \mathbf{x}'')$ and $\mathbf{X} = \mathbf{x} \Leftrightarrow \mathbf{X}' = \mathbf{x}' \wedge \mathbf{X}'' = \mathbf{x}''$, the conditioning event in (3.15) can be written as

$$\tau_t^*(U, \check{\mathbf{X}}, \mathbf{x}) := \big(\tau(U, \check{\mathbf{X}}, \mathbf{x}', \mathbf{x}''), \mathbf{X}'(t)\big) = (t, \mathbf{x}') =: t^*, \quad \mathbf{X}'' = \mathbf{x}'',$$

which is analogous to (3.15) with a pseudo-treatment τ_t^* in place of τ; in order to determine which other variables can be included in $\mathbf{X}''$ when $\mathbf{X}'(T)$ is included in $\mathbf{X} \equiv (\mathbf{X}', \mathbf{X}'')$, one can then apply the criterion to the graph that results from deleting the nodes of $\mathbf{X}'$ from the original graph, replacing τ and $\mathbf{X}'$ by $T_t^* := \tau_t^*(U, \check{\mathbf{X}}, \mathbf{X})$, and directing all the arrows pointing to τ or $\mathbf{X}'$ in the original graph to T_t^*.

In our example, because we can delete $\varphi_7(U_7, t, x_8) = x_7$ right away, the inclusion of $\mathbf{X}' = (X_7, X_8)$ in $\mathbf{X} \equiv (\mathbf{X}', \mathbf{X}'')$ leads to a conditioning event that includes

$$\tau_t^*(U, X_1, X_3, X_4, \mathbf{x}) := \big(\tau(U, X_3, X_4), \varphi_8(U_8, X_1, X_4, t)\big) = (t, x_8) =: t^*, \quad \mathbf{X}'' = \mathbf{x}'',$$

and to the graph shown in Fig. 3.4, to which Pearl's criterion may be applied: The path $T_t^* \leftarrow X_3 \rightarrow R$ does not permit **(ii)**, so **(i)** must hold and hence X_3 must be included in $\mathbf{X}$. If **(i)** is to hold with the path

$$T_t^* \leftarrow X_4 \leftarrow X_1 \rightarrow X_3 \leftarrow X_2 \rightarrow X_5 \rightarrow R$$

then at least one of X_1, X_2, X_4 or X_5 must be included in $\mathbf{X}$. But X_4 alone cannot deal with the path $T_t^* \leftarrow X_1 \rightarrow X_3 \leftarrow X_2 \rightarrow X_5 \rightarrow R$, so we conclude that if at least one of X_1, X_2 and X_5 joins X_3, X_7 and X_8 in $\mathbf{X}$ then the effect of T on R can be estimated from observed data.

Thus, for each subset $\mathbf{X}'$ of $\mathbf{X}$ that is a function of T, one can use Pearl's criterion to determine which subsets $\mathbf{X}''$ of $\mathbf{X}$ that are not functions of T permit the estimation of the effect of T on R by conditioning on $(\mathbf{X}', \mathbf{X}'')$.

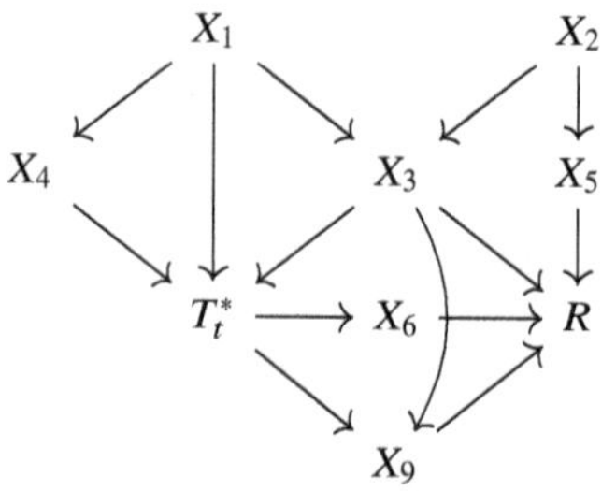

Fig. 3.4 Modified version of the graph of Fig. 3.3 to which Pearl's criterion may be applied to determine which variables in addition to X_7 and X_8 must be conditioned upon in order to allow the estimation of the effect of T on R.

(B) The situation may be different when the nodes that are functions of T are involved in the definition of R, as is the case with X_6 and X_9 in the model of Fig. 3.3, and one considers including them in **X**—which, as noted in Remark 3.2, is unnecessary.

Suppose we intend to condition on X_6, i.e. to include this variable in **X**. Then under the conditioning event in (3.15) we have

$$R = \rho(V, X_3, X_5, x_6, \varphi_9(U_9, X_3, t)), \quad \tau(U, X_3, X_4) = t, \quad \varphi_6(U_6, t) = x_6,$$

and we can apply the criterion to the graph obtained from that of Fig. 3.3 by deleting X_6, replacing T by

$$T_t^* := \tau_t^*(U, X_3, X_4) = \big(\tau(U, X_3, X_4), \varphi_6(U_6, t)\big),$$

and linking T_t^* directly to R. The conclusion is that including X_3 and at least one of X_1, X_2, X_4 and X_5 in addition to X_6 in **X** allows us to estimate the effect of T on R—or at least some of the effect of T on R. The problem is that by conditioning on $\{X_1, X_3, X_5, X_6\}$, for example, the effect is characterized by

$$\mathcal{L}\big(\rho(V, x_3, x_5, x_6, \varphi_9(U_9, x_3, t))\big) = \mathcal{L}\big(\rho(V, x_3, x_5, X_6, \varphi_9(U_9, x_3, t))\big|X_6 = x_6\big),$$

while by conditioning on $\{X_1, X_3, X_5\}$ the effect is characterized by

$$\mathcal{L}\big(\rho(V, x_3, x_5, \varphi_6(U_6, t), \varphi_9(U_9, x_3, t))\big),$$

and there is quite a difference between these two: while the second law may be said to apply to the whole population and describes the effect of treatment through two arguments of ρ, the first applies to a subpopulation in which part of the treatment effect is, so to speak, overruled by a fixed value of X_6; and whether one rather than the other is relevant depends on the problem in hand.

A more extreme situation is met with if one includes both X_6 and X_9 in **X**. Under the conditioning event in (3.15) we then have

$$R = \rho(V, X_3, X_5, x_6, x_9), \quad \tau(U, X_3, X_4) = t, \quad \varphi_6(U_6, t) = x_6, \quad \varphi_9(U_9, X_3, t) = x_9,$$

and if T is to exert any effect on R it is through a conditioning event consisting of the second and last of these equations. In the modified graph with T, X_6 and X_9 replaced by $T_t^* = \big(\tau(U, X_3, X_4), \varphi_9(U_9, X_3, t), \varphi_6(U_6, t)\big)$, only T_t^*, X_3 and X_5 point arrows at R, and we can apply the criterion to conclude that including X_3 and one of X_1, X_2, X_4 and X_5 in $\mathbf{X}$, in addition to X_6 and X_9, allows us to remove from (3.15) all the conditioning events that involve t. But, whatever the choice we make, the law that is obtained no longer involves t, which seems rather odd if we think—as we have been doing—of the treatment effect as the role played by an argument of a random function. Thus, if we condition on $\{X_1, X_3, X_6, X_9\}$, for example, the effect is characterized by $\mathcal{L}\big(\rho(V, x_3, \varphi_5(U_5, X_2), x_6, x_9)\big|\varphi_3(U_3, x_1, X_2) = x_3\big)$, while if we condition on $\{X_1, X_3, X_5, X_6, X_9\}$ it is characterized by $\mathcal{L}\big(\rho(V, x_3, x_5, x_6, x_9)\big)$, and not only are these different as they are constant in t.

In conclusion, although Pearl's criterion can be applied to models in which some of the X_j are functions of T, in models for which the response is a function of such X_j it may lead to characterizations of causal effects in which the role of T is partly or completely overruled by other variables.

3.3 More complicated causal effects

In some problems one needs to consider the effect on a response of *two or more treatments* associated with a sequence of situations; such problems presuppose a second-level model, and, because of the connection between the different treatments and situations, the specification and estimation of the joint effect of treatments cannot be reduced to an application of the methods of Chap. 2 with a single 'multiple treatment' and a single 'multiple situation'. An example of a general character will make clear what the difficulties and possibilities are.

Consider the model

$$\begin{cases} T = \tau(U, \mathbf{X}), \\ R = \rho(V, \mathbf{X}, T), \\ \mathbf{X}' = \xi(W, \mathbf{X}, T, R), \\ T' = \tau'(U', \mathbf{X}', T, R), \\ R' = \rho'(V', \mathbf{X}', T', T), \end{cases} \tag{3.16}$$

where conditionally on $\mathbf{X}$ (hence also unconditionally) the unobservable random variables U, U', V, V', W are independent and standard uniform and τ, τ', ξ, ρ and ρ' are some functions. This is a version of the deterministic model considered around (2.27) and (2.28) which may represent the application of two treatments, T and T', at different times under two situations, $\mathbf{X}$ and $\mathbf{X}'$, and the relationship of these variables with each other and with the corresponding responses, R and R': As indicated in Fig. 3.5, under an initial situation $\mathbf{X}$ a first treatment T is chosen and both situation and treatment influence the first response R; after some time, under a new situation $\mathbf{X}'$ influenced by $(\mathbf{X}, T, R)$, a new treatment T' is chosen, based on the most recent situation $\mathbf{X}'$, on the first response R and on the initial treatment T, and a final response R' is brought about by the new situation and the two treatments.

Remark 3.4 An equation such as $\mathbf{X} = \varphi(\eta)$, where φ is some function and η is a standard uniform independent of (U, U', V, V', W), could be added to (3.16), but it is unnecessary; similar observations apply to some of the models considered from now on. Evidently, the first two equations in (3.16), and the subgraph of nodes $\mathbf{X}$, T and R in Fig. 3.5, correspond to the basic model with unconfoundedness. Note that in this example the range of T' and the range of $\mathbf{X}'$, denoted here by $\mathcal{T}'$ and $\mathcal{X}'$, need not be the same as $\mathcal{T}$ and $\mathcal{X}$, respectively. □

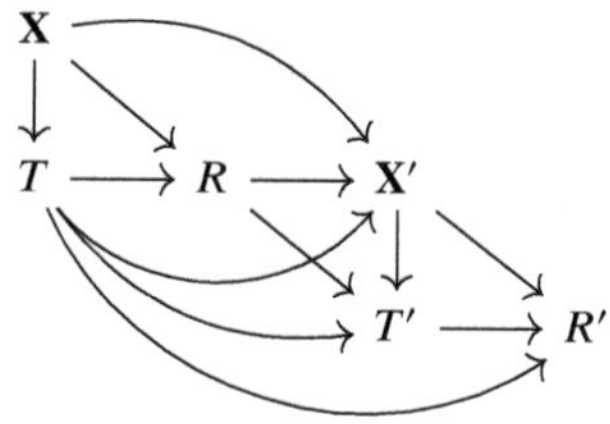

Fig. 3.5 Graph of the second-level causal model (3.16).

Just as in the case of the deterministic version of (3.16) mentioned around (2.27) and (2.28), one may consider the problem of determining the effect of the initial treatment T on the final response R': Successive substitutions yield

$$\begin{aligned} R' &= \rho'(V', \mathbf{X}', T', T) \\ &= \rho'(V', \xi[W, \mathbf{X}, T, R], \tau'[U', \mathbf{X}', T, R], T) \\ &= \rho'(V', \xi[W, \mathbf{X}, T, \rho\{V, \mathbf{X}, T\}], \tau'[U', \mathbf{X}', T, \rho\{V, \mathbf{X}, T\}], T) \\ &= \rho'(V', \xi[W, \mathbf{X}, T, \rho\{V, \mathbf{X}, T\}], \tau'[U', \xi[W, \mathbf{X}, T, R], T, \rho\{V, \mathbf{X}, T\}], T) \\ &=: \rho''(V'', \mathbf{X}, T), \end{aligned}$$

where V'' is a standard uniform random variable defined in terms of U', V, V' and W (e.g. by alternating the digits in the decimal expansions of these variables); and since these variables are independent of U conditionally on $\mathbf{X}$ we see that the basic model holds with unconfoundedness for a generic individual with data $(\mathbf{X}, T, R')$ and that the effect of T on R' can be investigated by the method of Chap. 2. Similarly, one may consider determining the effect of the second treatment T' on R' by conditioning on $(\mathbf{X}', T)$ (or on $(\mathbf{X}', T, R)$), since by the independence of U', V and V' the variables $R' = \rho'(V', \mathbf{X}', T', T)$ and $T' = \tau'(U', \mathbf{X}', T, R) = \tau'(U', \mathbf{X}', T, \rho\{V, \mathbf{X}, T\})$ are independent given $(\mathbf{X}', T)$ (and given $(\mathbf{X}', T, R)$); this, too, falls in the class of problems considered so far.

Investigating the joint effect of T and T' on the final response R' presents a different problem: In essence, it is about whether and how the law of the potential outcome

$$\rho'(V', \xi[W, \mathbf{X}, t, \rho\{V, \mathbf{X}, t\}], t', t) \tag{3.17}$$

varies with $(t,t') \in \mathcal{T} \times \mathcal{T}'$ and whether that law can be estimated from observed data, which need not be obvious because the potential outcome is unobservable. As posed, the question is incomplete because the law in question could, for example, be considered conditionally on $\{\mathbf{X} = \mathbf{x}\}$ for varying $\mathbf{x}$, but also unconditionally. We shall study the unconditional law of (3.17), which describes the effect of (T, T') on R' in a general sense, and in the end mention the conditional law.

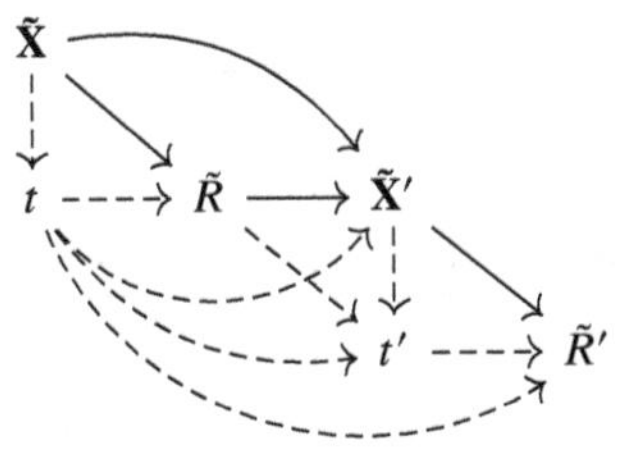

Fig. 3.6 'Intervention version' of the graph of Fig. (3.5); dashed-line arrows represent equations occurring in (3.16) but not in (3.18).

Note first that the potential outcome (3.17) can be seen as the result of forcing treatments T and T' to take the particular values t and t', discarding the first and penultimate equations in (3.16), and leaving $(\mathbf{X}, V, W, V')$ untouched, a procedure described by the 'intervention model'[4]

$$\begin{cases} \tilde{R} = \rho(\tilde{V}, \tilde{\mathbf{X}}, t), \\ \tilde{\mathbf{X}}' = \xi(\tilde{W}, \tilde{\mathbf{X}}, t, \tilde{R}), \\ \tilde{R}' = \rho'(\tilde{V}', \tilde{\mathbf{X}}', t', t), \end{cases} \tag{3.18}$$

where $(\tilde{\mathbf{X}}, \tilde{V}, \tilde{W}, \tilde{V}')$ has the same distribution as $(\mathbf{X}, V, W, V')$; it is represented in relation to the original model by the graph in Fig. 3.6.

Of course, the laws of $(\mathbf{X}, R, \mathbf{X}', R')$ and $(\tilde{\mathbf{X}}, \tilde{R}, \tilde{\mathbf{X}}', \tilde{R}')$ are generally different, but if in addition to (3.18) we introduce $\tilde{T} := \tau(\tilde{U}, \tilde{\mathbf{X}})$ and $\tilde{T}' := \tau'(\tilde{U}', \tilde{\mathbf{X}}', \tilde{T}, \tilde{R})$, with $(\tilde{U}, \tilde{U}')$ independent of $(\tilde{\mathbf{X}}, \tilde{V}, \tilde{W}, \tilde{V}')$ and with the same distribution as (U, U'), then

$$\mathbf{P}\big[(\tilde{\mathbf{X}}, \tilde{R}, \tilde{\mathbf{X}}', \tilde{R}') \in A, (\tilde{T}, \tilde{T}') = (t,t')\big] = \mathbf{P}\big[(\mathbf{X}, R, \mathbf{X}', R') \in A, (T, T') = (t,t')\big].$$

Using this observation and the independence of $\tilde{U}, \tilde{U}', \tilde{V}, \tilde{V}', \tilde{W}$ and $\tilde{\mathbf{X}}$, and assuming, in order to simplify the calculations, that the first response is discrete, we see that

$$\mathbf{P}(\tilde{R}' \le r') = \sum_{r,\mathbf{x},\mathbf{x}'} \mathbf{P}(\tilde{R}' \le r', \tilde{R} = r, \tilde{\mathbf{X}} = \mathbf{x}, \tilde{\mathbf{X}}' = \mathbf{x}')$$

[4] It is really a family of models indexed by (t, t'), but we ignore this in our notation. Again, this model can be seen as the result of an application of Pearl's '*do* operator' (cf. Remark 2.3).

$$
\begin{aligned}
&= \sum_{r,\mathbf{x},\mathbf{x}'} \mathbf{P}\big(\rho'(\tilde{V}',\mathbf{x}',t',t) \le r', \tilde{R} = r, \tilde{\mathbf{X}} = \mathbf{x}, \tilde{\mathbf{X}}' = \mathbf{x}'\big) \\
&= \sum_{r,\mathbf{x},\mathbf{x}'} \mathbf{P}\begin{pmatrix} \rho'(\tilde{V}',\mathbf{x}',t',t) \le r',\ \rho(\tilde{V},\mathbf{x},t) = r,\ \tilde{\mathbf{X}} = \mathbf{x}, \\ \xi(\tilde{W},\mathbf{x},t,\rho(\tilde{V},\mathbf{x},t)) = \mathbf{x}' \end{pmatrix} \\
&= \sum_{r,\mathbf{x},\mathbf{x}'} \frac{\mathbf{P}\begin{pmatrix} \rho'(\tilde{V}',\mathbf{x}',t',t)\le r',\rho(\tilde{V},\mathbf{x},t)=r,\tilde{\mathbf{X}}=\mathbf{x}, \\ \xi(\tilde{W},\mathbf{x},t,\rho(\tilde{V},\mathbf{x},t))=\mathbf{x}',\ \tau(\tilde{U},\mathbf{x})=t,\ \tau'(\tilde{U}',\mathbf{x}',t,r)=t' \end{pmatrix}}{\mathbf{P}\big(\tau(\tilde{U},\mathbf{x}) = t,\ \tau'(\tilde{U}',\mathbf{x}',t,r) = t'\big)} \\
&= \sum_{r,\mathbf{x},\mathbf{x}'} \frac{\mathbf{P}\big(\tilde{R}' \le r', \tilde{R} = r, \tilde{\mathbf{X}} = \mathbf{x}, \tilde{\mathbf{X}}' = \mathbf{x}', \tilde{T} = t,\ \tilde{T}' = t'\big)}{\mathbf{P}\big(\tau(\tilde{U},\mathbf{x}) = t,\ \tau'(\tilde{U}',\mathbf{x}',t,r) = t'\big)} \\
&= \sum_{r,\mathbf{x},\mathbf{x}'} \frac{\mathbf{P}\big(\tilde{R}' \le r' \big| \tilde{R} = r, \tilde{\mathbf{X}} = \mathbf{x}, \tilde{\mathbf{X}}' = \mathbf{x}', \tilde{T} = t,\ \tilde{T}' = t'\big)}{\varphi(\mathbf{x},\mathbf{x}',r,t,t')} \\
&= \sum_{r,\mathbf{x},\mathbf{x}'} \frac{\mathbf{P}\big(R' \le r' \big| R = r, \mathbf{X} = \mathbf{x}, \mathbf{X}' = \mathbf{x}', T = t,\ T' = t'\big)}{\varphi(\mathbf{x},\mathbf{x}',r,t,t')},
\end{aligned}
$$

where we wrote

$$
\begin{aligned}
\frac{1}{\varphi(\mathbf{x},\mathbf{x}',r,t,t')} &= \frac{\mathbf{P}\big(\tilde{R} = r, \tilde{\mathbf{X}} = \mathbf{x}, \tilde{\mathbf{X}}' = \mathbf{x}', \tilde{T} = t,\ \tilde{T}' = t'\big)}{\mathbf{P}\big(\tau(\tilde{U},\mathbf{x}) = t,\ \tau'(\tilde{U}',\mathbf{x}',t,r) = t'\big)} \\
&= \mathbf{P}\big(\tilde{R} = r, \tilde{\mathbf{X}} = \mathbf{x}, \tilde{\mathbf{X}}' = \mathbf{x}'\big).
\end{aligned}
$$

This sequence of identities results from an attempt to turn the probabilities $\mathbf{P}\big(\tilde{R}' \le r', \tilde{R} = r, \tilde{\mathbf{X}} = \mathbf{x}, \tilde{\mathbf{X}}' = \mathbf{x}'\big)$, which pertain to (3.18), into a product of probabilities pertaining to (3.16). It is carried out by 'completing' the events pertaining to (3.18) by the events $\{\tau(\tilde{U},\mathbf{x}) = t\}, \{\tau'(\tilde{U}',\mathbf{x}',t,r) = t'\}$, through multiplication and division by $\mathbf{P}\big(\tau(\tilde{U},\mathbf{x}) = t\big)$ and $\mathbf{P}\big(\tau'(\tilde{U}',\mathbf{x}',t,r) = t'\big)$. The procedure will be illustrated several times in Chap. 4.

Applying essentially the same procedure to $\mathbf{P}\big(\tilde{R} = r, \tilde{\mathbf{X}} = \mathbf{x}, \tilde{\mathbf{X}}' = \mathbf{x}'\big)$, we have

$$
\begin{aligned}
\frac{1}{\varphi(\mathbf{x},\mathbf{x}',r,t,t')} &= \frac{\mathbf{P}\big(\tilde{R} = r, \tilde{\mathbf{X}} = \mathbf{x}, \tilde{\mathbf{X}}' = \mathbf{x}', \tilde{T} = t\big)}{\mathbf{P}\big(\tau(\tilde{U},\mathbf{x}) = t\big)} \\
&= \frac{\mathbf{P}\begin{pmatrix} \rho(\tilde{V},\mathbf{x},t)=r,\tilde{\mathbf{X}}=\mathbf{x}, \\ \xi(\tilde{W},\mathbf{x},t,\rho(\tilde{V},\mathbf{x},t))=\mathbf{x}',\ \tau(\tilde{U},\mathbf{x})=t \end{pmatrix}}{\mathbf{P}\big(\tau(\tilde{U},\mathbf{x}) = t\big)} \\
&= \frac{\mathbf{P}\big(R = r, \mathbf{X} = \mathbf{x},\ \mathbf{X}' = \mathbf{x}', T = t\big)}{\mathbf{P}\big(\tau(U,\mathbf{x}) = t\big)} \\
&= \mathbf{P}\left(\mathbf{X}' = \mathbf{x}' \,\middle|\, \begin{matrix} \mathbf{X} = \mathbf{x}, T = t, \\ R = r \end{matrix}\right) \frac{\mathbf{P}\big(\mathbf{X} = \mathbf{x}, T = t, R = r\big)}{\mathbf{P}\big(\tau(U,\mathbf{x}) = t\big)} \\
&= \mathbf{P}\left(\mathbf{X}' = \mathbf{x}' \,\middle|\, \begin{matrix} \mathbf{X} = \mathbf{x}, T = t, \\ R = r \end{matrix}\right) \mathbf{P}\left(R = r \,\middle|\, \begin{matrix} \mathbf{X} = \mathbf{x}, \\ T = t \end{matrix}\right) \mathbf{P}(\mathbf{X} = \mathbf{x}).
\end{aligned}
$$

Putting the two results together we get

$$\mathbf{P}(\tilde{R}' \le r') = \sum_{r,\mathbf{x},\mathbf{x}'} \mathbf{P}(R' \le r' | \mathbf{X} = \mathbf{x}, T = t, R = r, \mathbf{X}' = \mathbf{x}', T' = t') \times \\ \mathbf{P}(\mathbf{X}' = \mathbf{x}' | \mathbf{X} = \mathbf{x}, T = t, R = r) \times \\ \mathbf{P}(R = r | \mathbf{X} = \mathbf{x}, T = t)\, \mathbf{P}(\mathbf{X} = \mathbf{x}). \tag{3.19}$$

In particular, (3.19) shows that the law of the potential outcome (3.17)—the law of the response $\tilde{R}' = \rho'(\tilde{V}', \tilde{\mathbf{X}}', t', t)$ of the intervention model (3.18)—can be estimated from data on $(\mathbf{X}, \mathbf{X}', T, T', R, R')$. It also shows that the joint effect of T and T' on the final response R' compounds the effect of T on the first response R conditionally on the first situation $\mathbf{X}$, the effect of $(\mathbf{X}, T, R)$ on the new situation $\mathbf{X}'$, and the effect of T' on R' conditionally on $(\mathbf{X}, T, R, \mathbf{X}')$.

This example illustrates two aspects of the analysis of a second-level model: the specification of the causal effect—which often amounts to choosing one among several meaningful measures of causal effect, like the unconditional law of (3.17)—and the investigation of whether the causal effect can be expressed in terms of quantities that can be estimated from observed data—a question that need not have a positive or obvious answer.

Regarding the interpretation and potential usefulness of a result such as (3.19), we may note that if the distribution of a potential outcome can indeed be estimated from observed data and, based on a large sample, is found to depend on the values of the treatment variables with high probability, then it is probable that a substantial number of individuals underlying the sample had responses which were nonconstant functions of the treatments. If this is the case, then whether one may or may not have convinced those individuals to opt for certain values of the treatments instead of those they chose at the time and whether one may or may not convince the same or similar individuals to opt for certain values of the treatments in the future are legitimate questions, but they do not diminish the evidence for an effect of the treatments.[5]

In the case of our example, knowing that

$$(t, t') \to \rho'(V', \xi[W, \mathbf{X}, t, \rho\{V, \mathbf{X}, t\}], t', t)$$

is nonconstant with positive probability essentially implies that

$$\mathbf{P}\{\rho'(V', \xi[W, \mathbf{X}, t, \rho\{V, \mathbf{X}, t\}], t', t) \in A \,|\mathbf{X} = \mathbf{x}\}$$

depends on (t, t') for some $\mathbf{x}$ and A. In a real-life problem this information could be used as a basis for prescribing or recommending particular values of treatment to individuals of a given population. In fact, it is easy to see from (3.19) that this last probability equals

[5] A similar observation has been made on pp. 32–32 in connection with the simpler problem of the effect of a single treatment.

$$\sum_{r,\mathbf{x}'} \mathbf{P}(R' \in A|\mathbf{X} = \mathbf{x}, T = t, R = r, \mathbf{X}' = \mathbf{x}', T' = t') \times$$
$$\mathbf{P}(\mathbf{X}' = \mathbf{x}'|\mathbf{X} = \mathbf{x}, T = t, R = r)\mathbf{P}(R = r|\mathbf{X} = \mathbf{x}, T = t);$$

since this too can be estimated from observed data, it could provide information about the most favourable directions in which to move t and t' in situations $\mathbf{x}$, and hence guide a future intervention on the population in question. Although in many cases the probabilities estimated from past data do not apply to future data, in some real-life problems the directions of the effects do, and sometimes it may be enough to get these right.

3.4 Notes

The terminology used here to describe levels of causality has been proposed by Cox and Wermuth [20].

Model (3.1) has been used as an example by Pearl and co-workers (e.g. [68], pp. 113–4, and [73], p. 64). In real-life problems, different variables in a set of equations such as (3.1) may represent measurements of the same quantity made at different times, the more recent ones being functions of the earlier ones.

The need for considering detailed models of cause and effect in real-life problems, rather than just the basic model of Chap. 2, has long been defended by Pearl (see pp. 708–9 of [66], pp. 129–130 of [68], or p. 5 of [69]); for a different opinion see Imbens and Rubin [45, 46]. Yet it must be admitted that if in relatively simple problems, such as those considered in Robins and Wasserman [82], setting up and justifying a second-level model is possible, in many observational studies of the type considered in Rosenbaum [83] and Imbens and Rubin [46] it may be simply impossible. In fields such as Economics and Sociology, identifying and estimating causal effects seems to be particularly difficult; see the valuable analyses of Freedman in the later chapters of Freedman [36] and in Chap. 5 of Freedman [37].

On the possible danger of creating confounding while attempting to remove it by conditioning on variables, see Pearl [69], Rubin [89] and Shrier [98].

Since realistic second-level models tend to be quite large, the method of checking one such model by testing all the conditional independencies implied by it should involve some form of multiple testing correction in order to control the number of incorrect rejections. If many tests are carried out then one thing to check is whether the histogram (or the probability plot) of the p-values (which should be 'weakly dependent') looks compatible with a uniform distribution, or with an appropriate null distribution.

In connection with the possibility of inferring the 'causal structure' behind data sets see pp. 243–54 of Freedman [37] and also Prob. 3.8 below.

For the origins of Pearl's criterion see pp. 17–8, 79–81 of Pearl [67], pp. 106, 114 of Pearl [68], and the earlier references provided in these sources. For other proofs of the criterion see Perković *et al.* [74]. A computer program is almost a necessity if one is to apply the criterion to large models; see Textor *et al.* [104].

As noted by Perković *et al.* ([74], p. 3), it is possible to determine which variables need to be conditioned upon in order to estimate an effect even when the relationships between some of the variables are not known exactly, although the measure of the effect may depend on what those relationships are; see Prob. 3.9 for an example.

Examples of the kind considered in Sect. 3.3 seem to have been studied first by Robins [80], but our reference for this section is Robins's commentary [81]. Equation (3.19) is a special case of a result due to Robins (see pp. 696–7 of Robins [81], pp. 102–3 of Pearl [67], and the references cited in these sources). The general result involves further iterations of the procedure, illustrated by (3.16), of defining a new situation, a new treatment and a new response based on earlier situations, treatments and responses; the corresponding identity—the penultimate in the theorem of Sect. 2 of Robins [81]—can be guessed from that of the special case. We note that the potential outcome (3.17) and $T' = \tau'(U', \mathbf{X}', T, R)$ are independent conditionally on the variables $T = \tau(U, \mathbf{X})$, $\mathbf{X}$, $R = \rho(V, \mathbf{X}, T)$ and $\mathbf{X}' = \xi(W, \mathbf{X}, T, R)$ that precede them; this corresponds to assumption (1) in Robins [81].

3.5 Problems

3.1 Individuals are invited to participate in a study of the effectiveness of a treatment. If they accept to participate then they are randomly assigned to the treatment group or to the 'control' group. If assigned to the treatment group an individual may accept treatment, decline and join the control group, or leave the study, and if assigned to the control group an individual may comply, demand treatment or leave the study; so the assignment is really an *invitation* to treatment. Of the individuals who refused the invitation or left the study no more is heard. After a certain period, a response which is expected to be affected by the treatment is measured on each of the remaining participants. Is it possible to say something about the effect of the treatment on the response with the data—the invitation to treatment, the reaction to the invitation, the value of the treatment and the response of each participant—from such a study?

In order to provide more structure to this question let $\mathbf{X}$ be a set of 'background characteristics' of a generic individual invited to participate in the study, A the individual's indicator of initial acceptance, I the random invitation to treatment, P the indicator of actual participation, T the actual treatment, with 0 denoting no treatment, R the response, and assume that

$$\begin{cases} \mathbf{X} = \xi(W_1), \\ A = \alpha(W_2, \mathbf{X}), \\ I = \iota(W_3, \mathbf{X}, A), \\ P = \pi(W_4, \mathbf{X}, A, I), \\ T = \tau(U, \mathbf{X}, A, I, P), \\ R = \rho(V, \mathbf{X}, A, I, P, T), \end{cases} \tag{3.20}$$

where A, I and P are binary, T and R are numerical, ξ is vector-valued, W_1, W_2, W_3, W_4, U and V are standard uniform random variables, and $\mathbf{P}(A = 1 = P) > 0$.

In what follows assume in addition that W_1, W_2, W_3, W_4 and U are independent. This appears realistic in applications where individuals are drawn randomly from some population and $\mathbf{X}$, A, I, P and T are realized sequentially. Indeed, the invitation to treatment is random, and conditionally on $\{\mathbf{X} = \mathbf{x}, \alpha(W_2, \mathbf{x}) = 1, \iota(W_3, \mathbf{x}, a) = i, \pi(W_4, \mathbf{x}, a, I) = 1\}$ the penultimate equation,

$$T = \tau(U, \mathbf{x}, 1, i, 1),$$

represents the drawing of a treatment among individuals with characteristics equal to $\mathbf{x}$ in the subpopulation of individuals who remain in the study after the invitation to treatment, which are the only ones providing data to the study; so it does not seem implausible that U should remain what it is upon the conditioning event.

On the other hand, in a real-life problem the independence of U and V may be problematic because in the subpopulation of individuals who stay in the study and have certain characteristics the response may be confounded by the actual treatment (as it may happen in any other population).

(a) Suppose that U and V are independent and independent of the other standard uniforms and that there exist $\mathbf{x}$ and $i \in \{0, 1\}$ such that $0 < \mathbf{P}(T = t|\mathbf{X} = \mathbf{x}, A = 1, I = i, P = 1) < 1$ for all t. Show that the effect of the treatment can be studied at least in a subpopulation by showing that

$$\mathcal{L}(R|\mathbf{X} = \mathbf{x}, A = 1, I = i, P = 1, T = t) = \mathcal{L}(\rho(V, \mathbf{x}, 1, i, 1, t)),$$

and that $\mathcal{L}(R|\mathbf{X} = \mathbf{x}, A = 1, P = 1, T = t)$ is an integrated version of this law.

(b) Suppose that U and V are dependent but independent of the other standard uniforms. In this case one may regard the invitation to treatment as an *intervention* on the population represented by the individuals recruited initially and propose to study the effect of that intervention in the subpopulation that accepts to remain in the study following the invitation, the assumption being that if the intervention has an effect then the treatment too has some effect (whatever that effect is). Show that this proposal is feasible if there exists $\mathbf{x}$ such that $0 < \mathbf{P}(I = i|\mathbf{X} = \mathbf{x}, A = 1, P = 1) < 1$ for $i = 0, 1$, by writing the response as

$$R = \tilde{\rho}(\tilde{W}, \mathbf{X}, A, I, P)$$

for some $\tilde{\rho}$, with $\tilde{W}$ standard uniform yielding (U, V) independently of W_1, W_2, W_3 and W_4 (cf. Prob. 1.4), and showing that

$$\mathcal{L}(R|\mathbf{X} = \mathbf{x}, A = 1, I = i, P = 1) = \mathcal{L}(\tilde{\rho}(\tilde{W}, \mathbf{x}, 1, i, 1)).$$

The estimation of this last law from data is an example of what is called an 'intention-to-treat analysis' ('intention' referring to the treatment indicated by randomization); see pp. 1–4, 15 of Freedman [36].

(c) Suppose the treatment is binary. Under the assumption of **(b)** it may sometimes be possible to say something about the effect of the treatment in a subpopulation, for example in the subpopulation of 'compliers'—individuals who not only remain in the study until the end but accept the treatment if indicated to them. Write $\tilde{\mathbf{P}}$ for the probabilities $\mathbf{P}$ conditional on $A = 1 = P$, so that for instance

$$\tilde{\mathbf{P}}(R \le r | I = i, T = t) = \mathbf{P}(R \le r | A = 1, I = i, P = 1, T = t),$$

and $\tilde{\mathbf{E}}$ for the expectation under $\tilde{\mathbf{P}}$. A natural measure of the treatment effect among compliers is the ratio

$$\epsilon(\mathbf{x}) := \frac{\tilde{\mathbf{E}}\big[\rho(V, \mathbf{X}, 1, I, 1, 1)\big|\mathbf{X} = \mathbf{x}, I = 1, T = 1\big]}{\tilde{\mathbf{E}}\big[\rho(V, \mathbf{X}, 1, I, 1, 0)\big|\mathbf{X} = \mathbf{x}, I = 1, T = 1\big]},$$

which compares the mean response among compliers when they are treated with the mean response when they are not treated. Although the expectation in the numerator pertains to an observable event, the one in the denominator does not; but we have

$$\mathbf{P}\left(\rho(V, \mathbf{X}, 1, I, 1, 0) \le r, \mathbf{X} = \mathbf{x}, A = 1 = P, I = 1, T = 1\right) =$$

$$\mathbf{P}\begin{pmatrix}\rho(V, \mathbf{X}, 1, I, 1, 0) \le r, \\ \mathbf{X} = \mathbf{x}, A = 1 = P, I = 1\end{pmatrix} - \mathbf{P}\begin{pmatrix}\rho(V, \mathbf{X}, 1, I, 1, 0) \le r, \\ \mathbf{X} = \mathbf{x}, A = 1 = P, I = 1, T = 0\end{pmatrix} =$$

$$\mathbf{P}(\rho(V, \mathbf{x}, 1, 1, 1, 0) \le r)\mathbf{P}\begin{pmatrix}\mathbf{X} = \mathbf{x}, I = 1, \\ A = 1 = P\end{pmatrix} - \mathbf{P}\begin{pmatrix}R \le r, \mathbf{X} = \mathbf{x}, A = 1 = P, \\ I = 1, T = 0\end{pmatrix},$$

and the last probability here also pertains to an observable event. Show that if

$$(I{=}0 \wedge P{=}1 \Rightarrow T{=}0) \wedge \mathcal{L}(\rho(V, \mathbf{x}, 1, 1, 1, 0)){=}\mathcal{L}(\rho(V, \mathbf{x}, 1, 0, 1, 0)) \ \forall \mathbf{x} \quad (3.21)$$

then

$$\epsilon(\mathbf{x}) = \frac{\tilde{\mathbf{E}}(R\,\mathbf{1}_{\{I=1, T=1\}}|\mathbf{X} = \mathbf{x})}{\tilde{\mathbf{E}}(R\,\mathbf{1}_{\{I=0, T=0\}}|\mathbf{X} = \mathbf{x})c(\mathbf{x}) - \tilde{\mathbf{E}}(R\,\mathbf{1}_{\{I=1, T=0\}}|\mathbf{X} = \mathbf{x})},$$

where

$$c(\mathbf{x}) := \frac{\tilde{\mathbf{P}}(I = 1|\mathbf{X} = \mathbf{x})}{\tilde{\mathbf{P}}(I = 0|\mathbf{X} = \mathbf{x})},$$

so that $\epsilon(\mathbf{x})$ can be estimated from data. Interpret the two conditions in (3.21) and consider whether they are always realistic.[6]

3.2 Suppose you want to define—or simulate data from—a model such as (3.1) but having a particular set of conditional distributions, namely a particular distribution function F_{X_1, X_2} for (X_1, X_2), a particular distribution function $x \to F_{X_3}^{(x_1, x_2)}(x) := \mathbf{P}(X_3 \le x | (X_1, X_2) = (x_1, x_2))$ for X_3 conditionally on (X_1, X_2), etc. Which functions $\varphi_1, \varphi_2, \ldots$ can you take for that purpose? What if the variables are required to have a nondegenerate multivariate normal distribution? *Suggestion:* Use Probs. 1.1, 1.2.

---

[6] For a sequel see Prob. 5.17.

3.3 How would you define—or simulate data from—the most general system of five random variables $X_1, \ldots, X_5$ in which X_2 is a function of X_1, X_3 a random function of (X_1, X_2), X_4 a random function of (X_1, X_2, X_3) and X_5 a random function of (X_1, X_2, X_3, X_5), the successive random functions being independent of each other and independent of the variables that precede them? Draw the graph that represents such a model. Which variables must one condition upon in order to determine the causal effect of X_4 on X_5? And the effect of X_1 on X_5?

3.4 Draw a graph representing the model of Prob. 2.10 and use Pearl's criterion to show that the effect of T on R can be determined by conditioning on a subset of $\mathbf{X}$ (cf. Prob. 2.10 (c)).

3.5 Use Pearl's criterion to show that in Prob. 2.12, in the special case mentioned at the end of (b), conditioning on $\mathbf{X}'_n$ suffices to quantify the effect of the treatment on the response.

3.6 Show that the second measure of causal effect in Example 3.2 is obtained by integrating the first. Why cannot the conditioning event be dropped from the second measure?

3.7 Verify the statements made in Remark 3.3.

3.8 Suppose that we were convinced that the real data followed the model represented by the graph of Fig. 3.1 but that in reality they followed another model represented by the same graph with an additional node X_7 linked directly to each of T, X_1, X_2 and R (without other additional edges or nodes).

(a) Would we have to include X_7 among the set of variables to be conditioned upon in order to determine the causal effect of T on R?

(b) If we had no idea about the existence of X_7 and no data on it we might still discover, based on a large sample on the other variables, that the model was not quite right; how? What if in the true model X_7 pointed no arrows to X_1 and X_2—would we still be able to find out that the assumed model was wrong? *Hint:* According to the assumed model, X_1 and X_2 should be independent.

3.9 Consider the model represented in Fig. 3.7, in which the dashed edges with no arrows represent lack of exact knowledge about the causal relation between some of the variables—whether X_1 influences X_2 or the other way around, etc. Depending on the direction of the missing arrows, conditioning on a given subset of the variables may or may not allow one to estimate the effect of T on R from data on $(X_1, \ldots, X_4, T, R)$.[7]

Use Pearl's criterion or direct calculation (as in Remark 3.3) to investigate which specifications of the arrows make it possible to estimate the effect of T on R by conditioning on (X_1, X_3). (Recall that when specifying the direction of the arrows one must avoid closed paths, such as $X_2 \to X_1 \to X_3 \to X_2$.)

[7] This exercise is based on Fig. 1 of Perković *et al.* [74].

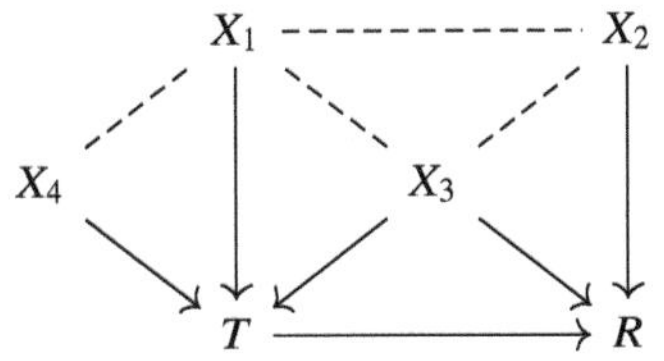

Fig. 3.7 Graph of a model in which the relations between some of the variables are unknown.

Show, in particular, that the specification represented in Fig. 3.8 does not allow the determination of the causal effect of T on R by conditioning on (X_1, X_3), and that the one represented in Fig. 3.9 does in the form of

$$\mathcal{L}\left(R \middle| \begin{matrix} X_1 = x_1, X_3 = x_3, \\ T = t \end{matrix}\right) = \mathcal{L}\big(\rho(V, \varphi_2(U_2, x_1), x_3, t) \big| \varphi_3(U_3, x_1 \varphi_2(U_2, x_1)) = x_3\big),$$

in notation familiar to us from this chapter.

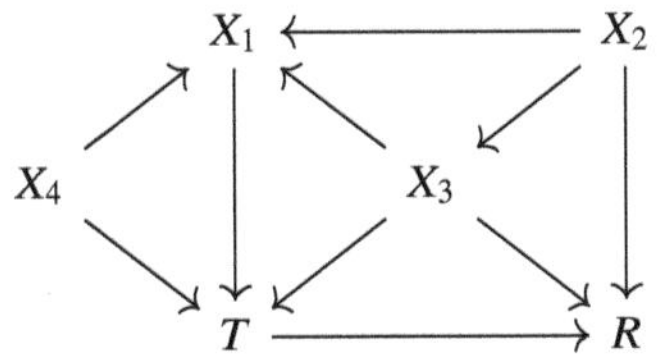

Fig. 3.8 A particular specification of the incomplete model represented in Fig. 3.7.

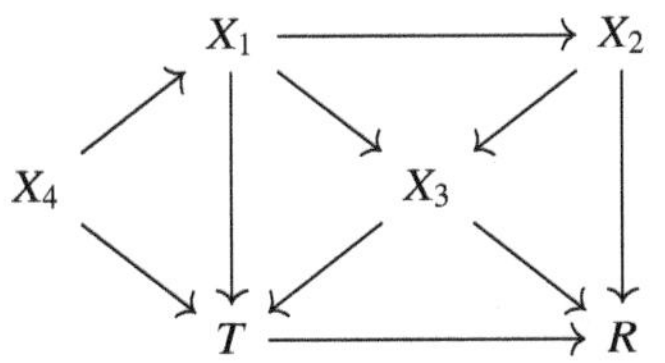

Fig. 3.9 A particular specification of the incomplete model represented in Fig. 3.7.

CHAPTER 4

Specification and identification of causal effects: some examples

The basic questions about a detailed model of cause and effect concern the specification of the effect of interest and the possibility of identifying or estimating it from observations. The causal effect is embodied in the probability law of a potential outcome, i.e. in the law of an unobservable random variable. Whether this law can be expressed by a functional of laws of observable random variables, hence whether it can be estimated through empirical versions of the latter laws, depends on which data are regarded as observable, and perhaps on how the observable data are sampled. This chapter examines such questions with examples considered by the various schools of statistical causality.

4.1 Introduction

Our purpose in this chapter is to study a number of problems on statistical causality which have been considered (as examples, mostly) in the literature, and to solve them by elementary probability. Apart from the description of case-control studies in terms of the basic model in Sect. 4.6, these problems involve calculations of the sort used in the derivation of (3.19). The sections that follow are independent and can be read in any order, though the first two sections treat similar problems and Sect. 4.4 could perhaps be read before Sect. 4.5.

4.2 Smoking and the genotype theory

Pearl considers the model

$$\begin{cases} X = \varphi_0(U_0), \\ Y = \varphi_1(X, U_1), \\ Z = \varphi_2(Y, U_2), \\ W = \varphi_3(X, Z, U_3), \end{cases} \tag{4.1}$$

where U_0, U_1, U_2, U_3 are independent standard uniforms, W, Y, Z are regarded as observable, X and the U_i as unobservable, and the φ_i as unknown real-valued functions. The model, summarized by Fig. 4.1, is intended to describe the combined effects on the development of lung cancer of smoking and of a putative genotype that is not only carcinogenic but also predisposes people to smoke; the random variables pertain to an individual randomly drawn from some population, X stands for the individual's genotype, Y for his level of smoking and Z, assumed to be discrete for simplicity, for the concentration of tar deposits in his lungs determined at some point in time, and W for the indicator of whether the individual develops lung cancer at a later time.

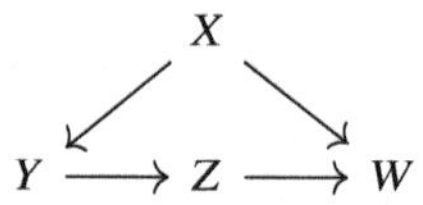

Fig. 4.1 Graph of model (4.1).

The question of interest is whether the effect of smoking on the development of lung cancer (the effect of Y on W) can be estimated despite the confounder of that effect (X) being unobservable; the answer is that it can, thanks to the knowledge about the concentration of tar in the lungs (Z), a more immediate cause of cancer. This model is probably not very realistic, but the question behind it is not trivial and could be of interest in similar problems.

Model (4.1) implies

$$W = \varphi_3\left(X, \varphi_2(Y, U_2), U_3\right), \quad Y = \varphi_1(X, U_1), \tag{4.2}$$

which we may identify as a version of the basic model with unconfoundedness: the 'treatment' Y is a function of X, the response W is a function of the treatment Y and of X, the latter confounds the effect of Y on the response, but since U_1 and U_2 are independent (conditionally on X as well as unconditionally) the treatment effect is fully characterized by $\mathcal{L}(W|X = x, Y = y)$ and may be quantified through functionals of it, such as

$$p_y(w) := \sum_x \mathbf{P}(W = w|X = x, Y = y)\mathbf{P}(X = x). \tag{4.3}$$

This can be estimated from data on (W, X, Y), but our assumption is that only data on (W, Y, Z) are available. In order to show that one can write $p_y(w)$ in terms of probabilities pertaining to W, Y and Z, we first put

$$m_z(w) := \sum_x \mathbf{P}(W = w|X = x, Z = z)\mathbf{P}(X = x)$$

$$= \sum_x \mathbf{P}[\varphi_3(x,z,U_3) = w | \varphi_0(U_0) = x, \varphi_2(\varphi_1(x,U_1),U_2) = z]\, \mathbf{P}(X = x)$$

$$= \sum_x \mathbf{P}(\varphi_3(x,z,U_3) = w)\mathbf{P}(X = x),$$

where in the last equality we make use of the independence of the U_i, and

$$l_y(w) := \sum_z m_z(w)\mathbf{P}(Z = z|Y = y), \tag{4.4}$$

and then note that

$$p_y(w) = \sum_x \mathbf{P}\left[\varphi_3(x, \varphi_2(y,U_2),U_3) = w \,|\, \varphi_0(U_0) = x, \varphi_1(x,U_1) = y\right] \mathbf{P}(X = x)$$

$$= \sum_x \mathbf{P}\left[\varphi_3(x, \varphi_2(y,U_2),U_3) = w\right] \mathbf{P}(X = x)$$

$$= \sum_{x,z} \mathbf{P}\left[\varphi_3(x, \varphi_2(y,U_2),U_3) = w, \varphi_2(y,U_2) = z\right] \mathbf{P}(X = x)$$

$$= \sum_z \mathbf{P}\left[\varphi_2(y,U_2) = z\right] \sum_x \mathbf{P}\left[\varphi_3(x,z,U_3) = w\right] \mathbf{P}(X = x)$$

$$= \sum_z \mathbf{P}\left[\varphi_2(y,U_2) = z\right] m_z(w) = \sum_z \mathbf{P}(Z = z|Y = y)\, m_z(w)$$

$$= l_y(w),$$

the penultimate step following from

$$\mathbf{P}(Z = z|Y = y) = \mathbf{P}[\varphi_2(y,U_2) = z|\varphi_1(\varphi_0(U_0),U_1) = y] = \mathbf{P}(\varphi_2(y,U_2) = z).$$

Thus the result will be proved if we prove that $m_z(w)$ can be written in terms of probabilities pertaining to W, Y and Z.

By the independence of X, U_1, U_2 and U_3, we have

$$\mathbf{P}(W = w|Y = y, Z = z) = \frac{\mathbf{P}\left[\varphi_3(X,z,U_3) = w, \varphi_1(X,U_1) = y, \varphi_2(y,U_2) = z\right]}{\mathbf{P}(Y = y, Z = z)}$$

$$= \sum_x \frac{\mathbf{P}\begin{pmatrix} \varphi_3(x,z,U_3)=w,\ \varphi_1(x,U_1)=y, \\ \varphi_2(y,U_2)=z, X=x \end{pmatrix}}{\mathbf{P}(\varphi_1(X,U_1) = y, \varphi_2(y,U_2) = z)}$$

$$= \sum_x \frac{\mathbf{P}(\varphi_3(x,z,U_3) = w)\mathbf{P}\begin{pmatrix} \varphi_1(x,U_1)=y,\ \varphi_2(y,U_2)=z, \\ X=x \end{pmatrix}}{\mathbf{P}(\varphi_1(X,U_1) = y)\mathbf{P}(\varphi_2(y,U_2) = z)}$$

$$= \sum_x \frac{\mathbf{P}(\varphi_3(x,z,U_3) = w)\mathbf{P}\begin{pmatrix} \varphi_1(X,U_1)=y, \\ X=x \end{pmatrix}}{\mathbf{P}(\varphi_1(X,U_1) = y)}$$

$$= \sum_x \mathbf{P}(\varphi_3(x, z, U_3) = w)\mathbf{P}(X = x|Y = y)$$

(incidentally, the last couple of steps show that X and Z are independent given Y). Since

$$\mathbf{P}(\varphi_3(x, z, U_3) = w) = \mathbf{P}(W = w|Z = z, X = x)$$

(an observation already used around the definition of $m_z(w)$), the preceding identity is equivalent to

$$\mathbf{P}(W = w|Y = y, Z = z) = \sum_x \mathbf{P}(W = w|Z = z, X = x)\mathbf{P}(X = x|Y = y).$$

Finally, integrating both sides here with respect to the distribution of Y we get

$$\sum_{y'} \mathbf{P}(W = w|Y = y', Z = z)\mathbf{P}(Y = y') = \sum_x \mathbf{P}(W = w|Z = z, X = x)\mathbf{P}(X = x)$$
$$= m_z(w),$$

and the first term involves only probabilities pertaining to W, Y and Z.

In conclusion, the effect of Y on W in the form of (4.3) can be estimated from data on (W, Y, Z) because

$$p_y(w) = \sum_{y',z} \mathbf{P}(W=w|Y=y', Z=z)\mathbf{P}(Y=y')\mathbf{P}(Z=z|Y=y) = l_y(w).$$

This result has been obtained by identifying the effect of smoking on cancer through the reduced model (4.2) and using the information of the full model (4.1) to derive an alternative expression—that of $l_y(w)$ in (4.4)—for a measure of that effect—namely $p_y(w)$. But $l_y(w)$ has its own causal interpretation within the full causal mechanism: Under (4.1), $m_z(1)$ quantifies the effect of a concentration z of deposited tar on the development of lung cancer irrespective of genotype, since, for example, $m_z(1) - m_{z'}(1)$ is equal to

$$\sum_x \{\mathbf{P}(W = 1|X = x, Z = z) - \mathbf{P}(W = 1|X = x, Z = z')\}\mathbf{P}(X = x),$$

the expected difference between the 'risk' of cancer at different concentrations z and z' of deposited tar for an individual/genotype randomly drawn from the population. The differences $m_z(1) - m_{z'}(1)$ are meaningful because under model (4.1) the development of lung cancer is purely a function of genotype and tar accumulation and of other, 'exogenous factors' embodied by U_3. On the other hand, tar deposits occur only through smoking (in particular they are not affected by genotype), so $\mathbf{P}(Z = z|Y = y)$ is a proper measure of the effect of smoking on tar accumulation. Thus (see (4.4)), $l_y(1)$ quantifies the effect of smoking on cancer by averaging the risk of cancer as consequence of tar accumulation with respect to the distribution of the tar accumulation that results from smoking at level y.

4.3 Eelworms, fumigants and crop yields

Another model of Pearl's is

$$\begin{cases} A = \varphi_0(\xi_0), B = \varphi_1(A, \xi_1), \\ U = \varphi_2(A, \xi_2), \\ X = \varphi_3(A, \xi_3), \\ V = \varphi_4(U, X, \xi_4), \\ W = \varphi_5(B, V, \xi_5), \\ Y = \varphi_6(X, V, W, \xi_6), \end{cases} \tag{4.5}$$

where the ξ_i are independent, U, V, W, X and Y are regarded as observable, all the other variables (including A and B) as unobservable, and the φ_i are regarded as unknown. This is intended to represent the effects of the population density of eelworms, the application of fumigants, and the predation of eelworms by birds and other animals, on the crop yield of oats, as indicated in Fig. 4.2. The amount of fumigants applied, X, is decided by the farmer based on intuition and past experience (last year's measurements, impressions, etc.), all of which constitute an unmeasurable and hence unobservable set of quantities A. Some parts of A at least have an effect on the estimate of the eelworm population obtained before fumigation, U, and on the predatory activity on eelworms, represented by B, which, being difficult to quantify, is regarded as unobservable. The eelworm population is estimated again as V some time after fumigation, and then again later in the season, when the effect of the fumigant is lost, as W. While V is affected by X, the only systematic effects on W are due to V and to B. Finally, the yield of oats, Y, is influenced by X, V and W.

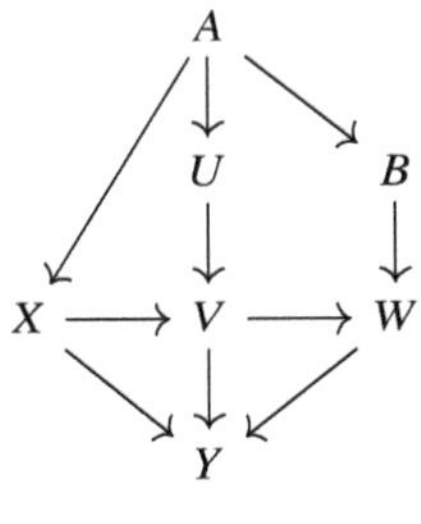

Fig. 4.2 Graph of model (4.5).

The question of interest is similar to that of Sect. 4.2: can one estimate the effect of X on Y—the effect of fumigation on crop yield—from observed data on (U, V, W, X, Y) (hence without making use of data on (A, B)), and, if so, how?

Note that (4.5) implies

$$\begin{aligned} Y &= \varphi_6(X, V, W, \xi_6) = \varphi_6(X, V, \varphi_5(B, V, \xi_5), \xi_6) \\ &= \varphi_6\left[X, \varphi_4(U, X, \xi_4), \varphi_5(B, \varphi_4(U, X, \xi_4), \xi_5), \xi_6\right] \end{aligned}$$

$$= \varphi_6\left[X, \varphi_4(\varphi_2(A,\xi_2), X, \xi_4), \varphi_5(B, \varphi_4(\varphi_2(A,\xi_2), X, \xi_4), \xi_5), \xi_6\right]$$
$$=: \rho(X, A, \xi),$$

say, the third argument of ρ being occupied by $\xi := (\xi_j)_{j\neq 3}$. Together with $X = \varphi_3(A, \xi_3)$ this may be identified with the basic model with unconfoundedness; the effect of X on Y is characterized by $\mathcal{L}(Y|A = a, X = x)$ and may be quantified through functionals of it, such as

$$\mu_x(y) := \sum_a \mathbf{P}(Y = y|X = x, A = a)\mathbf{P}(A = a). \tag{4.6}$$

Because data on A are unobservable, it is not clear that this can be estimated; we shall prove that it can through manipulations of the kind used in our previous example.

By the independence of the ξ_i,

$$\mathbf{P}\left(\varphi_6(x, v, w, \xi_6) = y \,\middle|\, \begin{matrix} X = x, W = w, \\ V = v, A = a \end{matrix}\right) = \mathbf{P}(\varphi_6(x, v, w, \xi_6) = y)$$

and

$$\mathbf{P}(\varphi_6(x, v, w, \xi_6) = y \,|X = x, W = w, V = v) = \mathbf{P}(\varphi_6(x, v, w, \xi_6) = y),$$

so

$$\mathbf{P}\left(Y = y \,\middle|\, \begin{matrix} X = x, W = w, \\ V = v, A = a \end{matrix}\right) = \mathbf{P}(Y = y|X = x, W = w, V = v).$$

Using this equality we get

$$\mathbf{P}(Y = y|X = x, A = a) = \sum_{v,w} \mathbf{P}(Y = y, W = w, V = v|X = x, A = a)$$

$$= \sum_{v,w} \frac{\mathbf{P}\left(Y = y \,\middle|\, \begin{smallmatrix} X=x, W=w, \\ V=v, A=a \end{smallmatrix}\right)\mathbf{P}\left(\begin{smallmatrix} X=x, W=w, \\ V=v, A=a \end{smallmatrix}\right)}{\mathbf{P}(X = x, A = a)}$$

$$= \sum_{v,w} \mathbf{P}\left(Y = y \,\middle|\, \begin{matrix} X = x, W = w, \\ V = v \end{matrix}\right) \frac{\mathbf{P}\left(\begin{smallmatrix} X=x, W=w, \\ V=v, A=a \end{smallmatrix}\right)}{\mathbf{P}(X = x, A = a)}.$$

Writing $\Psi(a, v, w, x)$ for the second term of the summand on the right we thus have the identity

$$\mathbf{P}(Y = y|X = x, A = a) = \sum_{v,w} \mathbf{P}\left(Y = y \,\middle|\, \begin{matrix} X = x, W = w, \\ V = v \end{matrix}\right) \Psi_1(a, v, w, x), \tag{4.7}$$

and we have to show that $\Psi_1(a, v, w, x)\mathbf{P}(A = a)$ can be estimated from observed data (the first term of the summand can).

Note that $\Psi_1(a, v, w, x) = \mathbf{P}(W = w, V = v|X = x, A = a)$; since this is formally analogous to the probability $\mathbf{P}(Y = y|X = x, A = a)$ we have started with, we may try and reapply the procedure just used: We have

$$\Psi_1(a,v,w,x) = \sum_u \mathbf{P}(W=w, V=v, U=u | X=x, A=a)$$

$$= \sum_u \mathbf{P}\left(W=w \,\middle|\, \begin{matrix} X=x, V=v, \\ U=u, A=a \end{matrix}\right) \mathbf{P}\left(V=v \,\middle|\, \begin{matrix} X=x, U=u, \\ A=a \end{matrix}\right) \times \mathbf{P}(U=u \,|X=x, A=a)$$

and, using as always the independence of the ξ_i,

$$\mathbf{P}(U=u\,|X=x, A=a) = \mathbf{P}(\varphi_2(a,\xi_2)=u),$$

$$\mathbf{P}(V=v\,|X=x, U=u, A=a) = \mathbf{P}(V=v\,|X=x, U=u),$$

$$\mathbf{P}(W=w\,|X=x, V=v, U=u, A=a) = \mathbf{P}(\varphi_5\{\varphi_1(a,\xi_1), v, \xi_5\}=w).$$

Thus

$$\Psi_1(a,v,w,x) = \sum_u \mathbf{P}\left(V=v \,\middle|\, \begin{matrix} X=x, \\ U=u \end{matrix}\right) \mathbf{P}\left(\begin{matrix} \varphi_5\{\varphi_1(a,\xi_1), v, \xi_5\}=w \\ \varphi_2(a,\xi_2)=u \end{matrix}\right)$$

$$= \sum_u \mathbf{P}\left(V=v \,\middle|\, \begin{matrix} X=x, \\ U=u \end{matrix}\right) \frac{\mathbf{P}\left(\begin{matrix} \varphi_5\{\varphi_1(a,\xi_1), v,\xi_5\}=w, \\ U=u, A=a \end{matrix}\right)}{\mathbf{P}(A=a)}.$$

To see how much progress we have made, let us multiply both sides of (4.7) by $\mathbf{P}(A=a)$ and sum over a; using the last equality we get (see (4.6))

$$\mu_x(y) = \sum_{u,v,w} \mathbf{P}\left(Y=y \,\middle|\, \begin{matrix} X=x, W=w, \\ V=v \end{matrix}\right) \mathbf{P}\left(V=v \,\middle|\, \begin{matrix} X=x, \\ U=u \end{matrix}\right) \sum_a \Psi_2(a,u,v,w),$$

where

$$\Psi_2(a,u,v,w) = \mathbf{P}(\varphi_5\{\varphi_1(a,\xi_1), v, \xi_5\}=w, U=u, A=a).$$

This last probability is the remaining problematic term. In it, $\varphi_5\{\varphi_1(a,\xi_1), v, \xi_5\}$ could almost be replaced by W, were it not for the absence of the event $\{V=v\}$. We cannot slip $\{\varphi_4(u,X,\xi_4)=v\}$ into our probabilities because $X=\varphi_3(A,\xi_3)$ depends on A; but the event $\{\varphi_4(u,x',\xi_4)=v, \varphi_3(a,\xi_3)=x'\}$ is allowed to appear there:

$$\Psi_2(a,u,v,w) = \sum_{x'} \mathbf{P}(\varphi_5\{\varphi_1(a,\xi_1), v, \xi_5\}=w, \varphi_3(a,\xi_3)=x', U=u, A=a)$$

$$= \sum_{x'} \frac{\mathbf{P}\left(\begin{matrix} \varphi_5\{\varphi_1(a,\xi_1), v, \xi_5\}=w,\ \varphi_4(u,x',\xi_4)=v, \varphi_3(a,\xi_3)=x', \\ U=u, A=a \end{matrix}\right)}{\mathbf{P}(\varphi_4(u,x',\xi_4)=v)}$$

$$= \sum_{x'} \frac{\mathbf{P}\left(\substack{\varphi_5\{\varphi_1(A,\xi_1),V,\xi_5\}=w,\ \varphi_4(U,x',\xi_4)=v,\varphi_3(A,\xi_3)=x',\\ U=u,A=a}\right)}{\mathbf{P}(\varphi_4(u,x',\xi_4)=v)}$$

$$= \sum_{x'} \frac{\mathbf{P}(X=x', W=w, V=v, U=u, A=a)}{\mathbf{P}(\varphi_4(u,x',\xi_4)=v)}.$$

Summing over a rids us of probabilities pertaining to A and leads us to

$$\mu_x(y) = \sum_{u,v,w} \mathbf{P}\left(Y=y \,\middle|\, \begin{matrix} X=x, W=w, \\ V=v \end{matrix}\right) \mathbf{P}\left(V=v \,\middle|\, \begin{matrix} X=x, \\ U=u \end{matrix}\right) \Psi_3(u,v,w),$$

where

$$\Psi_3(u,v,w) = \sum_{x'} \mathbf{P}(W=w\,|X=x', V=v, U=u)\, \frac{\mathbf{P}(X=x', V=v, U=u)}{\mathbf{P}(\varphi_4(u,x',\xi_4)=v)}.$$

Finally,

$$\frac{\mathbf{P}(X=x', V=v, U=u)}{\mathbf{P}(\varphi_4(u,x',\xi_4)=v)} = \frac{\mathbf{P}(\varphi_4(u,x',\xi_4)=v, \varphi_3(A,\xi_3)=x', \varphi_2(A,\xi_2)=u)}{\mathbf{P}(\varphi_4(u,x',\xi_4)=v)}.$$

$$= \mathbf{P}(\varphi_3(A,\xi_3)=x', \varphi_2(A,\xi_2)=u)$$

$$= \mathbf{P}(X=x', U=u),$$

so

$$\mu_x(y) = \sum_{v,w} \mathbf{P}\left(Y=y \,\middle|\, \begin{matrix} X=x, V=v, \\ W=w \end{matrix}\right) \sum_{u} \mathbf{P}\left(V=v \,\middle|\, \begin{matrix} X=x, \\ U=u \end{matrix}\right)$$

$$\times \sum_{x'} \mathbf{P}(W=w\,|V=v, X=x', U=u)\,\mathbf{P}(X=x', U=u),$$

and this indeed can be estimated from data on (U, V, W, X, Y).

Let us try to interpret this formula:

$$\mathbf{P}(Y=y|X=x, V=v, W=w) = \mathbf{P}(\varphi_6(x,v,w,\xi_6)=y)$$

characterizes a causal effect of X on Y, though it is an effect under the situation where V and W are fixed at numerical values v and w rather than being affected by X; it is the 'direct effect' of X on Y, as opposed to the 'indirect effect' that X has on Y through V and W, a distinction considered in more detail in the next section. The effect of X on V, which can be studied by considering the sub-model consisting of A, U, V, X, is obtained by conditioning on U (since A is unobservable):

$$\mathbf{P}(V=v|X=x, U=u) = \mathbf{P}(\varphi_4(u,x,\xi_4)=v).$$

The effect of V on W is studied from the model without Y, from which we see (using Pearl's criterion, for example) that it is obtained by conditioning on X and U:

$$\mathbf{P}\left(W = w \,\middle|\, \begin{matrix} V = v, X = x', \\ U = u \end{matrix}\right) = \mathbf{P}\left(\varphi_5(B, v, \xi_5) = w \,\middle|\, \begin{matrix} \varphi_3(A, \xi_3) = x', \\ \varphi_2(A, \xi_2) = u \end{matrix}\right).$$

Thus $\mu_x(y)$ averages the direct effect of X on Y over the product of its indirect effects via V and W integrated with respect to the distribution of U and X, the variables being conditioned upon for the determination of the indirect effects.

4.4 Direct effect of initial treatment in a two-stage treatment plan

Consider the model

$$\begin{cases} Y_1 = \lambda Y_2 + \alpha Y_4 + \delta U + \epsilon_1, \\ Y_2 = \nu Y_3 + \epsilon_2, \\ Y_3 = \theta Y_4 + \gamma U + \epsilon_3, \\ Y_4 = \epsilon_4, U = \epsilon_5, \end{cases} \tag{4.8}$$

where the ϵ_i are independent random variables with zero means, the Y_i are regarded as observable, U as unobservable, and $\alpha, \gamma, \delta, \lambda, \nu, \theta$ as unknown numbers.[1] This is a special case of a model used by Pearl [71] to comment on some work of Cox and Wermuth [108]:

$$\begin{cases} Y_1 = \varphi_1(Y_2, Y_4, U, \epsilon_1), \\ Y_2 = \varphi_2(Y_3, \epsilon_2), \\ Y_3 = \varphi_3(Y_4, U, \epsilon_3), \\ Y_4 = \varphi_4(\epsilon_4), U = \varphi_5(\epsilon_5), \end{cases} \tag{4.9}$$

where the φ_i are regarded as unknown real-valued functions. The two models are summarized by Fig. 4.3. They represent the situation of a two-stage treatment plan in which patients of unknown 'health status' U undergo a first treatment Y_4 and then a second treatment Y_2 dictated in part by the outcome Y_3 of an examination carried out some time after the first treatment; the final outcome Y_1, measured some time after the second treatment, is influenced by the first as well as by the second treatment, and also by health status.

Using the second and third equations in the first equation of (4.8), we see that by 'regressing' Y_1 on Y_4 one can estimate the effect of the initial treatment Y_4 on the final outcome Y_1, namely the parameter $\alpha + \lambda\nu\theta$, in a consistent and unbiased way.[2] Cox and Wermuth, however, were interested in the estimation of the *direct effect* of the first treatment. It seems clear enough that Y_4 affects Y_1 directly as well

[1] Sometimes, as in (4.8), we list the variables in the reverse order of their definition, which by this time should cause no confusion.

[2] On the other hand, regressing Y_1 on (Y_2, Y_4) yields a biased estimate of the first equation of (4.8), because U—which, being unobservable, is forced to contribute to the error term—is correlated with Y_2. This phenomenon is well-known from regression analysis (see for example our Prob. 2.9

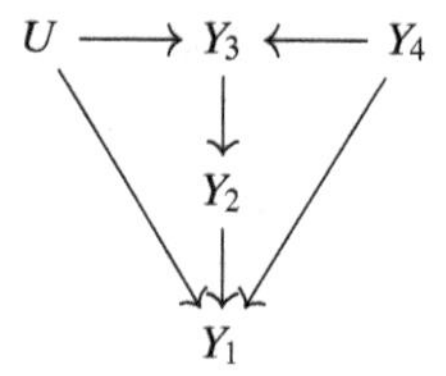

Fig. 4.3 Graph of models (4.8) and (4.9).

as indirectly through its influence on Y_3 and Y_2, that one may want to quantify one effect rather than the other, and that in (4.8) the direct effect should be α. But if the first three equations are $Y_1 = \alpha Y_2 Y_4 + \delta U + \epsilon_1$, $Y_2 = Y_3 + \epsilon_2$ and $Y_3 = Y_4 + \gamma U + \epsilon_3$, for example, then the situation may not be so clear. It is useful, therefore, to define the direct effect of Y_4 on Y_1 in connection with the more general model (4.9).

One of way of doing this is by integrating $\mathcal{L}(Y_1|U = u, Y_2 = y_2, Y_4 = y_4)$ with respect to the distribution of (U, Y_2) and considering the result as a function of y_4. If we write[3]

$$\mu_t := \sum_{u,y_1,y_2} y_1 \mathbf{P}(Y_1 = y_1|U = u, Y_2 = y_2, Y_4 = t)\mathbf{P}(U = u, Y_2 = y_2) \qquad (4.10)$$

then $\mu_{t'} - \mu_t$ for $t' \neq t$ quantifies the difference between the outcomes of two randomly chosen patients who happened to have the same health status and to have received the same second treatment but to have had different initial treatments. If in (4.8) we regard U as observable, the first equation constitutes a linear regression model of Y_1 on (U, Y_2, Y_4) (because this random vector is independent of ϵ_1), so $\mathbf{E}[Y_1|U = u, Y_2 = y_2, Y_4 = y_4] = \lambda y_2 + \alpha y_4 + \delta u$ and $\mu_{t'} - \mu_t = \alpha(t' - t)$, and this definition of direct effect can be identified with α. The definition applies more generally to (4.9), of course, but then $\mu_{t'} - \mu_t$ need not reduce to a single parameter.

Another way of defining the direct effect of Y_4 on Y_1 is to imagine that the second treatment Y_2 can be fixed at some value y_2, rather than being determined by the result Y_3 of the examination, without consequences for the way in which the other variables are generated. This situation, which need not be less realistic than the situation behind the models (4.8)–(4.9) (it is probably just as realistic if neither patients nor doctors have expectations about, or wish to influence, the second treatment), corresponds to the intervention model obtained from (4.9) by deleting the second equation and replacing the occurrences of Y_2 in the remaining equations by a numerical value y_2:

$$\begin{cases} \hat{Y}_1 = \varphi_1(y_2, \hat{Y}_4, \hat{U}, \hat{\epsilon}_1), \\ \hat{Y}_3 = \varphi_3(\hat{Y}_4, \hat{U}, \hat{\epsilon}_3), \\ \hat{Y}_4 = \varphi_4(\hat{\epsilon}_4), \; \hat{U} = \varphi_5(\hat{\epsilon}_5). \end{cases} \qquad (4.11)$$

---

or Freedman [36], question 14, pp. 56, 251) and finds a general explanation within the basic model: if φ_1 is identified with ρ and Y_2 with T, unconfoundedness fails unless one corrects for Y_4 and U.

[3] We assume that the ϵ_i are discrete random variables with mean zero, but the discussion extends in obvious ways to general random variables.

Under this model, the effect of $\hat{Y}_4$ on $\hat{Y}_1$ corresponds to a direct effect of Y_4 on Y_1 in (4.9) in the sense that the influence that Y_4 would have had on Y_1 through Y_2 has been cancelled by the replacement of Y_2 by a numerical value and the removal of the equation defining Y_2; and it is of course fully determined by

$$p_t^{(y_2)}(y) := \sum\nolimits_u \mathbf{P}(\hat{Y}_1 = y|\hat{U} = u, \hat{Y}_4 = t)\mathbf{P}(\hat{U} = u) = \mathbf{P}(\hat{Y}_1 = y|\hat{Y}_4 = t) \qquad (4.12)$$

(the expression on the right follows from the independence of $\hat{U}$ and $\hat{Y}_4$).

Let us write

$$v_t^{(y_2)} := \sum\nolimits_y y\, p_t^{(y_2)}(y); \qquad (4.13)$$

then $v_{t'} - v_t$ for $t' \neq t$ quantifies the difference between the outcomes of two randomly chosen patients who received the same fixed second treatment y_2, and who—as it happens—may or may not have had the same health status, but had different initial treatments. If in the analogous intervention version of (4.8) we regard $\hat{U}$ as observable, the first equation constitutes a linear regression of $\hat{Y}_1$ on $(\hat{U}, \hat{Y}_4)$ (with y_2 contributing to the intercept), so $\mathbf{E}[\hat{Y}_1|\hat{U} = u, \hat{Y}_4 = y_4] = \lambda y_2 + \alpha y_4 + \delta u$ and $v_{t'}^{(y_2)} - v_t^{(y_2)} = \alpha(t' - t)$, and this definition too can be identified with α.

Which definition of direct effect is of interest depends on the real-life problem in hand. The second definition is probably more useful in the analysis of clinical studies because the substitution of a numerical value for Y_2 in (4.8)–(4.9) does not constrain Y_3 in any way (since the second equation vanishes as well) whereas the probabilities involved in (4.10) do imply a constraint on Y_3 and consequently a constraint on Y_4, which constraints may not have operational meaning. That in the 'linear system' (4.8) the two definitions coincide may be of some interest, but we should notice the essential difference between them and try to define 'direct effect' operationally rather than through a parameter in a specific functional relationship.

Now Cox and Wermuth show that, despite the absence of data on U, the parameter α—the direct effect of the first treatment on the final outcome according to any of our two definitions—can be estimated indirectly by fitting a number of regression models implied by (4.8) to observable data. In his comment on their work, Pearl shows that the direct effect according to the second definition can be estimated from the observable data even under the more general model (4.9); in other words, (4.12)—hence (4.13), from which α is retrieved under (4.8)—can be written in terms of probabilities pertaining to Y_1, Y_2, Y_3 and Y_4 (in contrast, (4.10) does not, in general, seem to admit an alternative form free from probabilities pertaining to U). We are going to prove Pearl's result and then consider another situation—slightly different from the one behind the definition of direct effect—where the idea of an intervention model may be meaningful and perhaps practically useful.

Remark 4.1 Pearl observes that the result "extends the linear analysis" of Cox and Wermuth. This is certainly the case when the definition of interest is the second. It is also the case when the definition of interest is the first and, for example, (4.8) has its first equation replaced by $Y_1 = \varphi_1(Y_2, U, \epsilon_1) + \alpha Y_4$ for some φ_1; for under this model too both definitions of direct effect amount essentially to α and the result applies in the same way to the estimation of α. □

As noted by Pearl, the result applies slightly more generally to model (4.9) with the second equation replaced by $Y_2 = \varphi_2(Y_3, Y_4, \epsilon_2)$, which allows the initial treatment to influence the second treatment and may be represented by Fig. 4.3 with an additional arrow from Y_4 to Y_2; it says that

$$p_t^{(y_2)}(y) = \sum_{y_3} \mathbf{P}(Y_1 = y|Y_2 = y_2, Y_3 = y_3, Y_4 = t)\mathbf{P}(Y_3 = y_3|Y_4 = t) \tag{4.14}$$

and implies that the direct effect of the first treatment Y_4 on the final outcome Y_1 according to the second definition can be estimated from the observable data.

The proof is based on two observations: First,

$$\mathbf{P}\left(\varphi_3(y_4, \hat{U}, \hat{\epsilon}_3) = y_3 \,|\hat{Y}_4 = y_4\right) = \mathbf{P}\left(\varphi_3(y_4, U, \epsilon_3) = y_3 \,|Y_4 = y_4\right),$$

since the last three equations of each of (4.9) and (4.11) impose the same distribution on (U, Y_3, Y_4) and $(\hat{U}, \hat{Y}_3, \hat{Y}_4)$. Secondly,

$$\mathbf{P}\left(\varphi_1(y_2, y_4, \hat{U}, \hat{\epsilon}_1) = y_1 \,|\varphi_3(y_4, \hat{U}, \hat{\epsilon}_3) = y_3, \hat{Y}_4 = y_4\right) =$$
$$\mathbf{P}\left(\varphi_1(y_2, y_4, U, \epsilon_1) = y_1 \,|\varphi_3(y_4, U, \epsilon_3) = y_3, Y_4 = y_4\right) =$$
$$\mathbf{P}\left(\varphi_1(y_2, y_4, U, \epsilon_1) = y_1 \,|\varphi_2(y_3, y_4, \epsilon_2) = y_2, \varphi_3(y_4, U, \epsilon_3) = y_3, Y_4 = y_4\right) =$$
$$\mathbf{P}\left(\varphi_1(y_2, Y_4, U, \epsilon_1) = y_1 \,|\varphi_2(Y_3, Y_4, \epsilon_2) = y_2, \varphi_3(Y_4, U, \epsilon_3) = y_3, Y_4 = y_4\right) =$$
$$\mathbf{P}\left(Y_1 = y_1 \,|Y_2 = y_2, Y_3 = y_3, Y_4 = y_4\right),$$

because $(U, \epsilon_1, \epsilon_3, \epsilon_4)$ and $(\hat{U}, \hat{\epsilon}_1, \hat{\epsilon}_3, \hat{\epsilon}_4)$ have the same distribution and ϵ_1 and ϵ_2 are independent.

Using these identities and recalling the second identity in (4.12), we see that

$$p_{y_4}^{(y_2)}(y_1) = \mathbf{P}(\hat{Y}_1 = y_1|\hat{Y}_4 = y_4) = \sum_{y_3} \mathbf{P}(\hat{Y}_1 = y, \hat{Y}_3 = y_3|\hat{Y}_4 = y_4) =$$

$$\sum_{y_3} \mathbf{P}\left(\varphi_1(y_2, y_4, \hat{U}, \hat{\epsilon}_1) = y_1, \varphi_3(y_4, \hat{U}, \hat{\epsilon}_3) = y_3 \,|\hat{Y}_4 = y_4\right) =$$

$$\sum_{y_3} \frac{\mathbf{P}\begin{pmatrix} \varphi_1(y_2, y_4, \hat{U}, \hat{\epsilon}_1) = y_1, \\ \varphi_3(y_4, \hat{U}, \hat{\epsilon}_3) = y_3, \hat{Y}_4 = y_4 \end{pmatrix}}{\mathbf{P}(\varphi_3(y_4, \hat{U}, \hat{\epsilon}_3) = y_3, \hat{Y}_4 = y_4)} \frac{\mathbf{P}\left(\varphi_3(y_4, \hat{U}, \hat{\epsilon}_3) = y_3, \hat{Y}_4 = y_4\right)}{\mathbf{P}\left(\hat{Y}_4 = y_4\right)} =$$

$$\sum_{y_3} \mathbf{P}\left(\varphi_1(y_2, y_4, \hat{U}, \hat{\epsilon}_1) = y_1 \middle| \begin{matrix} \varphi_3(y_4, \hat{U}, \hat{\epsilon}_3) = y_3, \\ \hat{Y}_4 = y_4 \end{matrix}\right) \mathbf{P}(\varphi_3(y_4, \hat{U}, \hat{\epsilon}_3) = y_3|\hat{Y}_4 = y_4) =$$

$$\sum_{y_3} \mathbf{P}(Y_1 = y_1|Y_2 = y_2, Y_3 = y_3, Y_4 = y_4)\mathbf{P}(Y_3 = y_3|Y_4 = y_4),$$

which is the right-hand side of (4.14) with $(t, y) = (y_4, y_1)$.

4.4.1 A similar example

We consider a version of this example, vaguely inspired by diseases such as Lyme but rather academic, in order to show how simple distributional identities can in principle be useful for making predictions or drawing conclusions about real or imagined 'interventions'.

Among the individuals bitten by ticks that may carry a disease, some go to the doctor because they fear being infected (e.g. on the grounds of information they have about the disease or because of symptoms). The disease can be treated with antibiotics of two kinds: one is very effective but has undesirable side effects; the other is weaker, has few side effects, but is effective only at an early stage of the disease. The disease is difficult to diagnose, and diagnosis is based on a laborious, not immediately available and not infallible test. The only consensus among doctors about the treatment of a potentially infected individual coming for a first visit is that the test should be offered to him only after some time and provided certain symptoms persist or develop, and that if tested positive he should probably (thus: not certainly) be treated with the effective antibiotic and if not tested or tested negative should probably not be treated at all. Some doctors choose to treat the patient with the weak antibiotic on the first visit, but this is frowned upon by some of their colleagues who think that the risk of someone being infected is very small, that the symptoms reported by the patients are consistent with many other diseases, and that if a patient is really infected then the weak antibiotic will probably not help much anyway; consequently, whether a patient is assigned to the weak antibiotic or is asked to wait until he gets the chance of taking the test is dictated by the doctor rather than by the patient's health condition—implying that the assignment to the weak antibiotic is randomized. Eventually, a patient, depending on his infection status and general health condition and on the course of treatments he underwent, is declared free of the disease or else becomes chronically ill with it, the second possibility tending to occur if antibiotics are prescribed too late.

If we represent the largely unknown health condition (including the disease status) of a patient by U', the prescription of the weak antibiotic by Y_4' (equal to 1 if prescribed, equal to 0 otherwise), the result of the test by Y_3' (equal to 1 if the test is positive and to 0 if the test is negative or is not even offered), the prescription of the effective antibiotic by Y_2', and the patient's outcome by Y_1' (which can be thought of as binary, with 1 indicating chronically ill, or else as a number representing 'level of disability', 'survival time', or something similar), we may describe this situation by the model

$$\begin{cases} Y_1' = \varphi_1(Y_2', Y_4', U', \epsilon_1'), \\ Y_2' = \varphi_2(Y_3', Y_4', \epsilon_2'), \\ Y_3' = \varphi_3(Y_4', U', \epsilon_3'), \\ Y_4' = \varphi_4(\epsilon_4'), U' = \varphi_5(\epsilon_5'). \end{cases} \tag{4.15}$$

This is equivalent to (4.9) except for the appearance of Y_4' in the second equation, motivated by the influence that having or not having been treated with the first antibiotic may have on the prescription of the second antibiotic.

Suppose that years of practice reveal that a substantial number of patients (of whom some have undergone antibiotic treatment of one or both kinds and some have not) become chronically ill, and that even some of the patients who tested positive did not, for some reason, undergo antibiotic treatment. An intervention aimed at reducing the number of chronically ill patients might consist of putting every patient who tests positive on the effective antibiotic; and within such an intervention one could ask whether or not treating all patients initially with the weak antibiotic would have a further beneficial effect.

In obvious notation, we may describe the intervention by the model

$$\begin{cases} \check{Y}_1 = \varphi_1(\check{Y}_2, \check{Y}_4, \check{U}, \check{\epsilon}_1), \\ \check{Y}_2 = \check{Y}_3 + (1 - \check{Y}_3)\varphi_2(\check{Y}_3, \check{Y}_4, \check{\epsilon}_2), \\ \check{Y}_3 = \varphi_3(\check{Y}_4, \check{U}, \check{\epsilon}_3), \\ \check{Y}_4 = \varphi_4(\check{\epsilon}_4), \check{U} = \varphi_5(\check{\epsilon}_5). \end{cases} \tag{4.16}$$

If for instance $\check{Y}_1$ indicates whether the patient becomes chronically ill or not, then the question of interest is whether

$$\mathbf{E}(\check{Y}_1|\check{Y}_4 = 1) < \mathbf{E}(\check{Y}_1|\check{Y}_4 = 0). \tag{4.17}$$

If the probabilities $\mathbf{P}(\check{Y}_1 = y_1|\check{Y}_4 = y_4)$, which pertain to the intervention, can be estimated with data on (Y_1', Y_2', Y_3', Y_4') obtained from the earlier clinical practice, one may try to check (4.17) and consequently propose rules for an early prescription of the weak antibiotic.

That the probabilities in question can be estimated from data follows from

$$\mathbf{P}(\check{Y}_1 = y_1|\check{Y}_4 = y_4) = \mathbf{P}(\check{Y}_1 = y_1, \check{Y}_3 = 0|\check{Y}_4 = y_4) + \mathbf{P}(\check{Y}_1 = y_1, \check{Y}_3 = 1|\check{Y}_4 = y_4) =$$

$$\sum_{y_2=0,1} \mathbf{P}\left[\varphi_1(\check{Y}_2, y_4, \check{U}, \check{\epsilon}_1) = y_1, \check{Y}_2 = y_2, \check{Y}_3 = 0 \,\middle|\check{Y}_4 = y_4\right] +$$

$$\mathbf{P}\left[\varphi_1(1, y_4, \check{U}, \check{\epsilon}_1) = y_1, \check{Y}_3 = 1 \,\middle|\check{Y}_4 = y_4\right] =$$

$$\sum_{y_2=0,1} \mathbf{P}\left[\varphi_1(y_2, y_4, \check{U}, \check{\epsilon}_1) = y_1, \varphi_2(0, y_4, \check{\epsilon}_2) = y_2, \varphi_3(y_4, \check{U}, \check{\epsilon}_3) = 0 \,\middle|\check{Y}_4 = y_4\right] +$$

$$\frac{\mathbf{P}\left[\varphi_1(1, y_4, \check{U}, \check{\epsilon}_1) = y_1, \varphi_3(y_4, \check{U}, \check{\epsilon}_3) = 1, \check{Y}_4 = y_4\right]}{\mathbf{P}(\check{Y}_4 = y_4)} =$$

$$\sum_{y_2=0,1} \mathbf{P}\left[\varphi_1(y_2, y_4, U', \epsilon_1') = y_1, \varphi_2(0, y_4, \epsilon_2') = y_2, \varphi_3(y_4, U', \epsilon_3') = 0 \,\middle|Y_4' = y_4\right] +$$

$$\frac{\mathbf{P}\left(\substack{\varphi_1(1, y_4, U', \epsilon_1') = y_1, \varphi_2(1, y_4, \epsilon_2') = 1, \\ \varphi_3(y_4, U', \epsilon_3') = 1, Y_4' = y_4}\right)\mathbf{P}(\varphi_3(y_4, U', \epsilon_3') = 1)}{\mathbf{P}(\varphi_2(1, y_4, \epsilon_2') = 1, \varphi_3(y_4, U', \epsilon_3') = 1, Y_4' = y_4)} =$$

$$\sum_{y_2=0,1} \mathbf{P}\left[Y_1' = y_1, Y_2' = y_2, Y_3' = 0 \,\middle|Y_4' = y_4\right] +$$

$$\mathbf{P}\left[Y_1' = y_1 \,|Y_2' = 1, Y_3' = 1, Y_4' = y_4\right] \mathbf{P}(Y_3' = 1|Y_4' = y_4) =$$

$$\mathbf{P}(Y_1' = y_1, Y_3' = 0\,|Y_4' = y_4) + \mathbf{P}(Y_1' = y_1|Y_2' = 1, Y_3' = 1, Y_4' = y_4)\mathbf{P}(Y_3' = 1|Y_4' = y_4).$$

Evidently, the conclusions of this kind of analysis could be quite 'volatile': If an intervention guided by them were really carried out (rather than simply imagined for the purpose of producing and assessing predictions—e.g. checking (4.17)), the population of patients and the population of doctors might change considerably following the release of the news of the intervention and of the apparent need for it, and then the probabilities computed from past data would end up not applying in reality. But the most important thing in this case should be to get (4.17) (or a similar inequality) right, and that would perhaps dispense with correct estimates of 'post-intervention' probabilities.

Remark 4.2 We may observe that (4.16) is an 'intervention model' in a more general sense than the one considered so far—which is of replacing certain random variables in a system by numerical values and deleting the equations defining those variables—but it contains nothing essentially new. Note that the same result holds if we delete the penultimate equation from (4.16) and replace $\check{Y}_4$ by y_4 in the other equations. □

4.5 Sex discrimination in hiring

In order to study the effect of sex discrimination on the hiring of applicants for a certain job one may consider the model

$$\begin{cases} H = \varphi_1(B, Q, S, \epsilon_1), \\ Q = \varphi_2(B, S, \epsilon_2), \\ B = \varphi_3(S, \epsilon_3), \\ S = \varphi_4(\epsilon_4), \end{cases} \tag{4.18}$$

where H is a binary variable indicating whether a given applicant is hired or not, Q constitutes a summary of the applicant's qualifications, B a summary of the applicant's 'social and economic background', and S is the applicant's sex (equal to 1 if it is male, 0 if female), and as usual the ϵ_i are unobserved, independent standard uniforms and the φ_i are some unknown functions. A modicum of realism would perhaps require us to consider applicants of a certain age group applying for a first job and to include in B the background provided to the applicant by the family up to a certain age, and to take one such model for each kind of job, kind of company or institution, etc.; our purpose again is simply to illustrate some basic ideas and calculations.

As indicated by Fig. 4.4, sex can influence the fate of the applicant in a direct way but also in an indirect way through the applicant's qualifications and background, reflecting the possibility of men and women having different inclinations for certain subjects or kinds of education; it is sometimes said that the effect of S on H is 'mediated' by Q.

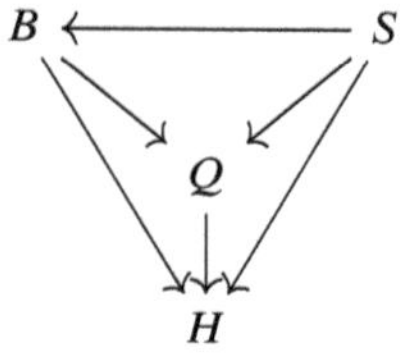

Fig. 4.4 Graph of model (4.18).

In a real or imagined intervention designed to assess the direct effect of sex on the applicant's success or to predict the result of anti-discriminatory measures, one may contemplate the situation where the hiring committee is not provided with any clue about the applicant's sex, or else may be given a false, fixed or randomized sex. This situation could be represented by the model

$$\begin{cases} H'_s = \varphi_1(B'_s, Q'_s, \sigma_s, \epsilon'_{1,s}), \\ Q'_s = \varphi_2(B'_s, s, \epsilon'_{2,s}), \\ B'_s = \varphi_3(s, \epsilon'_{3,s}), \end{cases} \tag{4.19}$$

where the variables are completely analogous to those of (4.18) except for σ_s, which is independent of the $\epsilon'_{i,s}$ and stands for the *assumption* made by the hiring committee about the applicant's sex, and for the indexing of all variables by s (to be set to 0 or 1) intended to concentrate the analysis on females or on males.[4]

The interest of such a study lies in comparing $\mathbf{P}(H'_s = h | B'_s = b, Q'_s = q)$, or a functional of it, with $\mathbf{P}(H = h | B = b, Q = q, S = s)$, or with the corresponding functional. When a large set of data on (B, H, Q, S) is available but no data have ever been collected on (B'_s, H'_s, Q'_s), the statistical question is whether the first probability can be estimated indirectly through estimates of the second probability.

In essence, the answer is affirmative: using the independence of σ_s, $\epsilon'_{1,s}$, $\epsilon'_{2,s}$ and $\epsilon'_{3,s}$, we see that

$$\begin{aligned}
\mathbf{P}(H'_s = h | B'_s = b, Q'_s = q) &= \mathbf{P}\left(\varphi_1(B'_s, Q'_s, \sigma_s, \epsilon'_{1,s}) = h \,\middle|\, \begin{matrix} \varphi_3(s, \epsilon'_{3,s}) = b, \\ \varphi_2(B'_s, s, \epsilon'_{2,s}) = q \end{matrix}\right) \\
&= \mathbf{P}\left(\varphi_1(b, q, \sigma_s, \epsilon'_{1,s}) = h \,\middle|\, \begin{matrix} \varphi_3(s, \epsilon'_{3,s}) = b, \\ \varphi_2(b, s, \epsilon'_{2,s}) = q \end{matrix}\right) \\
&= \mathbf{P}\left(\varphi_1(b, q, \sigma_s, \epsilon'_{1,s}) = h\right) \\
&= \sum_{s'} \mathbf{P}\left(\varphi_1(b, q, \sigma_s, \epsilon_{1,s}) = h, \sigma_s = s'\right)
\end{aligned}$$

[4] Because S is exogenous, one might include its analog S' in (4.19) and obtain equivalent results in the analysis that follows by conditioning on the event $\{S' = s\}$; cf. Remark 4.2.

$$= \sum_{s'} \frac{\mathbf{P}\left(\begin{smallmatrix}\varphi_1(b,q,\sigma_s,\epsilon_{1,s})=h,\, S=s',\\ \varphi_3(s',\epsilon_{3,s})=b,\varphi_2(b,s',\epsilon_{2,s})=q\end{smallmatrix}\right)\mathbf{P}(\sigma_s = s')}{\mathbf{P}\left(\varphi_3(s',\epsilon_{3,s}) = b, \varphi_2(b,s',\epsilon_{2,s}) = q, S = s'\right)}$$

$$= \sum_{s'} \mathbf{P}(H = h | B = b, Q = q, S = s')\mathbf{P}(\sigma_s = s'),$$

so the remaining question is whether $\mathbf{P}(\sigma_s = 0)$ is known or can be estimated or posited.

If the committee is unaware of the intervention and is given a randomized σ_s then $\mathbf{P}(\sigma_s = 0)$ is certainly known. Otherwise, σ_s cannot be fully controlled and its distribution will be unknown, but it may still be estimated in a small pilot intervention, or by simulation—e.g. by submitting to the committee a number of pseudo-applications mimicked from past applications and eliciting the committee's assumption about the putative applicant's sex. Positing values of $\mathbf{P}(\sigma_s = 0)$ is also possible, of course, and it may be useful for certain purposes.

For example, one might ask what happens if only women apply for jobs when the committee always assumes that the applicant is a man, i.e. what happens if $s = 0$ when $\sigma_0 = \sigma_1 = 1$. Averaged over background and qualifications in the population of women, the chance of women getting hired in that situation is

$$\sum_{b,q} \mathbf{P}(H_0' = h | B_0' = b, Q_0' = q)\mathbf{P}(B = b, Q = q | S = 0).$$

This can be contrasted with

$$\sum_{b,q} \mathbf{P}(H_1' = h | B_1' = b, Q_1' = q)\mathbf{P}(B = b, Q = q | S = 1),$$

which is the average chance of men getting hired in the same situation. In terms of expectations, the difference between these two measures is

$$\sum_{b,q} \mathbf{E}(H_0' | B_0' = b, Q_0' = q)\mathbf{P}(B = b, Q = q | S = 0) -$$

$$\mathbf{E}(H_1' | B_1' = b, Q_1' = q)\mathbf{P}(B = b, Q = q | S = 1) =$$

$$\sum_{b,q} \mathbf{E}(H | B = b, Q = q, S = 1)\left\{\mathbf{P}(B = b, Q = q | S = 0) - \mathbf{P}(B = b, Q = q | S = 1)\right\},$$

by the identity derived above with $\mathbf{P}(\sigma_0 = 1) = \mathbf{P}(\sigma_1 = 1) = 1$ (which incidentally shows that in this particular situation $\mathbf{P}(H_s' = h | B_s' = b, Q_s' = q)$ is independent of s, since we have fixed s at 0).

This last quantity can be estimated from data on (B, H, Q, S), and, despite being based on a very unlikely 'intervention', could perhaps serve as an indicator of sex discrimination.

4.6 Case-control studies

Let us consider the basic model with unconfoundedness, with the R_n and T_n binary and the $\mathbf{X}_n$ as well as the (U_n, V_n) independent and identically distributed, so that the vectors $(\mathbf{X}_n, T_n, R_n)$ too are independent and identically distributed, and think of R_n as an individual's status of a certain disease (equal to 1 if the individual is diseased), of T_n as the individual's indicator of exposure to an agent suspected of causing the disease (equal to 1 if the individual has been exposed), and of $\mathbf{X}_n$ as a set of background characteristics (year of birth, sex, occupation, etc.) which may influence the individual's exposure and disease statuses. For simplicity we assume that the $\mathbf{X}_n$ are discrete, so that the $(\mathbf{X}_n, T_n, R_n)$ too are discrete. As usual, we write $(\mathbf{X}, T, R)$ for a random vector with the same distribution as the $(\mathbf{X}_n, T_n, R_n)$.

In situations where the disease is rare it may be necessary to collect a very large random sample of individuals in order to obtain enough *cases*—i.e. diseased individuals—to provide evidence of the supposed effect of the exposure on the disease. Instead of collecting a random sample one may then think of obtaining a substantial number of cases from readily accessible subpopulations—typically patients in hospitals that treat the disease—and comparing them with respect to exposure with a more general subpopulation which, given the rarity of the disease, will consist mostly of nondiseased individuals. The intuition behind this proposal is that if the background characteristics of the cases are very similar to those of the nondiseased individuals and if the exposure indeed contributes to the disease then a greater rate of exposure must be found among the cases than among the nondiseased. But for it to work it is evident that great care is needed to ensure that the subpopulation of cases and the more general subpopulation are comparable enough with respect to everything except for frequency of disease and possibly frequency of exposure.

In a case-control study one tries to make the subpopulations comparable by matching each case with a 'control', namely an individual from a more general population who has the same or practically the same background characteristics as the case. In the language of the basic model, this kind of study can be described as follows: First, conditionally on the set $\{R_1, R_2, \ldots\}$ of responses of the whole population of interest, a sequence of cases is obtained by selecting random indices $I_1 < I_2 < \cdots$ from $\{n : R_n = 1\}$, and a sample of N cases is obtained from the indices in $C_N = \{I_1, I_2, \ldots, I_N\}$ by setting

$$(\mathbf{X}'_{2n-1}, T'_{2n-1}, R'_{2n-1}) \equiv (\mathbf{X}'_{2n-1}, T'_{2n-1}, 1) := (\mathbf{X}_{I_n}, T_{I_n}, 1), \quad n = 1, 2, \ldots, N.$$

Then, conditionally on $(\mathbf{X}_1, R_1), (\mathbf{X}_2, R_2), \ldots$ (the pairs of responses and background characteristics of the whole population), for each $\mathbf{x} \in \mathcal{X}$ a sequence of random indices $J_1(\mathbf{x}) < J_2(\mathbf{x}) < \cdots$ is selected from $\{n : \mathbf{X}_n = \mathbf{x}\} \setminus C_N$ independently of all the other variables and a corresponding sample of controls is obtained as

$$(\mathbf{X}'_{2n}, T'_{2n}, R'_{2n}) := (\mathbf{X}'_{2n-1}, T_{J_n(\mathbf{X}'_{2n-1})}, R_{J_n(\mathbf{X}'_{2n-1})}),$$

$n = 1, 2, \ldots N$.

Remark 4.3 Note that, thanks to the conditioning on the $(\mathbf{X}_n, R_n)$, the $J_n(\mathbf{x})$ may be chosen in such a way that $R'_{2n} = 0$ for all n, but also in such a way that $J_n(\mathbf{x}) > I_N$ for all $n = 1, 2, \ldots, N$ and all $\mathbf{x}$, so that $R'_{2n} = 1$ too may occur. While in the former case no diseased individuals are found among the controls, in the latter the number of diseased and exposed individuals with background characteristics $\mathbf{x}$ among the controls may very well follow the probabilities $\mathbf{P}(T = t, R = r|\mathbf{X} = \mathbf{x})$. Whatever the case, if not the whole sample of controls then at least a subset of it can be used to estimate $\mathbf{P}(T = t|R = 0, \mathbf{X} = \mathbf{x})$. □

The vectors $(\mathbf{X}'_1, T'_1, R'_1), (\mathbf{X}'_2, T'_2, R'_2), \ldots, (\mathbf{X}'_{2N}, T'_{2N}, R'_{2N})$ defined in this way represent the data observed in the study. They are not a random sample because in them each case is followed by its control, but more important is that their elements, being incompletely and selectively observed, do not have the same distribution as the $(\mathbf{X}_n, T_n, R_n)$, so it is not obvious that a case-control study can help us to investigate the effect of the exposure on the disease. To show that it can, note first that, thanks to the way in which $I_1, I_2, \ldots, I_N$ and $J_1(\mathbf{x}), J_2(\mathbf{x}), \ldots$ are generated, the data from the cases and the data from the controls, respectively, allow us to estimate the probabilities

$$p_{\mathbf{x}} := \mathbf{P}(T = 1|R = 1, \mathbf{X} = \mathbf{x}) \quad \text{and} \quad q_{\mathbf{x}} := \mathbf{P}(T = 1|R = 0, \mathbf{X} = \mathbf{x}), \tag{4.20}$$

at least for the $\mathbf{x} \in \mathcal{X}$ assumed by the observed $\mathbf{X}'_n$.

Of course, these probabilities are not our object of interest: we know that under our model the effect of the exposure on the disease is fully characterized by the probabilities $\mathbf{P}(R = 1|T = t, \mathbf{X} = \mathbf{x})$ as functions of t, that is by

$$\mathbf{P}(R = 1|T = 1, \mathbf{X} = \mathbf{x}) \quad \text{and} \quad \mathbf{P}(R = 1|T = 0, \mathbf{X} = \mathbf{x}), \tag{4.21}$$

or by some measure of discrepancy between them. But these probabilities can hardly be estimated from the data collected in a case-control study because in the set of cases the number of diseased individuals is fixed and the set of controls, which may have had its source of diseased individuals even more depleted by the preceding sampling of cases, will rarely have a diseased individual to show.

Surprisingly, there is a measure of discrepancy between the probabilities in (4.21) which can be expressed in terms of those in (4.20), and hence estimated from the data collected in a case-control study: the *odds ratio*, or, more precisely, the odds ratio conditional on $\{\mathbf{X} = \mathbf{x}\}$, defined by

$$\mathrm{e}(\mathbf{x}) = \frac{O(R = 1|T = 1, \mathbf{X} = \mathbf{x})}{O(R = 1|T = 0, \mathbf{X} = \mathbf{x})}, \tag{4.22}$$

where

$$O(R = 1|T = t, \mathbf{X} = \mathbf{x}) = \frac{\mathbf{P}(R = 1|T = t, \mathbf{X} = \mathbf{x})}{1 - \mathbf{P}(R = 1|T = t, \mathbf{X} = \mathbf{x})},$$

the *odds* of an individual with characteristics $\mathbf{x}$ being diseased given that he was exposed to treatment t, is an increasing function of $\mathbf{P}(R = 1|T = t, \mathbf{X} = \mathbf{x})$. If the exposure does not cause the disease, $O(R = 1|T = t, \mathbf{X} = \mathbf{x})$ is constant in t and hence $\mathrm{e}(\mathbf{x}) = 1$; otherwise, $\mathbf{P}(R = 1|T = 1, \mathbf{X} = \mathbf{x}) > \mathbf{P}(R = 1|T = 0, \mathbf{X} = \mathbf{x})$ and hence $\mathrm{e}(\mathbf{x}) > 1$. Since, as is easy to verify,

$$\mathrm{e}(\mathbf{x}) = \frac{O(T = 1|R = 1, \mathbf{X} = \mathbf{x})}{O(T = 1|R = 0, \mathbf{X} = \mathbf{x})} = \frac{p_{\mathbf{x}}(1 - q_{\mathbf{x}})}{q_{\mathbf{x}}(1 - p_{\mathbf{x}})}, \tag{4.23}$$

$\mathrm{e}(\mathbf{x})$ also compares the odds of an individual having been exposed given that he is diseased with the odds of an individual having being exposed given that he is not diseased. Thus, although it is (4.22) that characterizes the causal effect of the exposure on the disease, (4.23) shows that the effect can be estimated from estimates of the probabilities (4.20).

If we integrate $\mathrm{e}(\mathbf{x})$ with respect to $\mathbf{P}(\mathbf{X} \leq \mathbf{x}|R = 1)$ we get an overall measure of the effect of exposure on disease which in principle can be estimated from the observed data, namely $\mathbf{E}[\mathrm{e}(\mathbf{X})|R = 1]$, where the conditioning on $\{R = 1\}$ reflects the sampling scheme, by which the values of the $\mathbf{X}_n$ that turn up do so conditionally on that event.

4.6.1 Berkson's paradox

In the model behind the case-control study it is essential that the independent and identically distributed random vectors $(\mathbf{X}_1, T_1, R_1), (\mathbf{X}_2, T_2, R_2), \ldots$ from which the data $(\mathbf{X}_1', T_1', R_1'), (\mathbf{X}_2', T_2', R_2'), \ldots$ are drawn obey the basic model with unconfoundedness. A violation of this condition may invalidate the use of the odds ratio (4.22) as a measure of the effect of the treatment and pervert the study. For example, if, based on its numerical values and independently of the other vectors, each $(\mathbf{X}_n, T_n, R_n)$ is either discarded or included in a *source*—representing a hospital, say—from which the data for the study are to be collected, the random vectors in the source no longer follow the basic model and the correlation between their treatment and response variables given the background variables will typically differ from the correlation between T_n and R_n given $\mathbf{X}_n$; in particular, even if the treatment has no effect on the response, so that T_n and R_n are independent conditionally on $\mathbf{X}_n$, in the source the treatment and response are typically correlated and therefore collecting the data from the source can lead to a wrong conclusion. This type of theoretical mechanism was proposed by Berkson [9] in order to explain why the occurrence of one disease among hospitalized patients is sometimes associated with another disease without there being hints of a causal relation between the two in the population at large. Over the years, Berkson's 'paradox' has been discussed many times in connection with the spurious association between a disease and a *treatment* (exposure to a substance, a medical treatment, or the occurrence of another disease) caused by *selection bias* in case-control studies, of which the inclusion in a source, just described, is an example.

In order to give a concrete example of Berkson's paradox, suppose that, instead of the basic model with unconfoundedness, the $(\mathbf{X}_n, T_n, R_n)$ obey the model

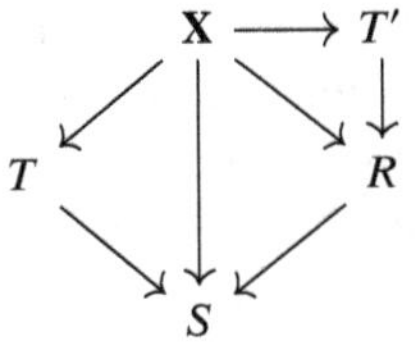

Fig. 4.5 Graph of model (4.24).

$$\begin{cases} \mathbf{X} = \xi(\epsilon), \\ T = \tau(U, \mathbf{X}), \\ T' = \tau'(U', \mathbf{X}), \\ R = \rho(V, \mathbf{X}, T'), \\ S = \sigma(W, \mathbf{X}, T, R), \end{cases} \tag{4.24}$$

for a generic vector $(\mathbf{X}, T, T', R, S)$, where ϵ, U, U', V, W are independent standard uniforms, ξ, τ, τ', ρ and σ are some functions, the binary variable T' represents an unobservable 'treatment' which causes the disease, the binary T represents the purported cause of the disease, and S is a binary variable indicating whether or not $(\mathbf{X}, T, R)$ falls into the source from which the data for the case-control study are to be drawn. In this model, summarized in Fig. 4.5, T and R are independent conditionally on $\mathbf{X}$ (the purported cause playing no role in the response), but, conditionally on $\mathbf{X}$ and on the event $\{S = 1\}$, which represents the inclusion of $(\mathbf{X}, T, R)$ in the source, they are generally dependent.

Indeed, we have

$$\mathbf{P}(T = t|\mathbf{X} = \mathbf{x}, R = r, S = s) = \mathbf{P}(T = t|\mathbf{X} = \mathbf{x}) \frac{\mathbf{P}\{\sigma(W, \mathbf{x}, t, r) = s\}}{\mathbf{P}\{\sigma\left[W, \mathbf{x}, \tau(U, \mathbf{x}), r\right] = s\}},$$

and in general the term on the right does not reduce to $\mathbf{P}(T = t|\mathbf{X} = \mathbf{x}, R = r)$, which in the case of (4.24) equals $\mathbf{P}(T = t|\mathbf{X} = \mathbf{x})$, since T is not involved in R; see Prob. 5.10. If σ is constant in its third argument, which means that the purported cause of the disease does not influence admission to the source and implies that there is no arrow from T to S in Fig. 4.5, then we do have

$$\mathbf{P}(T = t|\mathbf{X} = \mathbf{x}, R = r, S = s) = \mathbf{P}(T = t|\mathbf{X} = \mathbf{x}, R = r), \tag{4.25}$$

and if $\mathbf{P}(S = 1|\mathbf{X} = \mathbf{x}, R = r) > 0$ for $r = 0, 1$ we can conduct a valid case-control study by collecting data from the source, while if $\mathbf{P}(S = 1|\mathbf{X} = \mathbf{x}, R = 1) > 0$ but $R = 0 \Rightarrow S = 0$ (as when $\sigma(w, \mathbf{x}, t, r) = r \cdot \sigma_0(w, \mathbf{x})$, say) we can conduct a valid study by collecting data on the cases from the source and data on the controls from outside the source. Otherwise, T and R will normally be dependent conditionally on $\{\mathbf{X} = \mathbf{x}, S = 1\}$, and an estimate of the odds ratio involving data from the source will yield a distorted measure of the effect of the supposed exposure on the disease, without the true cause of the disease, T', ever being in sight.

As an exercise (Prob. 5.10), the reader may show that the situation is essentially the same when T too influences R, that is, when $R = \rho(V, \mathbf{X}, T', T)$ in (4.24), which corresponds to adding an arrow from T to R in Fig. 4.5; in particular, (4.25) holds if σ is constant in its third argument, but generally fails otherwise.

4.7 Determination of causal effects with instrumental variables

In problems where unconfoundedness fails in the basic model, the study of a causal effect in the full sense of the word is usually ruled out, but sometimes it is still possible to estimate a sort of causal effect. Angrist, Imbens and Rubin [2] considered a special situation where the use of an 'instrumental variable' permits the estimation of the causal effect in a subpopulation of the population of interest.

The model studied by them is

$$R = \rho(\varepsilon, T), \quad T = \tau(\delta, I), \tag{4.26}$$

where the functions ρ and τ are regarded as unknown, T and I are binary random variables, I is observable, (δ, ε) is an unobservable random vector, I and (δ, ε) are independent but δ and ε are dependent. The model arises in the study of the effect of a treatment, T, on a response, R, with the help of a third variable, I, called an *instrument*, which is not of direct interest and affects R only through T. It may apply to a real-life problem in which a patient may or may not be prescribed a treatment according to the value of I and (possibly even without getting it prescribed) may or may not decide to follow it according to the value of T, which may be unobservable; if the patient's disposition towards the treatment, represented by δ and τ, is correlated with his response, then δ and ε must be assumed correlated; even if the prescription of the treatment is randomized, implying the independence of I and (δ, ε), unconfoundedness fails, so it is not clear that the effect of the treatment, which normally is the question of interest (rather than the effect of prescription), can be studied.

In the real-life problem treated by Angrist, Imbens and Rubin the question of interest is whether serving in the military in times of war has detrimental effects on the health of individuals; R stands for the health outcome of a generic individual, I for the individual's 'draft status'—whether he was called to serve or not—and T for his indicator of military service—whether he actually served or not. During the periods covered by the data, recruitment had been determined by a lottery, 'low' lottery numbers leading to drafting. Individuals who were called to serve may or may not have joined the army, so the data on such an individual might be $(I, T) = (1, 0)$ or $(I, T) = (1, 1)$; and individuals who were not called may have volunteered and joined the military, so $(I, T) = (0, 1)$ too might be observed. Although in principle there might be individuals who would have refused to serve if recruited but accepted to serve if not recruited, their δ satisfying $0 = \tau(\delta, 1) < \tau(\delta, 0) = 1$, such individuals should be rare; accordingly, it will be assumed that τ *is monotonic in its second argument*, i.e. that $\tau(\delta, 1) \geq \tau(\delta, 0)$ for all δ.

In order to study the effect of the treatment on the response—to study how varying t changes the law of $\rho(\varepsilon, t)$—one might think of estimating

$$\mathcal{L}(R|T = t) = \mathcal{L}(\rho(\varepsilon, t)|\tau(\delta, I) = t)$$

or

$$\mathcal{L}(R|T = t, I = i) = \mathcal{L}(\rho(\varepsilon, t)|\tau(\delta, i) = t, I = i),$$

or functionals of them, from data on (I, T, R). But the dependence between δ and ε shows that the conditioning on $\{\tau(\delta, I) = t\}$ or on $\{\tau(\delta, i) = t, I = i\}$ implies that changes in t cause changes in $\rho(\varepsilon, t)$ also through the first argument of ρ, so neither of these probability laws (and indeed no estimable probability law) describes the effect of the treatment on the response.

Surprisingly, the presence of the instrument I in the second equation of (4.26) makes it possible to describe the effect of T on R in some, albeit incomplete, sense.[5] Indeed, Angrist, Imbens and Rubin have shown that the parameter

$$\theta := \frac{\mathbf{E}(R|I = 1) - \mathbf{E}(R|I = 0)}{\mathbf{E}(T|I = 1) - \mathbf{E}(T|I = 0)}, \tag{4.27}$$

which is defined whenever I and T are correlated, and which can be estimated from data on (I, T, R) by replacing expected values by sample averages, is a measure of the causal effect of T on R.

To see this, note that by the independence of I and δ and the monotonicity of τ, which implies

$$\tau(\delta, 1) - \tau(\delta, 0) \neq 0 \quad \Leftrightarrow \quad \tau(\delta, 1) - \tau(\delta, 0) = 1 \quad \Leftrightarrow \quad \tau(\delta, 1) = 1 \ \wedge \ \tau(\delta, 0) = 0,$$

the denominator of θ is

$$\begin{aligned} \mathbf{E}(T|I = 1) - \mathbf{E}(T|I = 0) &= \mathbf{E}[\tau(\delta, 1)|I = 1] - \mathbf{E}[\tau(\delta, 0)|I = 0] \\ &= \mathbf{E}[\tau(\delta, 1) - \tau(\delta, 0)] \\ &= \mathbf{P}[\tau(\delta, 1) - \tau(\delta, 0) = 1]. \end{aligned}$$

By the independence of I and (δ, ε) and by

$$\rho\left(\varepsilon, \tau(\delta, 1)\right) - \rho\left(\varepsilon, \tau(\delta, 0)\right) \neq 0 \quad \Rightarrow \quad \tau(\delta, 1) > \tau(\delta, 0),$$

the numerator is

$$\begin{aligned} \mathbf{E}(R|I = 1) - \mathbf{E}(R|I = 0) &= \mathbf{E}[\rho\left(\varepsilon, \tau(\delta, 1)\right)|I = 1] - \mathbf{E}[\rho\left(\varepsilon, \tau(\delta, 0)\right)|I = 0] \\ &= \mathbf{E}[\{\rho\left(\varepsilon, 1\right) - \rho\left(\varepsilon, 0\right)\} \mathbf{1}_{\{\tau(\delta,1)-\tau(\delta,0)=1\}}]. \end{aligned}$$

[5] The role of I is quite different from that of the vector of potential confounders in the basic model. In fact, one may include such a vector $\mathbf{X}$ in both equations of (4.26) and carry out a stratified version of the analysis that follows by conditioning on $\{\mathbf{X} = \mathbf{x}\}$.

Thus

$$\theta = \frac{\mathbf{E}[\{\rho(\varepsilon,1)-\rho(\varepsilon,0)\}\,\mathbf{1}_{\{\tau(\delta,1)-\tau(\delta,0)=1\}}]}{\mathbf{P}[\tau(\delta,1)-\tau(\delta,0)=1]}$$
$$= \mathbf{E}\left[\rho(\varepsilon,1)-\rho(\varepsilon,0)\,|\tau(\delta,1)-\tau(\delta,0)=1\right] \tag{4.28}$$

is the *average effect of treatment on response in the subpopulation of individuals for whom* $\tau(\delta,1)-\tau(\delta,0)=1$. (In [2], where I is really a suggestion of treatment rather than the treatment $T=\tau(\delta,I)$, such individuals are called 'compliers': they get treatment only if treatment is suggested to them.)

Although it is impossible to identify the subpopulation of individuals to which the causal effect represented by θ applies ($\tau(\delta,1)-\tau(\delta,0)$ being unobservable), evidence that $\theta \neq 0$ does provide evidence of a treatment effect.

Remark 4.4 The study of the *existence* of the treatment effect may be tackled more simply by redefining (4.26) in the form of the basic model with unconfoundedness, namely as

$$R = \tilde{\rho}(V,I) \equiv \rho(\varepsilon,\tau(\delta,I)), \quad I = \iota(U),$$

where ι is some function, U and V are independent standard uniforms, and V is made to yield δ and ε (assumed uniform without loss of generality) with the appropriate joint distribution (cf. Prob. 1.4). Regarding the instrument as a treatment, the treatment effect is characterized by $\mathcal{L}(R|I=i)=\mathcal{L}(\tilde{\rho}(V,i))$ for varying i, and if evidence is found of an effect of I on R then it follows from the correlation between I and $T=\tau(\delta,I)$ and the assumption that T exerts its influence on R through the second argument of ρ that the evidence is equally applicable to the effect of T on R.

This is true without assumptions about the range of R and T and without requiring that T be observable; moreover,

$$\mathbf{E}\left[R\,|I=i\right]-\mathbf{E}\left[R\,|I=i'\right]=\mathbf{E}\left[\rho(\varepsilon,\tau(\delta,i))-\rho(\varepsilon,\tau(\delta,i'))\right]=$$

$$\mathbf{E}\left[\rho(\varepsilon,\tau(\delta,i))-\rho(\varepsilon,\tau(\delta,i'))\,|\tau(\delta,i)\neq\tau(\delta,i')\right]\mathbf{P}\left[\tau(\delta,i)\neq\tau(\delta,i')\right],$$

so what one may consider to be the effect of the treatment, albeit in an unobservable subpopulation characterized by the condition $\tau(\delta,i)\neq\tau(\delta,i')$, is proportional to the effect of the instrument. The effect of the treatment itself in that population is unclear, but it is undeniable in the sense that for ω outside a set of probability zero it implies that $\rho(\varepsilon(\omega),t_1)\neq\rho(\varepsilon(\omega),t_0)$ for some distinct t_0, t_1 which may depend on ω. What is gained by making additional assumptions (such as τ being monotonic and R and T being binary) is the possibility of making the effect of the treatment appear more clearly and identifying the proportionality constant as something that can be estimated from observed data, in the case above as $\mathbf{P}\left[\tau(\delta,1)\neq\tau(\delta,0)\right]=\mathbf{E}(T|I=1)-\mathbf{E}(T|I=0)$. □

Remark 4.5 If T is discrete rather than binary, similar arguments show that

$$\theta = \frac{\mathbf{E}[\{\rho(\varepsilon,\tau(\delta,1))-\rho(\varepsilon,\tau(\delta,0))\}\,\mathbf{1}_{\{\tau(\delta,1)>\tau(\delta,0)\}}]}{\mathbf{E}[\{\tau(\delta,1)-\tau(\delta,0)\}\,\mathbf{1}_{\{\tau(\delta,1)>\tau(\delta,0)\}}]}$$

$$= \frac{\mathbf{E}\left[\rho\left(\varepsilon, \tau(\delta, 1)\right) - \rho\left(\varepsilon, \tau(\delta, 0)\right) | \tau(\delta, 1) > \tau(\delta, 0)\right]}{\mathbf{E}\left[\tau(\delta, 1) - \tau(\delta, 0) | \tau(\delta, 1) > \tau(\delta, 0)\right]}$$

(see also the next remark). This, too, is a measure of the effect of the treatment on the response in the subpopulation of individuals for whom $\tau(\delta, 1) > \tau(\delta, 0)$; it is more difficult to interpret than the first because one cannot estimate the last denominator, but, again, evidence that $\theta \neq 0$ provides evidence of a treatment effect. □

Remark 4.6 The analysis can be extended to model (4.26) in the case where I and T are discrete random variables with I taking the numerical values $i_0 < i_1 < i_2 < \cdots$ and τ such that $\tau(\delta, i_0) \leq \tau(\delta, i_k)$ for $k \geq 1$. For simplicity, assume that $\mathbf{P}[\tau(\delta, i_k) > \tau(\delta, i_0)] > 0$ for all k and consider the parameters

$$\theta_k := \frac{\mathbf{E}(R|I = i_k) - \mathbf{E}(R|I = i_0)}{\mathbf{E}(T|I = i_k) - \mathbf{E}(T|I = i_0)}$$

$$= \frac{\mathbf{E}[\rho\left(\varepsilon, \tau(\delta, i_k)\right) - \rho\left(\varepsilon, \tau(\delta, i_0)\right)]}{\mathbf{E}[\tau(\delta, i_k) - \tau(\delta, i_0)]}$$

$$= \frac{\mathbf{E}\left[\rho\left(\varepsilon, \tau(\delta, i_k)\right) - \rho\left(\varepsilon, \tau(\delta, i_0)\right) | \tau(\delta, i_k) > \tau(\delta, i_0)\right]}{\mathbf{E}\left[\tau(\delta, i_k) - \tau(\delta, i_0) | \tau(\delta, i_k) > \tau(\delta, i_0)\right]},$$

which can be estimated from data on (I, T, R) and measures the effect of treatment in the subpopulation of individuals for whom $\tau(\delta, i_k) > \tau(\delta, i_0)$. The numbers

$$p_k := \frac{\mathbf{P}(I = i_k)\left\{\mathbf{E}(T|I = i_k) - \mathbf{E}(T|I = i_0)\right\}}{\sum_{l \geq 1} \mathbf{P}(I = i_l)\left\{\mathbf{E}(T|I = i_l) - \mathbf{E}(T|I = i_0)\right\}}$$

define a probability distribution on $\mathbb{N}$ which can be estimated from data on (I, T), so we can define an overall measure of treatment effect by $\Theta := \sum_{k \geq 1} \theta_k p_k$, which turns out to be

$$\Theta = \frac{\mathbf{E}[\rho\left(\varepsilon, \tau(\delta, I)\right) - \rho\left(\varepsilon, \tau(\delta, i_0)\right) \mid \tau(\delta, I) > \tau(\delta, i_0)]}{\mathbf{E}[\tau(\delta, I) - \tau(\delta, i_0) \mid \tau(\delta, I) > \tau(\delta, i_0)]}.$$

Thus Θ, which can be estimated from data on (I, T, R), is a measure (though not a very clear one) of the effect of treatment on response in individuals for whom $\tau(\delta, I) > \tau(\delta, i_0)$. When T is binary it simplifies to a conditional expectation and compares with (4.28). □

4.7.1 Connection with linear regression

The idea of using 'instrumental variables' for estimating a causal effect seems to have originated in applications of linear regression models in which exogeneity—the independence between the error term and the explanatory variables—is doubtful (as may be the case when the data are known to follow a regression model quite closely but only data on a subset of the explanatory variables are available for estimating the model). In some of these applications the method of "instrumental-variable regression" or "two-stage least-squares", which also involves an instrument correlated with the explanatory variables, is used in place of least-squares to estimate the coefficient of the treatment. We conclude this section by explaining this method in the case of simple linear regression and by connecting it with that of Angrist, Imbens and Rubin.

Let the observed data (I, T, R) on a generic individual satisfy the following special case of model (4.26):

$$R = \alpha + \beta T + \varepsilon, \quad T = \tau(\delta, I). \tag{4.29}$$

Since ε and T are dependent (because δ and ε are dependent and the second equation constrains T by δ), the least-squares estimates of α and β are biased.[6] Nevertheless, a consistent estimate of β (the parameter of interest when studying the treatment effect) can be obtained under certain conditions.

Let $\mathbf{I}$, $\mathbf{T}$, $\mathbf{R}$, $\boldsymbol{\varepsilon}$ and $\boldsymbol{\delta}$, stand for column vectors containing the elements of a random sample $(I_1, T_1, R_1, \varepsilon_1, \delta_1)$, $(I_2, T_2, R_2, \varepsilon_2, \delta_2)$,. . .,$(I_n, T_n, R_n, \varepsilon_n, \delta_n)$, of vectors with the same distribution as $(I, T, R, \varepsilon, \delta)$, and 1 for a column vector of 1s, so that

$$\mathbf{R} = 1\alpha + \mathbf{T}\beta + \boldsymbol{\varepsilon}. \tag{4.30}$$

Assume that $\bar{\mathbf{I}}$, the average of $\mathbf{I}$, is zero; if this is not the case we can always subtract it from $\mathbf{I}$. Multiplying both sides of (4.30) on the left by $\mathbf{I}^{\mathrm{T}}$ we get

$$\mathbf{I}^{\mathrm{T}}\mathbf{R} = \mathbf{I}^{\mathrm{T}}\mathbf{T}\beta + \mathbf{I}^{\mathrm{T}}\boldsymbol{\varepsilon}. \tag{4.31}$$

Since on the one hand

$$\frac{1}{n}\mathbf{I}^{\mathrm{T}}\mathbf{T} = \frac{1}{n}\sum\nolimits_{j=1}^{n}(I_j - \bar{\mathbf{I}})(T_j - \bar{\mathbf{T}}) =: \mathrm{cov}(\mathbf{I}, \mathbf{T}),$$

$$\frac{1}{n}\mathbf{I}^{\mathrm{T}}\mathbf{R} = \frac{1}{n}\sum\nolimits_{j=1}^{n}(I_j - \bar{\mathbf{I}})(R_j - \bar{\mathbf{R}}) =: \mathrm{cov}(\mathbf{I}, \mathbf{R}),$$

and on the other hand, by the independence of I and ε, $n^{-1}\mathbf{I}^{\mathrm{T}}\boldsymbol{\varepsilon} \to 0$ (say in probability) as $n \to \infty$, we have from (4.31) that

$$\frac{1}{n}\mathbf{I}^{\mathrm{T}}\mathbf{R} - \frac{1}{n}\mathbf{I}^{\mathrm{T}}\mathbf{T}\beta = \mathrm{cov}(\mathbf{I}, \mathbf{R}) - \mathrm{cov}(\mathbf{I}, \mathbf{T})\beta \to 0.$$

But by the assumption that I and T are correlated, $\mathrm{cov}(\mathbf{I}, \mathbf{T})$ converges to some constant $\neq 0$, so $\mathrm{cov}(\mathbf{I}, \mathbf{R})/\mathrm{cov}(\mathbf{I}, \mathbf{T}) \to \beta$. In other words,

$$\check{\beta} := \frac{\mathrm{cov}(\mathbf{I}, \mathbf{R})}{\mathrm{cov}(\mathbf{I}, \mathbf{T})}$$

is a consistent estimator of β.

---

[6] See footnote 2 on p. 73.

Evidently, $\breve{\beta}$ is not the least-squares estimate; in fact, it is easy to check that when the I_j and T_j are binary $\breve{\beta}$ coincides with the estimator of θ mentioned after (4.27). But it is clear that when the I_j and T_j are binary and the monotonicity of τ holds the assumption in (4.29) of a linear relationship between response and treatment is not at all needed for the estimation of the treatment effect.

The appellation "two-stage least-squares" comes from $\breve{\beta}$ being obtainable in an alternative way by two applications of least-squares estimation in regression models. First, one *pretends* that I, T and δ satisfy the model

$$T = \tau(\delta, I) = a + b\,I + \delta,$$

where a and b are unknown parameters.[7] Applied to our random sample this becomes

$$\mathbf{T} = 1a + \mathbf{I}b + \boldsymbol{\delta}$$

and yields the usual least-squares estimators $\hat{a}$ and $\hat{b}$, the latter being

$$\hat{b} = \mathrm{cov}(\mathbf{I}, \mathbf{T})/\mathrm{var}(\mathbf{I}).$$

Then one plugs in the 'fitted values', or 'predicted values', $\hat{\mathbf{T}} = 1\hat{a} + \hat{\mathbf{I}}\,\hat{b}$ into (4.30) to get

$$\mathbf{R} = (\alpha + \hat{a}\,\beta)1 + \mathbf{I}\,\hat{b}\,\beta + \boldsymbol{\varepsilon}. \tag{4.32}$$

This identity is false (since the assumption is that (4.30) is true), but for large n it is approximately true. Even if it were true it would not specify a linear regression model because $\boldsymbol{\varepsilon}$ and $(\mathbf{I}, \hat{a}, \hat{b})$ are dependent (for instance, $\hat{b}$ is a function of $\mathbf{T}$, which depends on $\boldsymbol{\delta}$, and $\boldsymbol{\varepsilon}$ and $\boldsymbol{\delta}$ are dependent); but we can pretend it is true and, just as we have done above to get $\breve{\beta}$, multiply both sides of (4.32) on the left by $\mathbf{I}^{\mathrm{T}}$ and solve the resulting equation to get an estimator for β, namely

$$\tilde{\beta} := \frac{\mathrm{cov}(\mathbf{I}\hat{b}, \mathbf{R})}{\mathrm{cov}(\mathbf{I}\hat{b}, \mathbf{I}\hat{b})} = \frac{1}{\hat{b}}\frac{\mathrm{cov}(\mathbf{I}, \mathbf{R})}{\mathrm{var}(\mathbf{I})} = \frac{\mathrm{cov}(\mathbf{I}, \mathbf{R})}{\mathrm{cov}(\mathbf{I}, \mathbf{T})},$$

which is none other than $\breve{\beta}$.

A more transparent and concise version of this approach is possible when $T = a + bI + \delta$ holds in place of the second equation of (4.29); substitution of T in $R = \alpha + T\beta + \varepsilon$ gives

$$R = \alpha + a\,\beta + I\,b\,\beta + \delta\,\beta + \varepsilon =: A + I\,B + E$$

for a generic individual with observed data (I, T, R) and $E = \delta\,\beta + \varepsilon$. This equation is a linear regression model proper (I being independent of (δ, ε)), so we have two linear models, one regressing T on I and the other R on I, which give consistent estimates of b and $B = b\,\beta$, and hence a consistent estimate of $B/b = \beta$ which is easily verified to coincide with $\breve{\beta}$ (cf. pp. 85–6 of Pearl, Glymour and Jewel [73]).

[7] What could fail is not the exogeneity (I and δ being independent) but the postulated form of τ.

4.8 Notes

The examples of Sects. 4.2 and 4.3 have been used several times by Pearl (e.g. pp. 83–84 and pp. 66–68 of [67]).

Freedman (pp. 272–3 of [37]) criticized the model in the first example; in his opinion, the 'perverse theory' about smoking and lung cancer, according to which it is the genotype rather than smoking that causes cancer, is refuted by carefully designed population studies, not by statistical analyses of 'detailed' models such as (4.1). In the second example the yield of oats is not influenced by B, which, as noted by Freedman (pp. 273–4 of [37]), is one of the reasons why the model is not sufficiently realistic. Of course, none of the examples is meant to be completely realistic but rather to illustrate definitions and calculations. Yet one cannot help wondering how often models such as (4.1) and (4.5) can be seriously entertained in real-life problems. As far as we can see, creating such models as approximations to reality is precisely what one should aim for, as long as one strives to be clear and critical about them and accepts that the task is impossible in many problems and long and arduous in others.

As far as we know, Freedman (pp. 267–9 of [37]) was the first to prove Pearl's result of Sect. 4.2 by means of standard probability; our approach is less general because it assumes the variables to be discrete, but we make an explicit connection between the quantity to be estimated from data and the treatment effect—comparing the proofs is easy since we follow Freedman's notation rather than Pearl's. Pearl's result of Sect. 4.3 is stated on p. 67 of Pearl [67], but we have found no proof of it in the literature. Freedman (p. 270 of [37]) certainly solved the problem but provides only an indication of how to proceed (here too we follow Freedman's notation rather than Pearl's).

It appears that the method of compounding the direct and indirect effects of one variable on another, as surmised from the interpretation attempted at the end of Sect. 4.3, should help one to write down a formula for the overall effect, no matter how large and complicated the model is (cf. Pearl [67], p. 67).

Sect. 4.4 is based on Pearl's [71] comments on some work by Cox and Wermuth (Sect. 2.3 of [108]). In other works cited in Pearl [71], Cox and Wermuth seem to have suggested that definitions of effects based on 'interventions' (presumably on potential outcomes, such as the $\hat{Y}_1$ involved in the second definition) generally have no bearing on real-life situations; the comments in [71] are Pearl's response to such a suggestion. We have not been able to understand exactly what Cox and Wermuth mean in the texts quoted by Pearl, but it may be related to our Remark 2.3.

Models (4.8) and (4.9) are versions of models considered earlier by Robins and Wasserman [82]. Our second definition of direct effect in Sect. 4.4 is used in Sect. 3.7 of Pearl, Glymour and Jewel [73] in an essentially different problem.

Sect. 4.5 is based on pp. 75–78 and 114–116 of Pearl, Glymour and Jewel [73]. The last quantity in this section (or, more precisely, a slight variant of it) is referred to as the 'natural indirect effect' of sex on hiring 'mediated by' qualification and background on p. 115 of Pearl, Glymour and Jewel [73].

The definition and study of causal effects in situations where the response is influenced by a variable in a direct way but also in an indirect way through another variable, exemplified in Sects. 4.4 and 4.5, are sometimes referred to as 'mediation analysis'.

Case-control studies, formalized in Sect. 4.6, have long been studied and carried out in Epidemiology (e.g. Rothmann [87]). There are various types of such studies; in particular, several controls may be matched to each case; see, for example, pp. 7, 83–86 of Rosenbaum [83], where case-control studies are called 'case-referent studies'. The significance of the odds ratio in case-control studies was first noticed by Cornfield [18]. Subsection 4.6.1 describes one version of Berkson's paradox; for other versions, and for some of the history behind the paradox and its implications for real-life studies, see Snoep *et al.* [97].

Sect. 4.7 is based on Angrist, Imbens and Rubin [2]. Some real-life problems in which the method of instrumental variables is potentially useful are described in Shehan *et al.* [96].

For the connection between "instrumental-variable regression" ("two-stage least-squares") and the method of instrumental variables in a more general, but necessarily less clear, case of a regression model containing other explanatory variables besides the treatment variable, see Freedman's exposition on pp. 181–4 and 197–8 of [36]. Our purpose has been to clarify the relationship between two apparently unrelated approaches; the assumptions required by the version of (4.26) conditional on a set of background variables are more likely to be met in a real-life problem than those required by the regression model considered by Freedman [36].

CHAPTER 5

Some methods of statistical inference

Under the basic model with unconfoundedness the effect of the treatment on the response is characterized by $\mathcal{L}(R_n|\mathbf{X}_n = \mathbf{x}, T_n = t)$, so inference about it amounts to the estimation of this law for various values of $(\mathbf{x}, t)$ and to testing the hypothesis that the law is constant in t for some or all $\mathbf{x}$, both procedures involving stratification or matching on the values of the $\mathbf{X}_n$. In particular, inference about the effect of the treatment involves estimates of functionals of this law, such as $\mathbf{E}(R_n|\mathbf{X}_n = \mathbf{x}, T_n = t)$ or $\sum_{\mathbf{x}} \{\mathbf{E}(R_n|\mathbf{X}_n = \mathbf{x}, T_n = t) - \mathbf{E}(R_n|\mathbf{X}_n = \mathbf{x}, T_n = t')\}\, \mathbf{P}(\mathbf{X}_n = \mathbf{x})$, $t \neq t'$, and tests of association between R_n and T_n conditionally on $\mathbf{X}_n = \mathbf{x}$. In this chapter we study statistics for testing the association between the response and the treatment within and over strata or matched sets, which provide exact, nonparametric tests on the effect of treatment, and use the central limit theorem to obtain confidence intervals for $\mathcal{L}(R_n|\mathbf{X}_n = \mathbf{x}, T_n = t)$ and functionals of it. In addition, we consider methods of data analysis designed to probe the plausibility of the basic model with unconfoundedness.

5.1 Orientation

The methods described in this chapter are meant to apply to a set of random vectors $(\mathbf{X}_n, T_n, R_n)$, $n = 1, \ldots, N$, defined by the generalized versions of the equations of the basic model (2.1) mentioned just before Example 2.1, namely

$$R_n = \rho_n(V_n, \mathbf{X}_n, T_n), \quad T_n = \tau_n(U_n, \mathbf{X}_n), \quad n \in \mathbb{N}, \tag{5.1}$$

where ρ_n and τ_n are real-valued functions and, for each n, U_n and V_n are standard uniform random variables conditionally on $\mathbf{X}_n$, which set may or may not constitute a random sample and is supposed to represent data on 'units', 'subjects', 'individuals', etc., with labels or indices $1, 2, \ldots, N$. When the $(\mathbf{X}_n, T_n, R_n)$ are a random sample, the model essentially implies that ρ_n and τ_n are constant in n, as we have been assuming until now for simplicity, but in the more general problems considered in Sect. 5.2 this assumption is restrictive. Besides some procedures for data analysis outlined at the end, we study tests of hypotheses about the existence of treatment effect and procedures for estimating measures of treatment effect, and all of these are based on a particular arrangement of the $(\mathbf{X}_n, T_n, R_n)$ into *strata* or *matched sets* $I_{\mathbf{x}} := \{n : \mathbf{X}_n = \mathbf{x}\} \cap \{1, \ldots, N\}$ for $\mathbf{x} \in \mathcal{X}$ (the range of the $\mathbf{X}_n$), some of which may be empty. Any such arrangement is called a *stratification* or *set of matches*: in stratification it corresponds to a partition of $\mathcal{X}$, but in matching the sets may make up only a subset of the sample and their definition may be based on things other than the values of the background variables, such as the number of units and the relative proportions of treatment values (or of response values) in them.

Because the methods involve testing or estimation within the $I_{\mathbf{x}}$, in strict sense they are valid only when the *covariate vectors* $\mathbf{X}_n$ are discrete (or at any rate when their distributions have discrete components, for otherwise every $I_{\mathbf{x}}$ contains at most one element), which will be assumed throughout.[1] It will also be assumed, though mainly for simplicity, that the T_n are discrete. And then, as usual, when considering the estimation of such things as $(\mathbf{x}, t) \to \mathbf{E}[R_n | T_n = t, \mathbf{X}_n = \mathbf{x}]$ it will be assumed that, at least for some $\mathbf{x}$, $0 < \mathbf{P}(T_n = t | \mathbf{X}_n = \mathbf{x}) < 1$ for some t; this ensures that at least some of the $I_{\mathbf{x}}$ can have units with distinct treatments (cf. Remark 2.2).

If some of the covariates are continuous then the methods apply approximately, to the extent (usually unknown in real-life problems) that an expectation such as $\mathbf{E}[f(R_n) | T_n = t, \mathbf{X}_n \in \mathcal{N}_{\mathbf{x}}]$ approximates $\mathbf{E}[f(R_n) | T_n = t, \mathbf{X}_n = \mathbf{x}]$ when $\mathcal{N}_{\mathbf{x}}$ is a 'small' neighbourhood of $\mathbf{x}$. For example, when the $\mathbf{X}_n$ are a random sample and take values in $\mathbb{R}^d$ stratification and matching are not based on $I_{\mathbf{x}}$ but rather on $\{n : \mathbf{X}_n \in \mathcal{N}\} \cap \{1, \ldots, N\}$ with $\mathcal{N} = A_{1,k_1} \times A_{2,k_2} \times \cdots \times A_{d,k_d}$ for open, half-open or closed intervals $A_{j,1}, A_{j,2}, \ldots$ constituting a partition of the range of the j-th coordinate of $\mathbf{X}_n$; the intervals of such a partition should be as short as the amount of data allows them to be, and, as ideas of integration suggest, their length should be more or less constant and 'deterministically' fixed rather than varying and randomly determined (as will be the case if, for example, their endpoints correspond to sample quantiles).

5.2 Testing for a treatment effect

The first assumption required for a test on the treatment effect is a more stringent version of unconfoundedness and *will be assumed throughout this chapter*:

A0 For every $I = \{i_1, i_2, \ldots, i_n\} \subset \{1, 2, \ldots, N\}$, conditionally on $\mathbf{X}_{i_1}, \mathbf{X}_{i_2}, \ldots, \mathbf{X}_{i_n}$ being equal to $\mathbf{x} \in \mathcal{X}$, the vectors $(U_{i_1}, U_{i_2}, \ldots, U_{i_n})$ and $(V_{i_1}, V_{i_2}, \ldots, V_{i_n})$ are independent. Moreover, if $1 \leq K \leq n$, $\mathbf{x}_1, \mathbf{x}_2, \ldots, \mathbf{x}_K \in \mathcal{X}$ are *distinct*, and $I_1, I_2, \ldots, I_K$ form a partition of I, then for $(u_i)_{i \in I}, (v_i)_{i \in I} \in \mathbb{R}^n$,

$$\mathbf{P}(U_i \leq u_i, V_i \leq v_i, i \in I | \mathbf{X}_i = \mathbf{x}_j, i \in I_j, j = 1, 2, \ldots, K) =$$
$$\textstyle\prod_{j=1}^{K} \mathbf{P}(U_i \leq u_i, V_i \leq v_i, i \in I_j | \mathbf{X}_i = \mathbf{x}_j, i \in I_j).$$

[1] We shall often refer to the $\mathbf{X}_n$ as *covariate vectors* and to their coordinates as *covariates*, terms frequently used in statistics but not necessarily in connection with 'causality'.

The first part (which for $n = K = 1$ reduces to unconfoundedness) ensures that the conditioning on the $\mathbf{X}_i$ of several units does not destroy unconfoundedness or change the conditional distributions of U_i and V_i for each i. It is not implausible that in a real-life problem one would have to assume that the information that several units (as opposed to one unit in isolation) have covariate vectors equal to a given $\mathbf{x}$ could 'entangle' U_i and V_i: for example, the appearance of so many individuals with similar characteristics and therefore equally eligible for treatment might lead a doctor responsible for the treatment assignment to pick the individuals to be treated on the basis of a new covariate, which could act both on U_i and on V_i.

The second part of of the assumption is intended to make inferences from different strata/matched sets independent of each other; in particular, the conditioning event represented by one stratum implies nothing about the treatment assignments and the responses of other strata.

Note that **A0** implies (take $K = n = N$, $I_j = \{j\}$, and use unconfoundedness and the distribution of (U_i, V_i) conditionally on $\mathbf{X}_i$ for each i) that

$$\mathbf{P}(U_i \leq u_i, V_i \leq v_i, i = 1, \ldots, N | \mathbf{X}_i = \mathbf{x}_i, i = 1, 2, \ldots, N) = \prod_{i=1}^{N} u_i v_i$$

for $u_i, v_i \in [0, 1]$ and *distinct* $\mathbf{x}_1, \mathbf{x}_2, \ldots, \mathbf{x}_N \in \mathcal{X}$. It does not imply that the (U_i, V_i) of units i in the same stratum or matched set $\{i : \mathbf{X}_i = \mathbf{x}\}$ are independent and have uniform marginal distributions; in fact, we shall see below that it is important to allow the (U_i, V_i) of units in the same stratum to be dependent.

Write

$$f_I^{(\mathbf{x})}((t_i)_{i \in I}) = \mathbf{P}(T_i = t_i, i \in I | \mathbf{X}_i = \mathbf{x}, i \in I), \tag{5.2}$$

$(t_i)_{i \in I} = (t_{i_1}, t_{i_2}, \ldots, t_{i_n}) \in \mathcal{T}^n$; this is the probability function of the treatments assigned to an arbitrary set $I = \{i_1, i_2, \ldots, i_n\} \subset \{1, 2, \ldots, N\}$ of units, conditional on the event that their covariate vectors all take the value $\mathbf{x}$. By the first part of **A0** we have, for $(r_i)_{i \in I} \in \mathbb{R}^n$,

$$\mathbf{P}(T_i = t_i, R_i \leq r_i, i \in I | \mathbf{X}_i = \mathbf{x}, i \in I) =$$
$$f_I^{(\mathbf{x})}((t_i)_{i \in I}) \mathbf{P}(\rho_i(V_i, \mathbf{x}, t_i) \leq r_i, i \in I | \mathbf{X}_i = \mathbf{x}, i \in I). \tag{5.3}$$

The hypothesis of no treatment effect relative to the units $1, 2, \ldots, N$ means that all the functions $(v, \mathbf{x}, t) \to \rho_i(v, \mathbf{x}, t)$ are constant in t for each $(v, \mathbf{x})$; thus, under the null hypothesis, the last probability is a function of the r_i and $\mathbf{x}$ alone, and $(T_i)_{i \in I} \equiv (T_{i_1}, T_{i_2}, \ldots, T_{i_n})$ and $(R_i)_{i \in I} \equiv (R_{i_1}, R_{i_2}, \ldots, R_{i_n})$ are independent conditionally on $\{\mathbf{X}_i = \mathbf{x}, i \in I\}$ for all $\mathbf{x}$ and all $I = \{i_1, i_2, \ldots, i_n\} \subset \{1, 2, \ldots, N\}$.

It follows that we can test the null hypothesis within a stratum or matched set by testing the independence of the responses and treatments of the units in it; this is what the first part of **A0** gives us in addition to unconfoundedness (which in our earlier chapters furnished the same result for single units). What the second part does is to allow us to test the null hypothesis by testing the independence of $(T_i)_{i \in I}$ and $(R_i)_{i \in I}$ over several strata or matched sets I (which are disjoint by definition), separately as so many null hypotheses or together as a single null hypothesis.

The other assumptions that need to be introduced should provide us with a *null distribution* for testing the conditional independence of a vector of treatments and a vector of responses within a stratum/matched set. To see the need for further assumptions suppose that τ_i varies with i; for example, say that i represents the time when a unit arrives for treatment, that τ_i tends to become smaller on average with increasing i, and that this has nothing to do with the characteristics $\mathbf{X}_i$ of the successive units but rather with some trend in treatment policy. Then the basic model and **A0** still offer the theoretical possibility of drawing inferences about the treatment effect, but the possibility may not be realizable unless one possesses the correct model for τ_i. Suppose instead that $\tau_i = \tau$ for all i but the U_i are dependent, which may be appropriate when the number of treatments in $I = \{i \leq N : \mathbf{X}_i = \mathbf{x}\}$ is constrained, say by $\sum_{i\in I} \tau_i(U_i, \mathbf{x}) = m$; then the drawing of inferences requires the knowledge of the joint distribution of the U_i, which in principle can be anything.

The way to proceed depends on the study in hand. One is to assume that

A1 For every $I = \{i_1, i_2, \ldots, i_n\} \subset \{1, 2, \ldots, N\}$, conditionally on $\mathbf{X}_{i_1}, \mathbf{X}_{i_2}, \ldots, \mathbf{X}_{i_n}$ being equal to $\mathbf{x} \in \mathcal{X}$, the random variables $T_{i_1}, T_{i_2}, \ldots, T_{i_n}$ are *exchangeable* in the sense that for $t_{i_1}, t_{i_2}, \ldots, t_{i_n} \in \mathcal{T}$

$$\mathbf{P}(T_i = t_i, i \in I | \mathbf{X}_i = \mathbf{x}, i \in I)$$

is invariant to permutations of the coordinates of $(t_{i_1}, t_{i_2}, \ldots, t_{i_n})$.

Assumption **A1** means that the $f_I^{(\mathbf{x})}$ of (5.2) is a symmetric function: permutation of its arguments does not change its value. It corresponds to what in the works of Rosenbaum and Rubin (cf. Sect. 3.2 of Rosenbaum [83]) is called an observational study with *overt bias*, or *free of hidden bias*, and it holds if $\tau_i \equiv \tau$ and the U_i are exchangeable.

The dependence of $f_I^{(\mathbf{x})}$ on the set I could, for example, represent a group effect, or a group *size* effect, on the distribution of the treatments. As examples of $f_I^{(\mathbf{x})}((t_i)_{i\in I})$ with $I = \{1, 2, \ldots, n\}$ we may cite $p(\mathbf{x})^{\sum_{i=1}^n t_i}(1 - p(\mathbf{x}))^{\sum_{i=1}^n (1-t_i)}$ for $t_i \in \{0, 1\}$ and $0 < p(\mathbf{x}) < 1$ (binary independent treatments), $\mathrm{r}(\mathbf{x})!\, n^{-\mathrm{r}(\mathbf{x})}/(t_1!t_2! \cdots t_n!)$ for $t_i \in \{0, 1, \ldots, \mathrm{r}(\mathbf{x})\}$ and $\sum_{i=1}^n t_i = \mathrm{r}(\mathbf{x}) \in \mathbb{N}$ (dependent treatments with a number $\mathrm{r}(\mathbf{x})$ of treatment levels that depends on $\mathbf{x}$), and $e^{-n\lambda(\mathbf{x})}\lambda(\mathbf{x})^{\sum_{i=1}^n t_i}/(t_1!t_2! \cdots t_n!)$ for $t_i \in \mathbb{N}_0$ and $\lambda(\mathbf{x}) > 0$ (independent Poisson treatments with covariate-dependent parameter); but the particular form of $f_I^{(\mathbf{x})}$ is irrelevant: it is the symmetry of it that matters and must be justified in applications.

Example 5.1 Suppose that the T_i are binary, $T_i = 1$ indicating that the i-th unit has been 'treated', $T_i = 0$ that it has been kept as a 'control'. Regard the $(\mathbf{X}_i, T_i, R_i)$ pertaining to units with the same value of $\mathbf{X}_i$ as data on the same individual obtained at different times. Thus, for example, if $\mathbf{X}_1 = \mathbf{X}_2 = \cdots = \mathbf{X}_8 = \mathbf{x}$ and $\mathbf{X}_i \neq \mathbf{x}$ for $i > 8$ then $R_1, R_2, \ldots, R_8$ may represent the responses measured on one individual at eight different times and $T_1, T_2, \ldots, T_8$ the treatments assigned to that individual at those times; in this case $R_1, R_2, \ldots, R_8$ need not to be assumed independent, and neither need $T_1, T_2, \ldots, T_8$ since the individual's treatment assignment may for instance be constrained to include treatment exactly four times. If **A1** holds then $T_1, T_2, \ldots, T_8$ are exchangeable—so the assignment $(1,0,0,1,0,1,0,1)$ is as likely as $(1,1,1,0,0,0,1,0)$, $(1,0,0,0,0,1,0,0)$ is as likely as $(0,0,1,1,0,0,0,0)$, and so on—and the assignments of treatment to different individuals are made independently.

This example applies to Fisher's famous 'lady tasting tea' experiment, in which a lady who believes she can distinguish a cup of tea prepared by pouring milk first from one prepared by pouring milk last tests her discriminating powers by tasting and trying to identify the four cups prepared according to her favourite recipe out of eight cups served in random order. It also applies to a larger experiment involving several ladies of similar taste in independent tasting contests. □

Note that **A1** does not involve the responses; and in principle there is no reason why it should. For instance, conditioning on the event $\{\mathbf{X}_i = \mathbf{x}, R_i = r_i, i \in I\}$ for $r_1, \ldots, r_n \in \mathbb{R}$ (distinct or all equal) rather than on $\{\mathbf{X}_i = \mathbf{x}, i \in I\}$ might destroy the symmetry of the resulting probabilities. This should happen in a model representing an experiment in which units with covariates equal to $\mathbf{x}$ are first identified, then from among them n are selected in such a way that (with the purpose of attaining some sort of variety or representativeness, for example) each has a particular level of an additional variable known to influence the response (say unit i with i kilos), and finally treatment is randomized over the selected units. In such an experiment (which would be perfectly compatible with **A0**) the randomization justifies **A1**, but not the version of it that conditions further on the responses, since the magnitude of these should provide some clues about which units got which treatment, irrespective of whether or not the treatment affects the response (so long as the additional variable does).

But **A1** is not completely general, because it does not cover case-control studies (Sect. 4.6). For example, if the data on the first case are $(\mathbf{X}_1, T_1, R_1) \equiv (\mathbf{X}_1, T_1, 1)$ and the data on its 'control' are $(\mathbf{X}_1, T_2, R_2) \equiv (\mathbf{X}_1, T_2, 0)$ (which may be arranged, as noted in Remark 4.3), then

$$f^{(\mathbf{x})}_{\{1,2\}}(t_1, t_2) = \mathbf{P}(T_1 = t_1, T_2 = t_2 | \mathbf{X}_1 = \mathbf{X}_2 = \mathbf{x})$$

is not symmetrical in t_1 and t_2 whenever the treatment tends to increase the response, since then this probability is larger with $(t_1, t_2) = (1, 0)$ than with $(t_1, t_2) = (0, 1)$, so **A1** fails with $I = \{1, 2\}$. This happens because the labels of the variables contain information about the method of sampling and hence about the variables.

A version of **A1** that applies to case-control studies is

A1′ For every $I = \{i_1, i_2, \ldots, i_n\} \subset \{1, 2, \ldots, N\}$, conditionally on $\mathbf{X}_i = \mathbf{x} \in \mathcal{X}$ and $R_i = r \in \mathbb{R}$ for all $i \in I$, the random variables $T_{i_1}, T_{i_2}, \ldots, T_{i_n}$ are exchangeable in the sense that for $t_{i_1}, t_{i_2}, \ldots, t_{i_n} \in \mathcal{T}$

$$\mathbf{P}(T_i = t_i, i \in I | \mathbf{X}_i = \mathbf{x}, R_i = r, i \in I)$$

is invariant to permutations of the coordinates of $(t_{i_1}, t_{i_2}, \ldots, t_{i_n})$.

For example, if $I = \{i_1, i_2\}$ and the event $\{R_{i_1} = R_{i_2} = r\}$ has positive probability, which implies that the units are both controls or both cases,

$$\mathbf{P}(T_1 = t_{i_1}, T_2 = t_{i_2} | \mathbf{X}_{i_1} = \mathbf{X}_{i_2} = \mathbf{x}, R_{i_1} = R_{i_2} = r)$$

is symmetric in (t_{i_1}, t_{i_2}). On the other hand, if $\{R_{i_1} = R_{i_2} = r\}$ has probability zero, this probability may be defined arbitrarily and does not contradict **A1′**.

Under the null hypothesis of no treatment effect (and under **A0**, as always in the present chapter), **A1** and **A1′** are equivalent since by the conditional independence of the treatments and the responses, which implies

$$f_I^{(\mathbf{x})}((t_i)_{i\in I}) = \mathbf{P}(T_i = t_i, i \in I | \mathbf{X}_i = \mathbf{x}, R_i = r, i \in I), \tag{5.4}$$

the probabilities in the latter reduce to those in the former. Because of this equivalence under the null hypothesis, the methods of testing to be described next are the same under the two assumptions—they are based solely on the validity of (5.4) under the null and on the symmetrical character of $f_I^{(\mathbf{x})}$.

5.2.1 Testing within a stratum

A1 and **A1′** have both one consequence that can be used to test the null hypothesis of no treatment effect in a way that does not involve the probabilities (5.2), which in most applications must be regarded as unknown.

For fixed $n \in \mathbb{N}$ and $I = \{i_1, i_2, \ldots, i_n\} \subset \mathbb{N}$ let $C_I^{(t)}$ be the random variable that counts the number of T_i with $i \in I$ assuming the value $t \in \mathcal{T}$:

$$C_I^{(t)} = \sum_{i\in I} \mathbf{1}_{\{T_i = t\}} = \sum_{i \in I} \delta_{T_i, t}. \tag{5.5}$$

Fix a vector $(t_i^*)_{i\in I} = (t_{i_1}^*, t_{i_2}^*, \ldots, t_{i_n}^*)$ with coordinates in $\mathcal{T}$ and put

$$c_t = \sum_{i\in I} \delta_{t_i^*, t}, \quad t \in \mathcal{T};$$

c_t is the number of coordinates of $(t_i^*)_{i\in I}$ that are equal to t. Clearly, $n = \sum_{t\in\mathcal{T}} c_t$ and $c_t = 0$ except for t in a finite set, and any vector $(t_i)_{i\in I} = (t_{i_1}, t_{i_2}, \ldots, t_{i_n})$ obtained by permuting the coordinates of $(t_i^*)_{i\in I}$ satisfies

$$c_t = \sum_{i\in I} \delta_{t_i, t}, \quad t \in \mathcal{T}; \tag{5.6}$$

let $\mathcal{T}_I((c_t)_{t\in\mathcal{T}})$ denote the set of all such vectors.

Since under the null both **A1** and **A1′** say that the probabilities $f_I^{(\mathbf{x})}((t_i)_{i\in I})$ are invariant under permutations of their arguments, we have, on abbreviating summation over the indices in $(t_i)_{i\in I} \in \mathcal{T}_I((c_t)_{t\in\mathcal{T}})$ by $\sum_{\mathcal{T}_I((c_t)_{t\in\mathcal{T}})}$,

$$\mathbf{P}\left(C_I^{(t)} = c_t, t \in \mathcal{T} \middle| \mathbf{X}_i = \mathbf{x}, i \in I\right) = \sum\nolimits_{\mathcal{T}_I((c_t)_{t\in\mathcal{T}})} f_I^{(\mathbf{x})}((t_i)_{i\in I})$$

$$= \#\mathcal{T}_I((c_t)_{t\in\mathcal{T}}) \times f_I^{(\mathbf{x})}((t_i^*)_{i\in I})$$

$((t_i)_{i\in I} \in \mathcal{T}_I((c_t)_{t\in\mathcal{T}}))$, where as usual we write $\#A$ for the number of elements in a set A. Now consider the ratio of

$$\mathbf{P}\left(T_i = t_i, i \in I, C_I^{(t)} = c_t, t \in \mathcal{T} \middle| \mathbf{X}_i = \mathbf{x}, i \in I\right) = f_I^{(\mathbf{x})}((t_i^*)_{i\in I})$$

to the preceding probability:

$$\frac{\mathbf{P}\left(T_i = t_i, i \in I, C_I^{(t)} = c_t, t \in \mathcal{T} \middle| \mathbf{X}_i = \mathbf{x}, i \in I\right)}{\mathbf{P}\left(C_I^{(t)} = c_t, t \in \mathcal{T} \middle| \mathbf{X}_i = \mathbf{x}, i \in I\right)} = \frac{1}{\#\mathcal{T}_I((c_t)_{t\in\mathcal{T}})}.$$

The term on the left here is the probability that $T_i = t_i$ for $i \in I$ conditionally on the event that $\mathbf{X}_i = \mathbf{x}$ for $i \in I$ and $C_I^{(t)} = c_t$ for $t \in \mathcal{T}$, so we conclude that

$$\mathbf{P}\left(T_i = t_i, i \in I \middle| \mathbf{X}_i = \mathbf{x}, i \in I, C_I^{(t)} = c_t, t \in \mathcal{T}\right) = \frac{1}{\#\mathcal{T}_I((c_t)_{t\in\mathcal{T}})}$$

for each $(t_1, t_2, \ldots, t_n) \in \mathcal{T}_I((c_t)_{t\in\mathcal{T}})$. Since by definition of $\mathcal{T}_I((c_t)_{t\in\mathcal{T}})$ the number of elements in it corresponds to the number of distinct *strings* of length $n = \sum_{t\in\mathcal{T}} c_t$ made up of $c_{t_1^*}$ elements labelled t_1^*, $c_{t_2^*}$ elements labelled t_2^*, etc., we also see that

$$\#\mathcal{T}_I((c_t)_{t\in\mathcal{T}}) = \frac{n!}{\prod_{t\in\mathcal{T}}(c_t!)}.$$

The same argument (one need only replace the event $\{\mathbf{X}_i = \mathbf{x}, i \in I\}$ by $\{\mathbf{X}_i = \mathbf{x}, R_i = r_i, i \in I\}$ in the above) shows that *under the null hypothesis* we have, for all $(t_i)_I \in \mathcal{T}_I((c_t)_{t\in\mathcal{T}})$ and $(r_i)_{i\in I} = (r_{i_1}, r_{i_2}, \ldots, r_{i_n}) \in \mathbb{R}^n$,

$$\mathbf{P}\left(T_i = t_i, i \in I \middle| \mathbf{X}_i = \mathbf{x}, R_i = r_i, i \in I, C_I^{(t)} = c_t, t \in \mathcal{T}\right) = \frac{\prod_{t\in\mathcal{T}}(c_t!)}{n!}. \quad (5.7)$$

It is this result that allows us to develop tests of the hypothesis of no treatment effect by conditioning on the responses and covariates and on the frequencies of the various levels of treatment and then confronting the actual treatment assignments with those expected under the *conditional null distribution*, or *null probabilities*, (5.7). Such tests are *conditional* in the sense that it is by conditioning *further* (on the $C_I^{(t)}$ of (5.5)) that the null distribution of treatment assignments becomes completely specified.

Many tests are possible, since there are many ways of looking for discrepancies between the observed treatment assignments and the treatment assignments expected under the null, but when both the treatments and the responses are ordinal or binary it seems natural to choose a test based on the sample covariance between them. We will now elaborate on a one-sided version of such a test; for the two-sided version see Prob. 5.1.

Assume that the treatment can only manifest itself by a positive association with the response, increasing levels of treatment tending to cause increasing responses. Let $I \subset \{1, 2, \dots, N\}$ be a set of $n := \#I$ units in a stratum or matched set characterized by covariate vectors equal to the same $\mathbf{x} \in \mathcal{X}$. For simplicity, in what follows we refer to such a set I of indices as a *stratum*, keeping in mind that it may also denote a matched set.

The sample covariance associated with the units in I is

$$\mathcal{K}_I \equiv \mathcal{K}((T_i, R_i)_{i\in I}) := \frac{1}{n}\sum_{i\in I} R_i T_i - \bar{R}_I \bar{T}_I,$$

where $\bar{R}_I := n^{-1}\sum_{i\in I} R_i$ and $\bar{T}_I = n^{-1}\sum_{i\in I} T_i$. Let us condition on the event that the $(\mathbf{X}_i)_{i\in I}$ are all equal to $\mathbf{x}$, the responses $(R_i)_{i\in I}$ are equal to $(r_i)_{i\in I} \in \mathbb{R}^n$, and the frequencies of the treatments satisfy

$$c_t = C_I^{(t)} = \sum_{i\in I} \mathbf{1}_{\{T_i = t\}} \tag{5.8}$$

for some integers c_t such that $\sum_{t\in\mathcal{T}} c_t = n$. As we have seen, under the null the distribution of $(T_i)_{i\in I}$ conditional on this event is uniform on the set $\mathcal{T}\big((c_t)_{t\in\mathcal{T}}\big)$ consisting of the $n!/\prod_{t\in\mathcal{T}}(c_t!)$ points $(t_i)_{i\in I}$ satisfying $c_t = \sum_{i\in I}\delta_{t_i,t}$ for $t \in \mathcal{T}$. In principle, these points may be enumerated and the corresponding set

$$\big\{\mathcal{K}((t_i, r_i)_{i\in I}) : (t_i)_{i\in I} \in \mathcal{T}\big((c_t)_{t\in\mathcal{T}}\big)\big\} \tag{5.9}$$

calculated; this set contains all the values that the random variable $\mathcal{K}((T_i, R_i)_{i\in I})$ can assume under the event we are conditioning upon, and therefore its frequency distribution is the conditional distribution of $\mathcal{K}((T_i, R_i)_{i\in I})$.

If q_α is the *upper critical value* of probability $\alpha \in]0, 1[$ under this distribution, a one-sided test against an association between treatment and response consists of rejecting the null hypothesis—more precisely the null relative to stratum I—if and only if the sample covariance actually observed exceeds q_α. This test has the desired significance level α because q_α is the smallest number q satisfying

$$\frac{\#\big\{(t_i)_{i\in I} \in \mathcal{T}\big((c_t)_{t\in\mathcal{T}}\big) : \mathcal{K}((t_i, r_i)_{i\in I}) \geq q\big\}}{\#\mathcal{T}\big((c_t)_{t\in\mathcal{T}}\big)} \leq \alpha,$$

and under the null the left-hand side of this inequality with $q = q_\alpha$ equals

$$\mathbf{P}\Big(\mathcal{K}((T_i, R_i)_{i\in I}) \geq q_\alpha \Big| \mathbf{X}_i = \mathbf{x}, R_i = r_i, i \in I, (T_i)_{i\in I} \in \mathcal{T}\big((c_t)_{t\in\mathcal{T}}\big)\Big);$$

thus, under the null, this probability is $\leq \alpha$, and integrating out the r_i and the c_t yields

$$\mathbf{P}\Big(\mathcal{K}((T_i, R_i)_{i\in I}) \geq q_\alpha \Big| \mathbf{X}_i = \mathbf{x}, i \in I\Big) \leq \alpha.$$

The conditional test can be carried out in another, equivalent manner, namely by computing the *p-value* of the observed sample covariance:

$$p((T_i, R_i)_{i\in I}) := \frac{\#\{(t_i)_{i\in I} \in \mathcal{T}((c_t)_{t\in\mathcal{T}}) : \mathcal{K}((t_i, R_i)_{i\in I}) \geq \mathcal{K}((T_i, R_i)_{i\in I})\}}{\#\mathcal{T}((c_t)_{t\in\mathcal{T}})};$$

if this is $\leq \alpha$ then the null hypothesis is rejected at the level of α.

As noted around (5.5)–(5.6), the elements of the set $\mathcal{T}((c_t)_{t\in\mathcal{T}})$ are obtained by permuting the coordinates of any fixed vector $(t_i^*)_{i\in I}$ in it. Since (cf. 5.8) under our conditioning event $(T_i)_{i\in I}$ is one such vector, we see that all the elements of the set in (5.9) are obtained from

$$\mathcal{K}((T_{\Pi(i)}, R_i)_{i\in I}) \tag{5.10}$$

as Π runs over all the permutations of the coordinates of the vector $(i)_{i\in I}$ of indices of I. Consequently, the conditional distribution of the test statistic $\mathcal{K}((T_i, R_i)_{i\in I})$ under the null is the same as the distribution of the random variable defined by taking the Π in (5.10) as a *random permutation* of the vector $(i)_{i\in I}$ and regarding *both* $(R_i)_{i\in I}$ and $(T_i)_{i\in I}$ as constant vectors.

When n is large, the enumeration of all the Π yielding the values of (5.10), and hence the null distribution of our test statistic, its critical value and p-value, is normally impracticable. Nevertheless, simulation can be used to compute these objects approximately by generating pseudo-random permutations Π and computing the corresponding value of $\mathcal{K}((T_{\Pi(i)}, R_i)_{i\in I})$ a large number of times to get an estimate of the distribution of the statistic, from which an estimate of q_α and an estimate of $p((T_i, R_i)_{i\in I})$ are readily calculated (as the sample critical value of the values of (5.10) thus generated and as the proportion of times in which those values exceed $\mathcal{K}((T_i, R_i)_{i\in I})$, respectively). In some cases, when n is large, a central limit theorem provides sufficiently accurate estimates of the required critical values and p-values, and thus dispenses with computer simulation; see Remark 5.1, p. 113.

Because of the conspicuous role that permutations play in it, this and similar tests are usually referred to as *permutation tests*. And since the permutations typically correspond to a random assignment of treatments to subjects, in problems where the only objects which are regarded as random are the treatments—the responses under each treatment being regarded as fixed constants—they are also called *randomization tests*.

There are in fact many other tests, since as we have noticed already there are many ways of measuring the discrepancies between the observed treatment assignments and the treatment assignments expected under the null. Some tests are equivalent to the one based on $\mathcal{K}((T_i, R_i)_{i\in I})$: for example, the test based on the *inner product statistic*

$$\mathbf{R}_I \cdot \mathbf{T}_I := \sum_{i\in I} R_i T_i, \tag{5.11}$$

the inner product of the vector $\mathbf{R}_I := (R_i)_{i\in I}$ of responses with the vector $\mathbf{T} := (T_i)_{i\in I}$ of treatments, and the test based on the *correlation* between $\mathbf{R}_I$ and $\mathbf{T}_I$ (assuming this is defined), are equivalent to it.This is because conditionally on the R_i and on the frequencies of the T_i (as determined by (5.8)) the terms $\bar{R}_I\bar{T}_I$, $\sum_{i\in I}(R_i - \bar{R}_I)^2$ and

$\sum_{i \in I}(T_i - \bar{T}_I)^2$ involved in the covariance and in the correlation are fixed constants, so the conditional distributions of the three statistics differ from each other at most by a change of scale and location.

When the responses, the treatments, or both, are categorical (nominal) with more than two levels, the inner product statistic is not meaningful, but the same principle of testing and the same method of performing the test approximately by computer simulation apply to yield permutation versions of the Kruskal-Wallis, chi-square and similar tests. For example, when the response and treatment variables assume only a few values, the chi-square test of homogeneity conditional on the frequencies of the responses provides a test of the hypothesis of no treatment effect; its null distribution is obtained from the values of the statistic over all the two-dimensional tables generated by permuting the treatments, and at least when the (T_i, R_i) are a random sample it converges to the chi-square distribution of the classical test of homogeneity/independence (e.g. Sects. 13.3 and 13.4 of Rice [79]) as $n \to \infty$.

The examples that follow concern mostly ordinal (or at least binary) responses.

Example 5.2 Consider the situation of Example 5.1 with $K = 1$, so that there is a single stratum characterized by covariate vectors equal to a given $\mathbf{x} \in \mathcal{X}$. Let the stratum have size $n = 8$ and let there be exactly $m = 4$ treated units. The set $\mathcal{T}(4,4)$ of treatment assignments satisfying these restrictions consists of the

$$\#\mathcal{T}(4,4) = \frac{8!}{4!4!} = 70$$

vectors $(t_1, t_2, \ldots, t_8)$ with binary coordinates such that $c_0 = \sum_{i=1}^{8} \delta_{t_i,0} = 4$ and $c_1 = \sum_{i=1}^{8} \delta_{t_i,1} = 4$, or equivalently such that $\sum_{i=1}^{8} t_i = 4$. The conditional distribution of the treatment assignments under the null hypothesis is therefore

$$\mathbf{P}\left(T_i = t_i, i = 1, \ldots, 8 \,\middle|\, \mathbf{X}_i = \mathbf{x}, R_i = r_i, i = 1, \ldots, 8, \sum_{j=1}^{8} t_j = 4\right) = \frac{1}{70}$$

for $(t_1, t_2, \ldots, t_8) \in \mathcal{T}(4,4)$. The elements of $\mathcal{T}(4,4)$ are listed in Table 5.1, sorted by increasing values of a test statistic S to be introduced below.

In order to illustrate the workings of the test and its conditional character let us first consider two situations where the responses may take positive integer values: In the first, the vector of responses that turns up is $\mathbf{r} = (1, 2, 3, 4, 5, 6, 7, 8)$, and in the second it is $\mathbf{r}' = (1, 1, 1, 2, 2, 3, 6, 7)$. The corresponding inner product statistics are the random variables $S := \mathbf{r} \cdot \mathbf{T}$ and $S' := \mathbf{r}' \cdot \mathbf{T}$, which depend on the random vector $\mathbf{T} := (T_1, T_2, \ldots, T_8)$ of treatment assignments. The possible values of these random variables are given in Table 5.1 next to the 70 possible values of $\mathbf{T}$ (the elements of $\mathcal{T}(4,4)$). The conditional probability functions of S and S' under the null hypothesis,

$$g_S(s) := \mathbf{P}\left(S = s \,\middle|\, \mathbf{X}_i = \mathbf{x}, i = 1, \ldots, 8, \sum_{j=1}^{8} T_j = 4\right), \quad s = 10, 11, \ldots, 26,$$

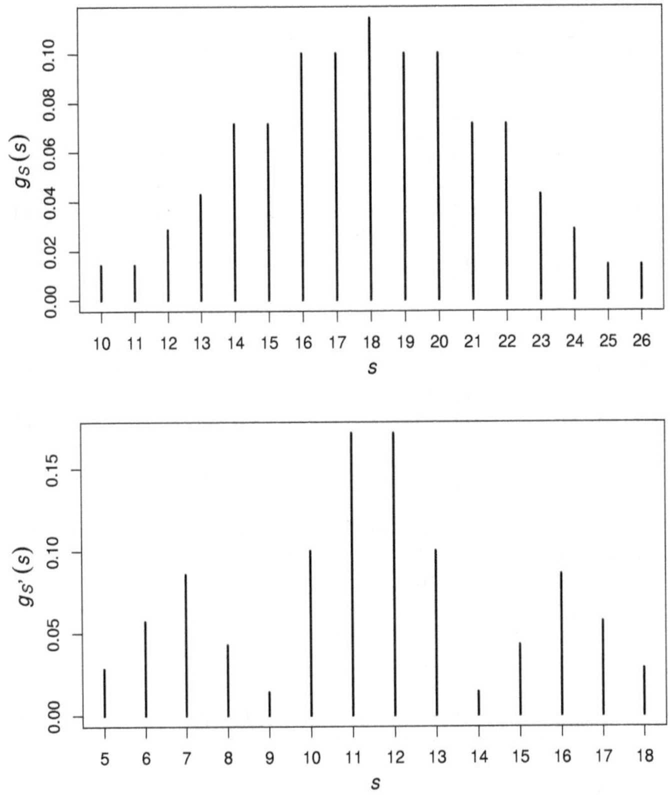

Fig. 5.1 Null conditional probability functions of the test statistics S and S'.

and

$$g_{S'}(s) := \mathbf{P}\left(S' = s \,\middle|\, \mathbf{X}_i = \mathbf{x}, i = 1, \ldots, 8, \sum\nolimits_{j=1}^{8} T_j = 4\right), \quad s = 5, 6, \ldots, 18,$$

are represented in Fig. 5.1; they are readily computed from Table 5.1 by aggregating and adding up the probabilities corresponding to the possible values of S and S'. It is seen that the conditional distribution can vary quite much with the responses.

To illustrate the use of the null distribution in carrying out the test, let the observed treatment assignment be $\mathbf{t} := (0, 0, 1, 0, 1, 0, 1, 1)$. Then the observed values of the inner product statistics are

$$s := \mathbf{r} \cdot \mathbf{t} = 3 + 5 + 7 + 8 = 23 \quad \text{and} \quad s' := \mathbf{r}' \cdot \mathbf{t} = 1 + 2 + 6 + 7 = 16.$$

Table 5.1 Enumeration of the elements of $\mathcal{T}(4,4)$ and of the corresponding values of the inner product statistics S and S'.

t_1	t_2	t_3	t_4	t_5	t_6	t_7	t_8	S	S'	t_1	t_2	t_3	t_4	t_5	t_6	t_7	t_8	S	S'
1	1	1	1	0	0	0	0	10	5	1	0	1	0	0	1	0	1	18	12
1	1	1	0	1	0	0	0	11	5	0	1	1	0	1	0	0	1	18	11
1	1	1	0	0	1	0	0	12	6	0	1	1	0	0	1	1	0	18	11
1	1	0	1	1	0	0	0	12	6	1	0	0	1	1	0	0	1	18	12
1	1	0	1	0	1	0	0	13	7	1	0	0	0	1	1	1	0	19	12
1	1	1	0	0	0	1	0	13	9	0	0	1	1	1	0	1	0	19	11
1	0	1	1	1	0	0	0	13	6	1	0	1	0	0	0	1	1	19	15
0	1	1	1	1	0	0	0	14	6	0	1	1	0	0	1	0	1	19	12
1	1	1	0	0	0	0	1	14	10	1	0	0	1	0	1	0	1	19	13
1	1	0	1	0	0	1	0	14	10	0	1	0	1	1	0	0	1	19	12
1	0	1	1	0	1	0	0	14	7	0	1	0	1	0	1	1	0	19	12
1	1	0	0	1	1	0	0	14	7	0	0	1	1	0	1	1	0	20	12
1	1	0	0	1	0	1	0	15	10	0	0	1	1	1	0	0	1	20	12
1	0	1	0	1	1	0	0	15	7	0	1	0	0	1	1	1	0	20	12
0	1	1	1	0	1	0	0	15	7	1	0	0	0	1	1	0	1	20	13
1	0	1	1	0	0	1	0	15	10	0	1	1	0	0	0	1	1	20	15
1	1	0	1	0	0	0	1	15	11	1	0	0	1	0	0	1	1	20	**16**
1	1	0	0	1	0	0	1	16	11	0	1	0	1	0	1	0	1	20	13
1	0	1	0	1	0	1	0	16	10	0	1	0	1	0	0	1	1	21	**16**
1	1	0	0	0	1	1	0	16	11	0	1	0	0	1	1	0	1	21	13
0	1	1	0	1	1	0	0	16	7	0	0	1	1	0	1	0	1	21	13
0	1	1	1	0	0	1	0	16	10	0	0	1	0	1	1	1	0	21	12
1	0	0	1	1	1	0	0	16	8	1	0	0	0	1	0	1	1	21	**16**
1	0	1	1	0	0	0	1	16	11	0	1	0	0	1	0	1	1	22	**16**
0	1	0	1	1	1	0	0	17	8	0	0	1	0	1	1	0	1	22	13
1	1	0	0	0	1	0	1	17	12	0	0	0	1	1	1	1	0	22	13
1	0	1	0	0	1	1	0	17	11	1	0	0	0	0	1	1	1	22	**17**
0	1	1	1	0	0	0	1	17	11	0	0	1	1	0	0	1	1	22	**16**
1	0	1	0	1	0	0	1	17	11	0	0	1	0	1	0	1	1	**23**	**16**
0	1	1	0	1	0	1	0	17	10	0	1	0	0	0	1	1	1	**23**	**17**
1	0	0	1	1	0	1	0	17	11	0	0	0	1	1	1	0	1	**23**	14
1	1	0	0	0	0	1	1	18	15	0	0	1	0	0	1	1	1	**24**	**17**
0	1	0	1	1	0	1	0	18	11	0	0	0	1	1	0	1	1	**24**	**17**
1	0	0	1	0	1	1	0	18	12	0	0	0	1	0	1	1	1	**25**	**18**
0	0	1	1	1	1	0	0	18	8	0	0	0	0	1	1	1	1	**26**	**18**

From the bottom of the last two columns in the table we see that seven treatment assignments yield a value of S at least as large as $s = 23$ and 12 assignments yield a value of S' at least as large as $s' = 16$. Thus, under the null the probability that S would yield a value at least as large as the one observed is $\frac{7}{70} = 0.1$ and the probability that S' would yield a value at least as large as the one observed is $\frac{12}{70} = 0.17$. These are the p-values of the conditional tests based on the inner product statistic in the two situations; for instance, in none of them is the null rejected at the level of 0.05.

As n increases, the size of the set $\mathcal{T}(c_0, c_1)$ increases at such a high rate—for instance, if we increased n from 8 to 10 and m from 4 to 5 in this illustration then the size of Table 5.1 would almost quadruple—that it becomes a problem to compute p-values by complete enumeration of its elements. The method mentioned just before

(5.11) as an alternative to enumeration is in this case equivalent to the simulation of randomly chosen rows from Table 5.1 and the selection of the corresponding values of S or S', which serve as random samples from their distributions.

Let us now consider the situation where the random vector $\mathbf{R}$ of responses is binary and its coordinates are constrained by the condition $\sum_{i=1}^{8} R_i = 4$, so that the set of possible outcomes for $\mathbf{R}$ coincides with $\mathcal{T}(4, 4)$. This corresponds to Fisher's tea tasting experiment, mentioned in Example 5.1; identifying the treatments with the ways of serving the tea and the responses with the lady's guesses at the cups prepared according to her favourite recipe, the null hypothesis corresponds to the assumption that the lady has no discriminating powers whatsoever. Here, the mechanics of the conditional test based on the statistic $\mathbf{R} \cdot \mathbf{T}$ are exactly the same as before, but in this case it is worth noting that the test reduces to Fisher's exact test for a 2×2 contingency table with all marginal totals fixed at 4. This follows by arranging the pairs (R_i, T_i) into a contingency table and observing that the number of pairs $(1, 1)$ in the table equals $\mathbf{R} \cdot \mathbf{T}$. Moreover, the conditional distribution of this statistic is hypergeometric, viz.

$$\binom{4}{s}\binom{4}{4-s}\bigg/\binom{8}{4}$$

for $s = 0, 1, 2, 3, 4$. Indeed, the *randomization* picks 4 out of 8 cups to treat and under the null hypothesis the lady picks 4 of the 8 at random; the expression above gives the probability that the lady picks exactly s cups of the 4 treated.

Fisher's exact test with general fixed margins is equivalent to the one based on the inner product statistic also when testing independence in one stratum with data from a case-control study (cf. the text around **A1′**). Under the sampling scheme of the present example but with a general number of 'trials' n (instead of our $n = 8$), it is asymptotically equivalent, as $n \to \infty$, to the test based on the normal approximation to the hypergeometric distribution which rejects the null for large $\mathbf{R} \cdot \mathbf{T}$; and its two-sided version is asymptotically equivalent to the chi-square test of independence. □

Example 5.3 Suppose the $(\mathbf{X}_i, T_i, R_i)$ are a stratified simple random sample from a finite population, and more specifically that n units with labels in I are sampled without replacement from the subpopulation characterized by covariate vectors equal to $\mathbf{x} \in \mathcal{X}$; then the treatment assignments within I are exchangeable and **A1** holds. The workings of the conditional test in I are exactly as in Example 5.2, except that the corresponding tables and probability functions may look less orderly when the treatments and the responses take arbitrary values. □

Example 5.4 Let the $(\mathbf{X}_i, T_i, R_i)$ be such that the (T_i, R_i) in $I = \{i : \mathbf{X}_i = \mathbf{x}\}$ are a random sample, the T_i are binary, and I happens to contain $m > 0$ 'treated' units (i.e. such that $T_i = 1$) and $n - m > 0$ 'untreated'. Again, the test can be carried out within I as in Example 5.2, but it can also be formulated in more familiar terms by noting that the distribution of the test statistic is obtained by permuting the responses randomly within a stratum just as by permuting the treatments, since applying Π to the indices of the R_i rather than to those of the T_is in (5.10) generates the same set of values, in keeping with the symmetric role played by the treatments and the responses in the null hypothesis (which states their conditional independence).

Indeed, since—the (T_i, R_i) being a random sample—conditionally on the $(\mathbf{X}_i, T_i)$ the responses in each treated group are independent and identically distributed and the same is true of the responses in each untreated group, the hypothesis of no treatment effect within the stratum is equivalent to the equality of the distributions in the two groups, so it may be tested by means of classical two-sample tests, such as the 'z-test' or, if the responses are continuous or at least take many different values, the Wilcoxon-Mann-Whitney test.

To see the connection with the z-test, note that if $\mathbf{R}_I$ is the vector of the responses in I and $\mathbf{T}_I$ that of the treatments then $S_1 := \mathbf{R}_I \cdot \mathbf{T}_I$ is the sum of the responses among the treated and, conditionally on the responses, it is equivalent—in the sense that it yields an equivalent conditional test—to

$$S_0 := \mathbf{R}_I \cdot \mathbf{1} - \mathbf{R}_I \cdot \mathbf{T}_I = \mathbf{R}_I \cdot (\mathbf{1} - \mathbf{T}_I),$$

where $\mathbf{1}$ is the n-vector with coordinates equal to 1. Because S_0 is the sum of the responses among the untreated, conditionally on the responses the difference between the average responses in the treated and untreated groups,

$$S := \frac{S_1}{m} - \frac{S_0}{n-m},$$

yields a test equivalent to the one provided both by S_1 and S_0. On the other hand, conditionally on the T_i and as a function of the R_i, S is a difference between independent sample averages of independent and identically distributed variables, and so, under the null, and provided m and $n - m$ are 'large', it is approximately normally distributed—with variance approximately equal to $\mathcal{S}_1^2/m + \mathcal{S}_0^2/(n-m)$, where $\mathcal{S}_1^2$ and $\mathcal{S}_0^2$ are the sample variances of the responses in the treated and untreated groups; see Prob. 5.2—and therefore provides an approximate test.

When the responses take many different values, instead of the responses in I one may take their *ranks* in the exact, conditional test based on $\mathbf{R}_I \cdot \mathbf{T}_I$; the ranks are not independent, of course, but the test does not require them to be.[2] On the other hand, with the ranks in place of the responses and conditionally on the treatments, $\mathbf{R}_I \cdot \mathbf{T}_I$ is the sum of the ranks among the treated and forms the basis of the Wilcoxon-Mann-Whitney test (Rice [79], p. 402). Moreover, under the null $\mathbf{R}_I \cdot \mathbf{T}_I$ is approximately normal with known mean and known variance (they are equal to $m(m+n+1)/2$ and $mn(m+n+1)/12$ when the R_i are continuous), so our conditional test with the ranks as responses is asymptotically equivalent (as $\min\{n-m, m\} \to \infty$) to a classical nonparametric test.

It is worth noting that the inner product statistic applies to dichotomous treatments as long as they are labelled as binary; for then its formulation as a statistic comparing two samples also makes it clear how the case of a categorical treatment with three or more levels can be dealt with by performing permutation versions of the Kruskal-Wallis (e.g. Rice [79], p. 453), Anderson-Darling (Scholz and Stephens [92]) and other tests for comparing independent samples. □

[2] As usual, the ranks of a set of responses with the same value are defined as the average of the ranks they would have if they had differed from each other.

5.2.2 Testing over strata

Testing within strata is often of interest because the magnitude and direction of the treatment effect may vary according to the values of the covariates and certain values may correspond to interesting subpopulations. Thus, one often carries out multiple tests, one within each of a number of strata, with the purpose of identifying the strata that show the stronger effects; and accounting for the multiple testing is easy thanks to the independence of tests from different strata: for example, in order to ensure that the probability of at least one incorrect rejection of the null is $\leq \alpha$ when testing for the treatment effect in K strata, one performs each test at the level of $1 - (1 - \alpha)^{1/K}$.

On the other hand, in many applications the strata reflect a range of 'representative conditions' or subpopulations, and it may be of interest to test for an effect of the treatment on the response by pooling evidence across that range. In other applications each stratum/matched set corresponds to a cluster of units related to each other in some sense (e.g. they are siblings or belong to a household) and the clusters are sampled randomly from a population (e.g. of families with two or more children); then the interest lies in the effect of the treatment in that population of clusters rather than in its effect within a particular cluster. The simplest and perhaps most natural way of combining evidence of a treatment effect over strata or matched sets is to test the null hypothesis that the treatment has no effect in any of the strata with a statistic that is a sum or weighted sum of statistics computed per stratum and whose distribution under the null is determined from the joint distribution of *all* treatment assignments. By **A0** and **A1** (**A1′** in case-control studies), the latter is determined by a product of probability functions like (5.7), one probability function per stratum, and the probability function of the statistic is the convolution of those probability functions. Moreover, the distribution of the statistic can be estimated by simulating independent random permutations of the treatments within strata and computing and adding up the statistics associated with those permutations.

For example, if the data are ordinal and they are arranged in K strata or matched sets $I_1, \ldots, I_K$, with I_k characterized by covariate vectors equal to $\mathbf{x}_k$, one may use the inner product statistic

$$S := \mathbf{R} \cdot \mathbf{T} \equiv \sum_{k=1}^{K} \mathbf{R}_{I_k} \cdot \mathbf{T}_{I_k}, \tag{5.12}$$

the inner product of the vectors $\mathbf{T}$ and $\mathbf{R}$ obtained by concatenating the K vectors $\mathbf{T}_{I_k}$ of treatments and the corresponding vectors $\mathbf{R}_{I_k}$ of responses, in order to test for the treatment effect over strata. The conditional distribution of S under the null is the same as that of the random variable

$$\mathcal{S} := \sum_{k=1}^{K} \sum_{i \in I_k} R_i T_{\Pi_k(i)}, \tag{5.13}$$

where Π_k is a random permutation of the indices in I_k, the Π_k are independent, and the R_i and T_i are regarded as fixed constants.

The following examples concern ordinal or dichotomous response and treatment variables and the inner product statistic; tests for categorical variables are described at the end of this subsection.

Example 5.5 Let us extend Example 5.2 to the case of several tea tasting experiments mentioned at the end of Example 5.1. Let the K values of the covariate vectors represent K ladies and write $S^{(k)}$ for the inner product statistic computed with the data from the k-th lady; $S^{(k)}$ can be used to test whether the k-th lady possesses any discriminating power. Suppose first that each lady has come forward with the claim that she can discriminate between the two ways of preparing tea. A legitimate point of view would be to test the null hypothesis that *none of the ladies has discriminating powers*. In this case the rejection of the null would indicate that at least one of the ladies has discriminating powers, and if rejection occurred then we would probably want to identify the ladies more likely to be the discriminating ones—which could be done by performing the K conditional tests separately and picking the ladies with smaller p-values (taking $s = 4$ in the probabilities at the end of Example 5.2 shows that the smallest p-value attainable in each test is $1/70$). To test the null, one could use the inner product of the vector of treatments obtained by concatenating the vectors of treatments of the K ladies with the vector of responses obtained by concatenating their vectors of responses:

$$S := \sum_{k=1}^{K} S^{(k)}.$$

Under the null, the $S^{(k)}$ have the same distribution—the hypergeometric distribution of Example 5.2—and are independent, so the null distribution of S can, in principle, be tabulated and the p-value of the test computed from it; alternatively the p-value can be estimated by simulation, simulating independent sets of K observations from the hypergeometric distribution and adding them to get observations from the null distribution of S.

Whatever the result of this test, any inference derived from it would have to refer to the particular group of ladies who made the claim, and perhaps even more specifically to the group of ladies *at the time* that the test was carried out.

Now suppose that the ladies have been sampled randomly from some population. A legitimate null hypothesis in this case would reflect the sampling mechanism, and would therefore state that a lady randomly sampled from the population has no discriminating powers. The population being finite, this would be equivalent to the hypothesis that *no lady in the population has discriminating powers*. The test based on S, exactly with the same mechanics, would serve here as well, but the inferences from it, even if based on a small fraction of the population, would concern the whole population (not just the ladies who turned up in the sample). □

Example 5.6 In Example 5.4 let there be K strata, each of the form $I_k = \{i : \mathbf{X}_i = \mathbf{x}_k\}$ and containing $m_k > 0$ treated and $n_k - m_k > 0$ untreated units, and suppose that the random samples of the (T_i, R_i) from different strata are independent conditionally on the $\mathbf{X}_i$. Let $S^{(k)}$ stand for the difference between the average response among the treated and the average response among the untreated in I_k, and $\tilde{S}^{(k)}$ for the sum of the ranks of the responses among the treated in I_k, which is meaningful if the responses in I_k take 'many' different values. The null hypothesis states that the distributions of the responses of treated and untreated units are the same in each stratum, and to test it we may consider

$$S := \sum_{k=1}^{K} S^{(k)} \quad \text{and} \quad \tilde{S} := \sum_{k=1}^{K} \tilde{S}^{(k)}.$$

Again, if the result of the test is found to be significant then the p-values of the tests per stratum may be examined in order to identify the strata in which the treatment effect is likely to be stronger. The mechanics of the tests based on S and $\tilde{S}$ are analogous to those of the test in Example 5.5. If m_k and $n_k - m_k$ are 'large' for all k one may use the limit null distributions of normalized versions of S and $\tilde{S}$, which are normal (see Remark 5.1), to get approximate p-values.

Evidently, S and $\tilde{S}$ are not the only possible test statistics. When the responses take many different values, $\tilde{S}$ may be replaced by a statistic proposed by Hodges and Lehmann which sometimes—namely when K is large compared to N—is more powerful (i.e. better at detecting departures from the null) than $\tilde{S}$.[3] The Hodges-Lehmann statistic has the same inner product form as $\tilde{S}$, but the responses on which it is based are defined differently: first, the responses in each stratum have their averages (or medians) subtracted from them, and then they are pooled into a single set and ranked; finally, the resulting ranks, assigned back to the appropriate strata and treatments, are used as the responses in place of the original responses. For the rest, the mechanics of the test are as before. □

Example 5.7 Let the treatments and the responses be binary—with a response of 1 standing for a positive outcome of the treatment, say—and interpret the possible values of the covariate vectors, $\mathbf{x}_1, \mathbf{x}_2, \ldots, \mathbf{x}_K$, as labels of *pairs* of individuals—siblings, identical twins, patients matched on personal characteristics, etc., so that there are $N = 2K$ individuals in total. Assume that only one individual of each pair is chosen at random to be treated, independently of the choices in other pairs, so that within each stratum both treatment assignments have probability $1/2$ and under the null they still have probability $1/2$ conditionally on the responses.

Evidently, it is not possible to test for a treatment effect within a stratum. To test for the effect of treatment on the set of pairs of individuals, or on a population they may represent, we may consider the number of positive outcomes among treated individuals,

$$S := \sum_{k:T_k=1} R_k = \sum_{k=1}^{K} T_k R_k,$$

or the difference between the proportion of positive outcomes among the treated and the proportion of positive outcomes among the untreated,

[3] For more information on this test see Sects. 2 and 3 of Hodges and Lehmann [44] or pp. 138-41 of Lehmann [53].

$$D := \frac{1}{K}\sum_{k=1}^{K} T_k R_k - \frac{1}{K}\sum_{k=1}^{K}(1 - T_k)R_k.$$

Since $K \cdot D = 2\sum_{k=1}^{K} T_k R_k - \sum_{k=1}^{K} R_k = 2 \cdot S - \sum_{k=1}^{K} R_k$, conditionally on the responses the distributions of S and D differ only by a change of scale and location and hence yield equivalent tests. In fact, both tests are equivalent to McNemar's test (Rice [79], p. 493). To see this, recall that in McNemar's test the K pairs of responses corresponding to the paired individuals are arranged in the table

		Response in control		
		0	1	Totals
Response in	0	$n_{1,1}$	$n_{1,2}$	$n_{1\cdot}$
treated individual	1	$n_{2,1}$	$n_{2,2}$	$n_{2\cdot}$
	Totals	$n_{\cdot 1}$	$n_{\cdot 2}$	K

where $n_{1,1}$ stands for the number of pairs whose treated and control (non-treated) individuals have both a negative outcome, $n_{1,2}$ for the number of pairs whose treated patient has a negative outcome and whose control has a positive outcome, etc., so that $n_{2\cdot}$ is the number of treated individuals with a positive outcome and $n_{\cdot 2}$ the number of control individuals with a positive outcome. McNemar's statistic (pp. 492–3 of Rice [79]) is equivalent to $n_{1,2} - n_{2,1}$, and clearly $K \cdot D = n_{2\cdot} - n_{\cdot 2} = n_{2,1} - n_{1,2}$.

The test based on D can be carried out by tabulating the null distribution of the statistic or by estimating the p-value by simulation; asymptotically (as $K \to \infty$), when *the pairs* of individuals are randomly sampled, it is equivalent to the McNemar test as this is usually applied, appealing to a normal or chi-square approximation. □

Example 5.8 Consider the same situation as in Example 5.7 except that the responses are continuous or at least take many different values. For convenience let the k-th stratum consist of the individuals with labels $2k - 1$ and $2k$ ($k = 1, \ldots, K$). Let $D_1^*, \ldots, D_K^*$ be the ranks of the absolute differences $D_k = |R_{2k} - R_{2k-1}|$ between the pairs of responses in the strata and consider the statistic

$$S := \sum_{k=1}^{K} D_k^* \left\{\mathbf{1}_{\{R_{2k} > R_{2k-1}\}} T_{2k} + \mathbf{1}_{\{R_{2k} < R_{2k-1}\}} T_{2k-1}\right\}.$$

The k-th term in this sum is the inner product

$$\left(D_k^* \mathbf{1}_{\{R_{2k} > R_{2k-1}\}}, D_k^* \mathbf{1}_{\{R_{2k} < R_{2k-1}\}}\right) \cdot (T_{2k}, T_{2k-1}),$$

which is equal to D_k^* if the response of the treated individual is greater than that of the untreated, and 0 otherwise. Hence S is the sum of the ranks of the absolute differences from strata in which the response of the treated individual is greater than that of the untreated, and so it is the well-known Wilcoxon signed rank statistic, a classical statistic for testing for a difference between two treatments based on paired samples (cf. Sect. 11.3.2 of Rice [79]).

The Hodges-Lehmann statistic, defined in the second half of Example 5.6, may be used for the same problem. In the same notation, subtraction by the means within the strata yields new responses $(R_{2k-1} - R_{2k})/2$, $(R_{2k} - R_{2k-1})/2$, $k = 1, \ldots, K$; denoting the ranks of these by $R_1^*, \ldots, R_{2K}^*$, the Hodges-Lehmann statistic is

$$S^* := \sum_{i=1}^{N} R_i^* T_i = \sum_{k=1}^{K} \left(R_{2k-1}^*, R_{2k}^*\right) \cdot (T_{2k-1}, T_{2k}).$$

It turns out that S^* and S are asymptotically equivalent (p. 141 of Lehmann [53]). □

Example 5.9 Consider a case-control study (Sect. 4.6) in which a diseased individual is matched with a single healthy 'control' sharing the same covariate vector. The data from such a study are sometimes arranged in a contingency table, even though this omits information on the matched sets:

		Exposure		
		1	0	Totals
Disease	1	$n_{1,1}$	$n_{1,2}$	$n_{1\cdot}$
	0	$n_{2,1}$	$n_{2,2}$	$n_{2\cdot}$
	Totals	$n_{\cdot 1}$	$n_{\cdot 2}$	N

A natural statistic for testing the null hypothesis is the difference between the proportions of individuals exposed among the diseased and nondiseased individuals,

$$D := \frac{\sum_{i=1}^{N} T_i R_i}{\sum_{i=1}^{N} R_i} - \frac{\sum_{i=1}^{N} T_i (1 - R_i)}{\sum_{i=1}^{N} (1 - R_i)} = \frac{n_{1,1}}{n_{1\cdot}} - \frac{n_{2,1}}{n_{2\cdot}},$$

which again is equivalent to the inner product statistic $\sum_i T_i R_i$. The approximate test for comparing two probabilities cannot replace the test based on D, since the T_i are permuted within matched sets and, because of the sampling plan of the study, the summands of $n_{1,1}/n_{1\cdot}$ and $n_{2,1}/n_{2\cdot}$ need not be identically distributed. □

Remark 5.1 We have seen that the conditional null distribution of the inner product statistic S of (5.12) is the same as the distribution of the random variable $\mathcal{S}$ of (5.13). Because everything in $\mathcal{S}$ is constant apart from the random permutations Π_k, it is relatively easy to show (Prob. 5.3) that the conditional mean and variance of S are

$$\mathbf{E}(S|\mathcal{D}) = \sum_k n_k \bar{R}_k \bar{T}_k \quad \text{and} \quad \mathrm{Var}(S|\mathcal{D}) = \sum_k \frac{n_k^2}{n_k - 1} v_k^2 V_k^2, \tag{5.14}$$

where $\mathcal{D}$ denotes the observed data, $n_k := \# I_k$,

$$\bar{R}_k := \frac{1}{n_k} \sum_{i \in I_k} R_i, \quad \bar{T}_k := \frac{1}{n_k} \sum_{i \in I_k} T_i,$$

$$v_k^2 := \frac{1}{n_k} \sum_{i \in I_k} (R_i - \bar{R}_k)^2, \quad V_k^2 := \frac{1}{n_k} \sum_{i \in I_k} (T_i - \bar{T}_k)^2.$$

Moreover, the distribution of $(S - \mathbf{E}(S|\mathcal{D}))/\sqrt{\mathrm{Var}(S|\mathcal{D})}$ approaches the standard normal distribution as $\min_k n_k \to \infty$, a more difficult result that may be studied in Chap. 12 of Ferguson's book [30]. □

Let us now consider the situation where the treatment and the response variables are categorical. As mentioned just before Example 5.2, we can carry out the exact, conditional version of the chi-square test in one stratum in order to test for an effect of the treatment on the response in that stratum. So a permutation test for the effect of the treatment—in fact for the hypothesis that the treatment has no effect in any stratum—may be based on the sum of chi-square statistics over all the strata, whose null distribution is analogous to that of the sum of inner product statistics described before Example 5.5 and can be estimated by simulation or, at least when the data in each stratum are a random sample, replaced by the asymptotic distribution (which is chi-square with degrees of freedom equal to the sum of the degrees of freedom of the statistics being added). Although this test is useful, there is another test which compounds the discrepancies between the observed and expected frequencies over the strata and which therefore is more powerful when the effect of the treatment on the response is qualitatively similar in different strata. This is the Cochran-Mantel-Haenszel test, which we describe next.

For simplicity consider first the case in which the data $(T_1, R_1), \ldots, (T_N, R_N)$ belong to the same stratum. If the values assumed by the response and the treatment are labelled as $1, \ldots, I$ and $1, \ldots, J$, respectively, the data can be summarized in an array of I rows and J columns whose entries are the *observed frequencies*

$$N_{ij} := \sum_{m=1}^{N} \mathbf{1}_{\{R_m=i, T_m=j\}}. \tag{5.15}$$

Write $\mathcal{D}$ for the set of marginal frequencies

$$N_{i\cdot} := \sum_{j=1}^{J} N_{ij}, \quad N_{\cdot j} := \sum_{i=1}^{I} N_{ij}, \quad N_{\cdot\cdot} \equiv N = \sum_{i,j} N_{ij}.$$

Conditionally on $\mathcal{D}$ and under the hypothesis of no treatment effect, the distribution of N_{ij} is the same as that of the number of responses equal to i contained in a random sample of size $N_{\cdot j}$ drawn without replacement from a population consisting of $N_{i\cdot}$ responses equal to i and $N - N_{i\cdot}$ responses different from i. Indeed, by attaching the treatment label j to the $N_{\cdot j}$ responses drawn in this way, we reproduce the random assignment of all the treatments of label j among the N responses, independently of which of the remaining $N - N_{\cdot j}$ treatments different from j are assigned to the remaining $N - N_{\cdot j}$ responses. Thus N_{ij} has the hypergeometric probability function

$$\mathbf{P}(N_{ij} = n|\mathcal{D}) = \binom{N_{i\cdot}}{n}\binom{N - N_{i\cdot}}{N_{\cdot j} - n}\Big/\binom{N}{N_{\cdot j}}$$

and (see Prob. 5.4) its mean and variance are

$$\mathbf{E}(N_{ij}|\mathcal{D}) = N_{\cdot j}\frac{N_{i\cdot}}{N} \quad \text{and} \quad \mathrm{Var}(N_{ij}|\mathcal{D}) = N_{\cdot j}\frac{N_{i\cdot}}{N}\frac{(N - N_{i\cdot})}{N}\frac{(N - N_{\cdot j})}{N-1}. \tag{5.16}$$

The Cochran-Mantel-Haenszel statistic compares the observed frequencies N_{ij} with the *expected frequencies* $\mathbf{E}(N_{ij}|\mathcal{D})$, taking account of $\mathrm{Cov}(N_{ij}, N_{kl}|\mathcal{D})$. Before giving expressions for this covariance we motivate and define the statistic.

With more than one stratum we indicate the variables and quantities pertaining to stratum s by adding a superscript to them and writing

$$N_{ij}^{(s)},\ N^{(s)} \equiv N_{\cdot\cdot}^{(s)},\ \mathbf{E}\big(N_{ij}^{(s)}|\mathcal{D}^{(s)}\big),\ \mathrm{Cov}\big(N_{ij}^{(s)}, N_{kl}^{(s)}|\mathcal{D}^{(s)}\big),\ \text{etc.}$$

The vector

$$\mathbf{N}_s := \left(N_{ij}^{(s)}\right)_{i=1,\dots,I-1,j=1,\dots,J-1}$$

contains all the frequencies except the N_{iJ} and N_{Ij}, so under the null hypothesis and when $N^{(s)}$ is large one may expect its distribution conditional on $\mathcal{D}^{(s)}$ to be non-degenerate and approximately normal with mean $\mathbf{E}[\mathbf{N}_s|\mathcal{D}^{(s)}]$ and covariance matrix $\mathrm{Cov}(\mathbf{N}_s|\mathcal{D}^{(s)})$.[4] Since $\mathbf{N}_s$ and $\mathbf{N}_{s'}$ are independent for $s \neq s'$, conditionally on $\bar{\mathcal{D}} := \bigcup_s \mathcal{D}^{(s)}$ and under the null hypothesis the vector $\mathbf{N} := \sum_s \mathbf{N}_s$, which adds the $(I-1)\times(J-1)$ frequencies over the strata, should be approximately normal with

$$\mathbf{E}(\mathbf{N}|\bar{\mathcal{D}}) = \sum\nolimits_s \mathbf{E}[\mathbf{N}_s|\mathcal{D}^{(s)}] \quad \text{and} \quad \mathrm{Cov}(\mathbf{N}|\bar{\mathcal{D}}) = \sum\nolimits_s \mathrm{Cov}(\mathbf{N}_s|\mathcal{D}^{(s)})$$

as mean and covariance matrix. Finally, these observations lead to

$$\big(\mathbf{N} - \mathbf{E}(\mathbf{N}|\bar{\mathcal{D}})\big)\,\mathrm{Cov}(\mathbf{N}|\bar{\mathcal{D}})^{-1}\,\big(\mathbf{N} - \mathbf{E}(\mathbf{N}|\bar{\mathcal{D}})\big)^{\mathrm{T}},$$

the Cochran-Mantel-Haenszel statistic, which under the hypothesis of no treatment effect converges to a chi-square random variable with $(I-1)\times(J-1)$ degrees of freedom as the sample size tends to infinity and the number of strata is held fixed.[5]

Evidently, the normal approximation being relied upon here may be poor even for large N, but the exact permutation test can always be carried out by simulation, and the 'standardization' through $\mathbf{E}(\mathbf{N}|\bar{\mathcal{D}})$ and $\mathrm{Cov}(\mathbf{N}|\bar{\mathcal{D}})$ remains sensible with small samples; with small N, however, some of the categories of the responses or treatments may have to be merged if they are not represented in some of the strata.

In order to determine the elements $\mathrm{Cov}\big(N_{ij}^{(s)}, N_{kl}^{(s)}|\mathcal{D}^{(s)}\big)$ of $\mathrm{Cov}(\mathbf{N}|\bar{\mathcal{D}})$, we revert to our simpler notation, dropping the superscripts, and consider first the case $i \neq k$ and $j = l$. Conditionally on $\mathcal{D}$ and under the null, the joint distribution of N_{ij} and N_{kj} is the same as the distribution of the numbers of responses equal to i and k, respectively, contained in a random sample of size $N_{\cdot j}$ drawn without replacement

[4] This is because linear combinations of proportions like $N_{ij}^{(s)}/N^{(s)}$ are sample averages to which the central limit theorem for sampling without replacement applies as $N^{(s)} \to \infty$; see pp. 199-200 of Rice [79], pp. 392–3 of Billingsley [11], or pp. 76, 83–4 of Ferguson [30].

[5] The definition of the statistic is motivated by the well-known result about normal random vectors which says that if a d-dimensional row vector $\mathbf{N}$ is normal with mean vector μ and full-rank covariance matrix Σ then $(\mathbf{N}-\mu)\Sigma^{-1}(\mathbf{N}-\mu)^{\mathrm{T}}$ is a chi-square random variable with d degrees of freedom; see for example p. 56 of Ferguson [30] for this property and pp. 383–5 of Billingsley [11] for the basic properties of normal random vectors.

from a population consisting of $N_{i\cdot}$ responses equal to i, $N_{k\cdot}$ responses equal to k, and $N - N_{i\cdot} - N_{k\cdot}$ responses different from i and k (the corresponding assignment of treatments being achieved, for example, by attaching the labels of treatment j to the first $N_{\cdot j}$ responses drawn). So (N_{ij}, N_{kj}) has the hypergeometric probability function

$$\mathbf{P}(N_{ij} = n, N_{kj} = n' | \mathcal{D}) = \binom{N_{i\cdot}}{n}\binom{N_{k\cdot}}{n'}\binom{N - N_{i\cdot} - N_{k\cdot}}{N_{\cdot j} - n - n'} \Big/ \binom{N}{N_{\cdot j}}$$

and (see Prob. 5.4) covariance

$$\mathrm{Cov}(N_{ij}, N_{kj} | \mathcal{D}) = -N_{\cdot j}\frac{(N - N_{\cdot j})}{N - 1}\frac{N_{i\cdot}}{N}\frac{N_{k\cdot}}{N}. \tag{5.17}$$

If $i = k$ and $j \neq l$, then it follows by symmetry (the distribution of (N_{ij}, N_{il}) corresponding to the same sampling scheme with the roles of the responses and the treatments interchanged) that

$$\mathrm{Cov}(N_{ij}, N_{il} | \mathcal{D}) = -N_{i\cdot}\frac{(N - N_{i\cdot})}{N - 1}\frac{N_{\cdot j}}{N}\frac{N_{\cdot l}}{N}.$$

Finally, if $i \neq k$ and $j \neq l$ then (N_{ij}, N_{kl}) has a more complicated distribution, but (see Prob. 5.4) its covariance can be shown to be

$$\mathrm{Cov}(N_{ij}, N_{kl} | \mathcal{D}) = \frac{N_{i\cdot}N_{k\cdot}}{N - 1}\frac{N_{\cdot j}}{N}\frac{N_{\cdot l}}{N}. \tag{5.18}$$

5.2.3 Multiple responses

An effect may have several aspects or manifest itself in several ways, so in real-life problems there may be more than one variable which can be taken as a response. If the treatment is ordinal and there are multiple ordinal response variables (or categorical ones with two levels) and one knows the direction of the association between the treatment and each response to be expected should the treatment have an effect, then sometimes one can transform the responses in such a way that they are all *aligned*—i.e. have the same direction of association with the treatment—and have comparable scales. In this situation one can use a statistic that is the sum of the inner-product statistics based on the various responses in order to test for a treatment effect. When each response taken on its own provides little or no evidence for an effect of the treatment—because the sample is small, the measurement of the responses is inaccurate or the effects are weak—such a statistic may be more expressive and provide the required evidence, even if, as is normally the case, the responses are correlated. On the other hand, specifying the various responses and the transformations required in order to align them before the data become available constitutes a daring test of a hypothesis elaborated through subject matter knowledge.

The main difficulty in combining evidence from multiple responses lies in finding transformations which render the responses comparable. When the responses take many values, one can use the sum of Hodge-Lehmann statistics (cf. Example 5.6, second paragraph)—or of Wilcoxon signed rank statistics (Example 5.8), if the strata correspond to paired units—computed separately per response in order to test for a treatment effect; even when some responses are more varied than others, the use of ranks should make for comparable responses. When the responses are categorical one can always try to use the sum of Cochran-Mantel-Haenszel statistics, described just before this subsection, though it may be difficult to align the responses.

Whatever the statistic used to combine the statistics on the various responses, the computation of its p-value by simulation is essentially the same as before (cf. p. 103), with the important difference that each permutation of the treatments is used to generate the values of all the statistics corresponding to the various responses and hence of their sum (instead of the value of a single statistic corresponding to a single response).

5.2.4 The effect of the treatment on the treated

So far we have said nothing about the estimation of a treatment effect. This problem is treated in Sect. 5.3, but under conditions that are more restricted than those needed for testing, and with methods that require large samples, so it is worth considering whether one can say something about the magnitude of an effect under comparatively weak assumptions akin to those of the present section. It turns out that when treatment is binary it is possible to obtain reasonably reliable approximate confidence intervals for the *effect of the treatment on the treated*, also known as *attributable effect*, which is an unobservable random variable rather than a parameter.[6]

We begin by considering the case of a single stratum of n units, and assume that the responses are given by

$$R_j = B_j + e_j T_j, \quad j = 1, 2, \ldots, n, \tag{5.19}$$

where the B_j and e_j are random variables (possibly degenerate ones) with arbitrary distributions, the distribution of the treatments T_j conditionally on the (B_j, e_j) (and hence independently of them) is that of n random drawings without replacement from a set of m 1s and $n - m$ 0s, so that $\sum_{j=1}^{n} T_j = m$ and, writing $p = m/n = 1 - q$,

$$\mathbf{E}(T_j) = p, \quad \mathbf{E}(1 - T_j) = q, \quad \mathrm{Var}(T_j) = pq, \quad \text{and} \quad \mathrm{Cov}(T_j, T_k) = -\frac{pq}{n-1};$$

the B_j stand for unobservable 'baseline values' or—when the corresponding T_j are 0—observable responses on non-treated units, and the e_j for the unobservable effects of the treatment.

---

[6] This is a somewhat special topic which may be omitted or read later; it is assumed only in two of the problems.

It is easy to see (Prob. 5.5 (a)) that this model is equivalent to the most general version of the basic model satisfying **A0** and **A1** in which the treatments are binary and have the given distribution. This is of course true even if both the B_j and the e_j are dependent and dependent on each other, which we allow in this subsection (in contrast to Sect. 5.3, where the $(B_1, e_1), \ldots, (B_n, e_n)$ would have to be a random sample).

The quantity we are interested in, the average effect of the treatment on the treated, is the unobservable random variable

$$\bar{\mathbf{e}}_1 := \frac{1}{m}\sum_{j=1}^{n} e_j T_j.$$

In order to obtain an approximate confidence interval for it we need to introduce further notation and argue a little. We write $\mathbf{R} = (R_1, \ldots, R_n)$ and similarly use $\mathbf{B}$, $\mathbf{e}$ and $\mathbf{T}$ for vectors of the B_j, e_j and T_j. If $\mathbf{x} = (x_1, \ldots, x_d)$, with each x_j a number or a random variable, and $\mathbf{y}$ is another such vector, we write

$$\bar{\mathbf{x}} = \frac{1}{d}\sum_{j=1}^{d} x_j, \quad v(\mathbf{x}) = \frac{1}{d}\sum_{j=1}^{d}(x_j - \bar{\mathbf{x}})^2, \quad c(\mathbf{x}, \mathbf{y}) = \frac{1}{d}\sum_{j=1}^{d}(x_j - \bar{\mathbf{x}})(y_j - \bar{\mathbf{y}}).$$

Finally, we write $\mathbf{r}_t$ $(t = 0, 1)$ for the vector of R_j such that $T_j = t$ (with coordinates ordered by the index label, say), accordingly write $\bar{\mathbf{r}}_t$ for its average, $v(\mathbf{r}_t)$ for its variance, and define $\mathbf{b}_t$, $\bar{\mathbf{b}}_t$, $v(\mathbf{b}_t)$, $\mathbf{e}_t$, $\bar{\mathbf{e}}_t$ and $v(\mathbf{e}_t)$, similarly.

To help fixing ideas we note that $\bar{\mathbf{b}}_0 = \bar{\mathbf{r}}_0$ and $v(\mathbf{b}_0) = v(\mathbf{r}_0)$, that these, like $\bar{\mathbf{r}}_1$ and $v(\mathbf{r}_1)$, represent observable random variables, unlike $\bar{\mathbf{b}}_1$ and $v(\mathbf{b}_1)$ which are unobservable, and that

$$\bar{\mathbf{e}} = \mathbf{E}(\bar{\mathbf{e}}_1|\mathbf{e}), \quad \hat{\mathbf{e}} := \bar{\mathbf{r}}_1 - \bar{\mathbf{r}}_0 = \frac{c(\mathbf{R}, \mathbf{T})}{pq} = \frac{c(\mathbf{R}, \mathbf{T})}{v(\mathbf{T})}.$$

The quantity here on the right is observable and serves as an estimator (or predictor) of the unobservable $\bar{\mathbf{e}}_1$. In fact (Prob. 5.5 (b)),

$$\hat{\mathbf{e}} - \bar{\mathbf{e}}_1 = \bar{\mathbf{b}}_1 - \bar{\mathbf{b}}_0 = \frac{1}{pq}\frac{1}{n}\sum_{j=1}^{n}(B_j - \bar{\mathbf{B}})T_j, \tag{5.20}$$

so

$$\mathbf{E}(\hat{\mathbf{e}} - \bar{\mathbf{e}}_1|\mathbf{B}) = 0 \quad \text{and} \quad \mathrm{Var}(\hat{\mathbf{e}} - \bar{\mathbf{e}}_1|\mathbf{B}) = \frac{v(\mathbf{B})}{pq(n-1)}; \tag{5.21}$$

in particular, $\hat{\mathbf{e}}$ is an unbiased estimator (or predictor) of $\bar{\mathbf{e}}_1$ and $\hat{\mathbf{e}} - \bar{\mathbf{e}}_1$ converges in probability to zero provided $\mathbf{E}\{v(\mathbf{B})\}/n \to 0$ as $n \to \infty$.

In view of these expressions, it is perhaps natural to base a confidence interval for $\bar{\mathbf{e}}_1$ on

$$\sqrt{pq(n-1)}\,\frac{\hat{\mathbf{e}} - \bar{\mathbf{e}}_1}{\sqrt{v(\mathbf{B})}}. \tag{5.22}$$

For, by the central limit theorem,[7] the distribution of this variable conditionally on $\mathbf{B}$ and $v(\mathbf{B}) > 0$ converges to the standard normal distribution as $n \to \infty$, $m/n \to c \neq 0, 1$, so if the unobservable $v(\mathbf{B})$ could be estimated consistently by some observable $\hat{v}(\mathbf{B})$ then one might substitute $\hat{v}(\mathbf{B})$ for $v(\mathbf{B})$ in (5.22) without affecting the asymptotic distribution. And if, in addition, the coverage of the interval conditionally on $\mathbf{B}$ were known to be satisfactory for all realizations of $\mathbf{B}$, one would expect the unconditional coverage to be satisfactory too. Unfortunately, conditionally on $\mathbf{B}$ and $v(\mathbf{B}) > 0$ the desired coverage of an interval of nominal confidence $1-\alpha$ based on (5.22) with $v(\mathbf{B})$ replaced by $\hat{v}(\mathbf{B})$, namely

$$|\hat{\mathbf{e}} - \bar{\mathbf{e}}_1| \leq z_{1-\frac{\alpha}{2}} \sqrt{\frac{\hat{v}(\mathbf{B})}{pq(n-1)}},$$

approaches

$$2\,\Phi\left(z_{1-\frac{\alpha}{2}} \sqrt{\frac{\hat{v}(\mathbf{B})}{v(\mathbf{B})}}\right),$$

where Φ is the standard normal distribution function and $z_{1-\alpha/2}$ its quantile of probability $1-\alpha/2$, and, no matter how large n is, at least when $n-m$ is 'small' the last quantity is $< 1-\alpha$ on average because $\hat{v}(\mathbf{B})$ tends to be smaller than $v(\mathbf{B})$. As illustrated in Prob. 5.5 (c), this is often the case if we choose $\hat{v}(\mathbf{B})$ to be $v(\mathbf{b}_0) = v(\mathbf{r}_0)$ (the variance of the observable part of $\mathbf{B}$), which seems the obvious thing to do.

The role of $v(\mathbf{B})$ in (5.22) can be diminished somewhat if we use the identity

$$v(\mathbf{B}) = p\,v(\mathbf{b}_1) + q\,v(\mathbf{r}_0) + p\,q\,(\hat{\mathbf{e}} - \bar{\mathbf{e}}_1)^2 \tag{5.23}$$

and note that

$$|\hat{\mathbf{e}} - \bar{\mathbf{e}}_1| \leq z_{1-\frac{\alpha}{2}} \sqrt{\frac{v(\mathbf{B})}{(n-1)pq}} \quad \Leftrightarrow \quad |\hat{\mathbf{e}} - \bar{\mathbf{e}}_1| \leq z^{(n)}_{1-\frac{\alpha}{2}} \sqrt{\frac{p\,v(\mathbf{b}_1) + q\,v(\mathbf{r}_0)}{(n-1)pq}}, \tag{5.24}$$

where

$$z^{(n)}_{1-\frac{\alpha}{2}} := \frac{z_{1-\alpha/2}}{\sqrt{1 - \frac{z^2_{1-\alpha/2}}{n-1}}}. \tag{5.25}$$

Since $v(\mathbf{r}_0)$ is observable, in order to produce an interval for $\bar{\mathbf{e}}_1$ from the last inequality we now need to find an estimate of $v(\mathbf{b}_1)$.

Suppose that the e_j are constant, say $e_j = \epsilon$ for all j. In this case

$$v(\mathbf{b}_1) = \frac{1}{m}\sum_{j=1}^{n} \left\{(B_j + \epsilon) - (\bar{\mathbf{b}}_1 + \epsilon)\right\}^2 T_j = \frac{1}{m}\sum_{j=1}^{n} (R_j - \bar{\mathbf{r}}_1)^2 T_j = v(\mathbf{r}_1),$$

---

[7] More precisely, the central limit theorem for random sampling without replacement from a finite population; see pp. 199–200 of Rice [79], pp. 392–3 of Billingsley [11], or pp. 76, 83–4 of Ferguson [30].

so $v(\mathbf{b}_1)$ is observable and

$$\hat{\mathbf{e}} \pm z^{(n)}_{1-\frac{\alpha}{2}} \sqrt{\frac{p\,v(\mathbf{r}_1) + q\,v(\mathbf{r}_0)}{(n-1)pq}} \tag{5.26}$$

is a confidence interval for $\bar{\mathbf{e}}_1 = \epsilon$ with the correct asymptotic coverage.

In the more interesting situation where the e_j are different, substituting $v(\mathbf{r}_0)$ for $v(\mathbf{b}_1)$ in (5.24) gives the interval

$$\hat{\mathbf{e}} \pm z^{(n)}_{1-\frac{\alpha}{2}} \sqrt{\frac{v(\mathbf{r}_0)}{(n-1)pq}}, \tag{5.27}$$

which, as we have seen, is generally too narrow and for that reason will be referred to as the *liberal interval* in the sequel.

The presence of $v(\mathbf{r}_1)$ in (5.26), in turn, appears excessive because it tends to add the variability of the e_j to that of the B_j. If this is so, then (5.26) is overly conservative, and one may think of combining it with the liberal interval in order to get an improved interval. For example, we observe that (5.26) is wider than the liberal interval if and only if

$$v(\mathbf{r}_0) < v(\mathbf{r}_1) = v(\mathbf{b}_1) + v(\mathbf{e}_1) + 2c(v(\mathbf{b}_1), v(\mathbf{e}_1)).$$

Because $v(\mathbf{b}_1) \approx v(\mathbf{b}_0) = v(\mathbf{r}_0)$, this inequality will tend to hold, and (5.26) will tend to be wider than (5.27), whenever the treatment effects vary and are uncorrelated or positively correlated with the baseline values. On the other hand, it is evident that using the liberal interval whenever $v(\mathbf{r}_1) < v(\mathbf{r}_0)$ (which indicates that the e_j and the B_j are negatively correlated) and (5.26) whenever $v(\mathbf{r}_1) \geq v(\mathbf{r}_0)$ produces a 'mixed interval' which can only improve upon the liberal one.

These considerations motivate the use of the generally *conservative interval* with endpoints

$$\hat{\mathbf{e}} \pm z^{(n)}_{1-\frac{\alpha}{2}} \sqrt{\frac{p \max\{v(\mathbf{r}_0), v(\mathbf{r}_1)\} + q\,v(\mathbf{r}_0)}{(n-1)pq}}. \tag{5.28}$$

It appears from simulation studies (Prob. 5.5 (d)) that when the treatment effects are not negatively correlated with the baseline values the conservative interval of nominal level 95% is indeed generally conservative, but its average length is not much greater than that of the best possible but unobservable interval (5.24); when the treatment effects are negatively correlated with the baseline values the interval is generally liberal, but less so than the liberal interval. If the treatment effects are zero or if their magnitude is small relative to that of the baseline values then $v(\mathbf{r}_1) \approx v(\mathbf{b}_1)$ and hence the coverage of the interval will not be in great excess; the coverage will be excessive only if $v(\mathbf{r}_1)$ is typically bigger than $v(\mathbf{b}_1)$, which requires that the treatment effects be variable and positively correlated with the baseline values.

For an example of application of (5.28) to a real-life problem see Prob. 5.6, where another, even more conservative replacement for $v(\mathbf{B})$ may be suggested.

Generalizing the model around (5.19) to the case of several strata requires mainly a change of notation. Let K be the number of strata, n_k and m_k the numbers of units and treated units in stratum k, and write $R_{j,k} = B_{j,k} + e_{j,k}T_{j,k}$ for the response on unit j of stratum k, where $T_{j,k}$, $B_{j,k}$ and $e_{j,k}$ have obvious meanings. As before, the $(B_{1,k}, e_{1,k}), \ldots, (B_{n_k,k}, e_{n_k,k})$ have arbitrary distributions, but conditionally on them $T_{1,k}, \ldots, T_{n_k,k}$ are like n_k random drawings without replacement from a set of m_k 1s and $n_k - m_k$ 0s, and treatments from different strata are independent. Finally, we use notation analogous to that introduced after (5.19), such as $\mathbf{B}^{(k)} = (B_{1,k}, \ldots, B_{n_k,k})$, $\mathbf{r}_t^{(k)}$ $(t = 0, 1)$, etc.

The quantities of interest are now weighted averages of those defined earlier for a single stratum, and they are indicated by adding a subscript or superscript:

$$\bar{\mathbf{e}}_1 := \sum_{k=1}^{K} w_k \bar{\mathbf{e}}_1^{(k)}, \quad \bar{\mathbf{e}}_1^{(k)} := \frac{1}{m_k} \sum_{j=1}^{n_k} e_{j,k} T_{j,k},$$

$$\hat{\mathbf{e}} := \sum_{k=1}^{K} w_k \hat{\mathbf{e}}_k, \quad \hat{\mathbf{e}}_k := \bar{\mathbf{r}}_1^{(k)} - \bar{\mathbf{r}}_0^{(k)} = \frac{c(\mathbf{R}^{(k)}, \mathbf{T}^{(k)})}{p_k q_k}.$$

The 'weights' w_k are non-negative and add to 1; for example, they may be taken as $w_k = 1/K$ when there is no natural ranking of the strata, or in proportion to the stratum sizes, say equal to the numbers

$$v_k := \frac{n_k - 1}{N - K}, \quad \text{where} \quad N := \sum_{k=1}^{K} n_k, \tag{5.29}$$

which we introduce for convenience. For large $\min\{n_1, \ldots, n_K\}$, the random variable $\hat{\mathbf{e}} - \bar{\mathbf{e}}_1 = \sum_{k=1}^{K} w_k \big(\hat{\mathbf{e}}_k - \bar{\mathbf{e}}_1^{(k)}\big)$ is, conditionally on the $B_{j,k}$ and $v(\mathbf{B}^{(k)}) > 0$ for all k, approximately normally distributed with mean 0 and variance

$$\operatorname{Var}\big(\hat{\mathbf{e}} - \bar{\mathbf{e}}_1 \big| \mathbf{B}^{(1)}, \ldots, \mathbf{B}^{(K)}\big) = \frac{1}{N - K} \sum_{k=1}^{K} \frac{w_k^2}{v_k p_k q_k} v\big(\mathbf{B}^{(k)}\big)$$

Thus, using

$$v(\mathbf{B}^{(k)}) = p_k v\big(\mathbf{b}_1^{(k)}\big) + q_k v\big(\mathbf{r}_0^{(k)}\big) + p_k q_k \big(\hat{\mathbf{e}}_k - \bar{\mathbf{e}}_1^{(k)}\big)^2$$

(cf. (5.23)) and (5.29), we see that the conditional probability of the event

$$|\hat{\mathbf{e}} - \bar{\mathbf{e}}_1|^2 \leq \frac{z_{1-\alpha/2}^2}{N - K} \left\{ \sum_{k=1}^{K} \frac{w_s^2\{p_k v(\mathbf{b}_1^{(k)}) + q_k v(\mathbf{r}_0^{(k)})\}}{v_k p_k q_k} + \sum_{k=1}^{K} \frac{\big(w_k\{\hat{\mathbf{e}}_k - \bar{\mathbf{e}}_1^{(k)}\}\big)^2}{v_k} \right\}$$

is $1 - \alpha$ approximately. This inequality involves the summands of $\hat{\mathbf{e}} - \bar{\mathbf{e}}_1$ on the right-hand side; in order to turn it into an interval for $\bar{\mathbf{e}}_1$ we write $x_k = w_k\big(\hat{\mathbf{e}}_k - \bar{\mathbf{e}}_1^{(k)}\big)$ and maximize $\big(\sum_{k=1}^{K} x_k\big)^2$ subject to the constraint

$$\Big(\sum_{k=1}^{K} x_k\Big)^2 = \frac{z_{1-\alpha/2}^2}{N - K} \left\{ \sum_{k=1}^{K} \frac{w_k^2\{p_k v(\mathbf{b}_1^{(k)}) + q_k v(\mathbf{r}_0^{(k)})\}}{v_k p_k q_k} + \sum_{k=1}^{K} \frac{x_k^2}{v_k} \right\}.$$

If the maximum of the term on the left is M, that maximum is attained in particular at $x_k = \sqrt{M} v_k$; moreover, $\sum_{k=1}^K x_k^2/v_k = M = \left(\sum_{k=1}^K x_k\right)^2$ at such x_k. Substituting $\left(\sum_{k=1}^K x_k\right)^2$ and $\sum_{k=1}^K x_k^2/v_k$ by M, solving the equation for $\sqrt{M}$ and recalling (5.29), we find that the endpoints of the interval for $\bar{\mathbf{e}}_1$ are

$$\hat{\mathbf{e}} \pm z_{1-\frac{\alpha}{2}}^{(N-K+1)} \sqrt{\sum_{k=1}^K \frac{w_k^2\{p_k v(\mathbf{b}_1^{(k)}) + q_k v(\mathbf{r}_0^{(k)})\}}{(n_k - 1) p_k q_k}}, \tag{5.30}$$

which therefore constitute an interval for $\bar{\mathbf{e}}_1$ of nominal confidence $1-\alpha$. As before, because the $v(\mathbf{r}_0^{(k)})$ are observable, in order to produce an observable interval for $\bar{\mathbf{e}}_1$ from this we need to find estimates of the $v(\mathbf{b}_1^{(k)})$. (The definition of the quantile in this interval is given by (5.25) with $N - K + 1$ in place of n.)

Replacing $v(\mathbf{b}_1^{(k)})$ by $v(\mathbf{r}_1^{(k)})$ or by $v(\mathbf{r}_0^{(k)})$ in (5.30), depending on whether it is assumed that $e_{j,k} = \epsilon_k$, say, for all $j = 1, 2, \ldots, n_k$, or that the $e_{j,k}$ differ within strata, yields approximate confidence intervals analogous to (5.26) and (5.27). Finally, a generally *conservative interval* generalizing (5.28) is

$$\hat{\mathbf{e}} \pm z_{1-\frac{\alpha}{2}}^{(N-K+1)} \sqrt{\sum_{k=1}^K \frac{w_k^2\{p_k \max\{v(\mathbf{r}_0^{(k)}), v(\mathbf{r}_1^{(k)})\} + q_k v(\mathbf{r}_0^{(k)})\}}{(n_k - 1) p_k q_k}}. \tag{5.31}$$

5.3 Estimation of a treatment effect

It is often of interest to estimate the magnitude of the effect of the treatment on the response rather than just testing for the existence of one. This more ambitious task requires stronger assumptions, and also much more data, than those used for testing. Since the treatment effect is characterized by $(\mathbf{x}, t) \to \mathcal{L}(R_n|\mathbf{X}_n = \mathbf{x}, T_n = t)$, the most basic thing one can think of estimating when this function does not depend on n is perhaps

$$\mu_{\mathbf{x},t} := \mathbf{E}(R_n|\mathbf{X}_n = \mathbf{x}, T_n = t);$$

the following assumption makes it possible to estimate this and similar quantities:

A2 For every $I = \{i_1, i_2, \ldots, i_n\} \subset \{1, 2, \ldots, N\}$, $\mathbf{x} \in \mathcal{X}$ and $t \in \mathcal{T}$, conditionally on $\mathbf{X}_i = \mathbf{x}$ and $T_i = t$ for all $i \in I$, $R_{i_1}, R_{i_2}, \ldots, R_{i_n}$ are a random sample.

Throughout this section we write

$$J_{\mathbf{x},t} = \{j \le N : \mathbf{X}_j = \mathbf{x}, T_j = t\}, \quad m_{\mathbf{x},t} = \#J_{\mathbf{x},t},$$

so that $I_{\mathbf{x}} = \bigcup_{t\in\mathcal{T}} J_{\mathbf{x},t}$ and $N = \sum_{\mathbf{x}\in\mathcal{X}} \#I_{\mathbf{x}} = \sum_{\mathbf{x}\in\mathcal{X}, t\in\mathcal{T}} m_{\mathbf{x},t}$, and put

$$\hat{\mu}_{\mathbf{x},t} = \frac{1}{m_{\mathbf{x},t}} \sum_{i\in J_{\mathbf{x},t}} R_i, \quad \hat{\sigma}_{\mathbf{x},t}^2 = \frac{1}{m_{\mathbf{x},t} - 1} \sum_{i\in J_{\mathbf{x},t}} (R_i - \hat{\mu}_{\mathbf{x},t})^2,$$

where empty sums are zero by definition. Then $\hat{\mu}_{\mathbf{x},t}$ is an unbiased estimator of $\mu_{\mathbf{x},t}$ conditionally on $m_{\mathbf{x},t} \geq 1$, and if $\mathrm{Var}(R_n \leq r|\mathbf{X}_n = \mathbf{x}, T_n = t) < \infty$ then for 'large' $m_{\mathbf{x},t}$ the distribution of

$$\frac{\hat{\mu}_{\mathbf{x},t} - \mu_{\mathbf{x},t}}{\hat{\sigma}_{\mathbf{x},t}/\sqrt{m_{\mathbf{x},t}}} \tag{5.32}$$

is approximately standard normal, so that

$$[\hat{\mu}_{\mathbf{x},t} - z_{1-\alpha/2}\hat{\sigma}_{\mathbf{x},t}/\sqrt{m_{\mathbf{x},t}}, \hat{\mu}_{\mathbf{x},t} + z_{1-\alpha/2}\hat{\sigma}_{\mathbf{x},t}/\sqrt{m_{\mathbf{x},t}}],$$

where $z_p = \Phi^{-1}(p)$, is a confidence interval for $\mu_{\mathbf{x},t}$ with level approximately equal to $1 - \alpha$ $(0 < \alpha < 1)$. More generally, given $t_1, t_2, \ldots, t_k \in \mathcal{T}$, by the independence of the responses in $J_{\mathbf{x},t_1}, J_{\mathbf{x},t_2}, \ldots, J_{\mathbf{x},t_k}$,

$$\left\{(\mu_1, \ldots, \mu_k) \in \mathbb{R}^k : |\mu_j - \hat{\mu}_{\mathbf{x},t_j}| \leq z_{1-\alpha^*/2}\hat{\sigma}_{\mathbf{x},t_j}/\sqrt{m_{\mathbf{x},t_j}},\ j = 1, \ldots, k\right\},$$

where $\alpha^* := 1 - (1 - \alpha)^{1/k}$, is a joint confidence region for $(\mu_{\mathbf{x},t_1}, \ldots, \mu_{\mathbf{x},t_k})$ with level approximately equal to $1 - \alpha$.

Also of interest are the differences

$$\delta_{\mathbf{x},t,t'} := \mu_{\mathbf{x},t} - \mu_{\mathbf{x},t'} \quad (t \neq t');$$

approximate confidence intervals for them can be based on the distribution of

$$\frac{\hat{\mu}_{\mathbf{x},t} - \hat{\mu}_{\mathbf{x},t'} - \delta_{\mathbf{x},t,t'}}{\sqrt{\hat{\sigma}^2_{\mathbf{x},t}/m_{\mathbf{x},t} + \hat{\sigma}^2_{\mathbf{x},t'}/m_{\mathbf{x},t'}}}, \tag{5.33}$$

which (see (Prob. 5.2)) for 'large' $m_{\mathbf{x},t}$ and $m_{\mathbf{x},t'}$ is approximately standard normal. Approximate confidence regions for $(\delta_{\mathbf{x}_1,t,t'}, \delta_{\mathbf{x}_2,t,t'}, \ldots, \delta_{\mathbf{x}_k,t,t'})$ can be obtained similarly to those for $(\mu_{\mathbf{x},t_1}, \ldots, \mu_{\mathbf{x},t_k})$.

These confidence intervals and regions can be unreliable when the distributions of the responses within the $J_{\mathbf{x},t}$ are skewed or the $m_{\mathbf{x},t}$ are small. The bootstrap method, applied within the strata/matched sets (formed independently of the data, as indicated at the end of Sect. 5.1) to imitate the sampling distributions of the ratios (5.32) and (5.33), provides alternatives to them. In the version described in Example 2.3 of Beran [8], possibly 'smoothed' as in the variant proposed by Polansky [75] in the case of a mean, the bootstrap is accurate in many cases, even with $m_{\mathbf{x},t}$ and $m_{\mathbf{x},t'}$ as small as five (a more usual version of the bootstrap is outlined below).

Estimates of an 'overall' effect of the treatment on the response are sometimes of interest, especially when T_n takes only two values. The parameter

$$\epsilon_{t,t'} := \sum\nolimits_{\mathbf{x}} \{\mu_{\mathbf{x},t} - \mu_{\mathbf{x},t'}\}\, w(\mathbf{x}), \tag{5.34}$$

where $w(\mathbf{x})$ may be $\mathbf{P}(\mathbf{X}_n = \mathbf{x})$ in the case where $\mathbf{X}_1, \mathbf{X}_2, \ldots, \mathbf{X}_N$ are a random sample, or another probability thought to be relevant for the problem in hand, is a general measure of the effect of the treatment. This can be estimated by replacing

$\mu_{\mathbf{x},t}$, $\mu_{\mathbf{x},t'}$ and $w(\mathbf{x})$ by their estimates, but there are various ways of doing this and at least as many ways of estimating the associated standard errors and producing approximate confidence intervals. In order to say something more definite we assume that $(\mathbf{X}_1, T_1, R_1), \ldots, (\mathbf{X}_N, T_N, R_N)$ are a random sample with the same distribution as a generic random vector $(\mathbf{X}, T, R)$ and that the $I_\mathbf{x}$ are strata.

Fix a small integer $K \geq 1$ and let E_K be the event that $m_{\mathbf{x},t}, m_{\mathbf{x},t'} \geq K$ if and only if $\mathbf{x} \in \mathcal{X}_K$ for a subset $\mathcal{X}_K$ of $\mathcal{X}$. Consider

$$\hat{\epsilon}^{(K)}_{t,t'} := \sum\nolimits_{\mathbf{x}\in\mathcal{X}_K} \{\hat{\mu}_{\mathbf{x},t} - \hat{\mu}_{\mathbf{x},t'}\}\, \hat{w}_K(\mathbf{x}), \tag{5.35}$$

where $\hat{w}_K(\mathbf{x}) := \#I_\mathbf{x}/n_K$, $n_K := \sum_{\mathbf{x}\in\mathcal{X}_K} \#I_\mathbf{x}$, and sums over an empty $\mathcal{X}_K$ are zero by definition. Conditionally on E_K this is generally biased as an estimator of $\epsilon_{t,t'}$, but if $\mathcal{X}$ and $\mathcal{T}$ are finite (and $\mathbf{P}(T = t|\mathbf{X} = \mathbf{x}), \mathbf{P}(T = t'|\mathbf{X} = \mathbf{x}) > 0$ for all $\mathbf{x}$, as implied by the definition of $\epsilon_{t,t'}$) then $\mathcal{X}_K = \mathcal{X}$ for large enough N and $\hat{\epsilon}^{(K)}_{t,t'}$ converges almost surely to $\epsilon_{t,t'}$ for any fixed K. Convergence certainly holds under weaker conditions, so $\hat{\epsilon}^{(K)}_{t,t'}$ is generally a sensible estimator. It is always biased if $\mathcal{X}$ is infinite, especially if the $\mathbf{X}_n$ are not discrete, since then the correct version of (5.35) is an integral and $\hat{\mu}_{\mathbf{x},t}$ estimates $\mathbf{E}[R|T = t, \mathbf{X} \in \mathcal{N}_\mathbf{x}]$ for a neighbourhood $\mathcal{N}_\mathbf{x}$ of $\mathbf{x}$ rather than $\mu_{\mathbf{x},t}$. Its bias increases with K, but its variance decreases (because $m_{\mathbf{x},t}$ and $m_{\mathbf{x},t'}$ increase with K). In addition, if the $\mathbf{X}_n$ are not discrete then its bias is greater when the stratification is coarser—i.e. when the $\mathcal{N}_\mathbf{x}$ implied by the $I_\mathbf{x}$ are larger—and its variance is smaller. Note that in (5.35) the $\hat{w}_K(\mathbf{x})$ are meant to estimate the probabilities $\mathbf{P}(\mathbf{X} = \mathbf{x})$, but the denominator in them is not N but n_K, which is the number of units that remain in $\bigcup_{\mathbf{x}\in\mathcal{X}_K} I_\mathbf{x}$; in general, (5.35) necessarily involves fewer strata and hence fewer data (the fewer the larger K is) than those available, and the $\hat{w}_K(\mathbf{x})$ are scaled so as to yield a probability distribution on $\mathcal{X}_K$.

Remark 5.2 Part of the bias of (5.35) comes from the mismatch between $\mathcal{X}_K$ and $\mathcal{X}$, which in turn comes partly from $J_{\mathbf{x},t}$ or $J_{\mathbf{x},t'}$ containing fewer than K units for some $\mathbf{x}$. This is because our estimators of $\mu_{\mathbf{x},t}$ are of the most basic kind and they are not defined unless $\#J_{\mathbf{x},t} \geq K$. In principle one may use estimators of $\mu_{\mathbf{x},t}$ in (5.35) which are defined for all t whenever $\mathbf{x}$ is represented in the sample: since $\mu_{\mathbf{x},t}$ is a conditional expectation, there are many high-powered estimators of it from the area of 'statistical prediction' (e.g. Devroye, Györfi and Lugosi [24], Breiman [13], Györfi *et al.* [41]) which have this property, which are accurate from the point of view of prediction criteria and generally consistent. The corresponding estimators of (5.34) obtained from them use the whole or practically the whole sample and their variance can be estimated by means of 'cross-validation' methods (e.g. Mosteller and Tukey [61]). On the other hand, because their definition extends beyond the range of the observed data, they are more biased than our $\hat{\mu}_{\mathbf{x},t}$—and indeed suffer from their own sort of bias because they allow combinations of treatments and covariates which may be implausible. □

One way of estimating $\mathrm{Var}(\hat{\epsilon}^{(K)}_{t,t'}|E_K)$ and obtaining a confidence interval for $\epsilon^{(K)}_{t,t'}$ is to treat the $\hat{w}_K(\mathbf{x})$ as constants, acting as if

$$v_1(\hat{\epsilon}_{t,t'}^{(K)}) := \sum\nolimits_{\mathbf{x}\in\mathcal{X}_K} \left(\frac{\hat{\sigma}_{\mathbf{x},t}^2}{m_{\mathbf{x},t}} + \frac{\hat{\sigma}_{\mathbf{x},t'}^2}{m_{\mathbf{x},t'}} \right) \hat{w}_K(\mathbf{x})^2 \quad \text{and} \quad \hat{\epsilon}_{t,t'}^{(K)} \pm z_{1-\alpha/2}\sqrt{v_1(\hat{\epsilon}_{t,t'}^{(K)})}$$

provided an accurate estimate and sufficient coverage respectively. Still, $v_1(\hat{\epsilon}_{t,t'}^{(K)})$ is accurate only when the $I_{\mathbf{x}}$ are formed from a fixed, finite population and $\epsilon_{t,t'}$ pertains to that population, or when $\mathcal{X}$ is finite; otherwise it can be too small and the associated interval too narrow—besides being off-centre, which is unavoidable since the point estimate is generally biased.

Another way, which should be safer when $\mathcal{X}$ is infinite and especially when the $\mathbf{X}_i$ are continuous, is based on the identity

$$\text{Var}(\hat{\epsilon}_{t,t'}^{(K)}|E_K) = \sum\nolimits_{\mathbf{x},\mathbf{y}\in\mathcal{X}_K} \text{Cov}\left(\hat{\delta}_{\mathbf{x}}\hat{w}_K(\mathbf{x}), \hat{\delta}_{\mathbf{y}}\hat{w}_K(\mathbf{y})\middle| E_K\right),$$

where we write $\hat{\delta}_{\mathbf{x}} = \hat{\mu}_{\mathbf{x},t} - \hat{\mu}_{\mathbf{x},t'}$ for short, and on an approximation of the covariances being summed. This yields

$$v_2(\hat{\epsilon}_{t,t'}^{(K)}) := \sum\nolimits_{\mathbf{x}\in\mathcal{X}_K} \left(\frac{\hat{\sigma}_{\mathbf{x},t}^2}{m_{\mathbf{x},t}} + \frac{\hat{\sigma}_{\mathbf{x},t'}^2}{m_{\mathbf{x},t'}} \right) \hat{w}_K(\mathbf{x})^2 + \hat{\delta}_{\mathbf{x}}^2 \frac{\hat{w}_K(\mathbf{x})(1-\hat{w}_K(\mathbf{x}))}{n_K} \tag{5.36}$$

as an approximate upper bound on $\text{Var}(\hat{\epsilon}_{t,t'}^{(K)}|E_K)$. Although this expression may be arrived at by intuition, its justification appears intrinsically complicated; a detailed account of it is provided in Prob. 5.7, where it is also seen that the bigger K is the more (5.36) will tend to overestimate $\text{Var}(\hat{\epsilon}_{t,t'}^{(K)}|E_K)$.

These approximations are meant really for $K > 1$, since they rely on estimates of the $\sigma_{\mathbf{x},t}^2$. A third approximation, which may be used even for $K = 1$, is based on the bootstrap method (p. 271 of Rice [79]), namely on the sampling distribution of the random variable

$$\sum\nolimits_{\mathbf{x}\in\mathcal{X}_K^*} \{\hat{\mu}_{\mathbf{x},t}^* - \hat{\mu}_{\mathbf{x},t'}^*\}\, \hat{w}_K^*(\mathbf{x}) - \hat{\epsilon}_{t,t'}^{(K)}, \tag{5.37}$$

where $\mathcal{X}_K^*$ and the $\hat{\mu}_{\mathbf{x},t}^*$ and $\hat{w}_K^*(\mathbf{x})$ are obtained from a random sample

$$(\mathbf{X}_1^*, T_1^*, R_1^*), \ldots, (\mathbf{X}_N^*, T_N^*, R_N^*)$$

from the empirical distribution of $(\mathbf{X}_1, T_1, R_1), \ldots, (\mathbf{X}_N, T_N, R_N)$. By simulating a large number of such samples one simulates a random sample with the distribution of (5.37) with which one can try to estimate the distribution of $\hat{\epsilon}_{t,t'}^{(K)} - \epsilon_{t,t'}^{(K)}$ and hence estimate its variance as well as its quantiles, and the latter serve to set a confidence interval for $\epsilon_{t,t'}^{(K)}$, which, unlike the intervals just given, need not be centred around the point estimate. This method must work well at least when N is very large, $\mathcal{X}$ and $\mathcal{T}$ are finite and the R_i are continuous, and one hopes that under more general conditions it will yield intervals that are at least as accurate as those based on our second approximation, though in practice one usually has no guarantee that an approximation is sufficiently accurate. At any rate, a safer interval is obtained by taking the smaller and larger of the lower and upper endpoints of our three intervals.

Note that in performing the bootstrap the stratification is kept fixed (having been fixed independently of the data, as indicated at the end of Sect. 5.1), and that in each bootstrap sample some of the strata may get fewer or more observations than they do in the original sample, and yet others may be discarded; this produces variability in the $\hat{\omega}_K^*(\mathbf{x})$, which one hopes will mimic the sampling variability of the $\hat{\omega}_K(\mathbf{x})$.

Whatever the estimators and intervals used for the overall effect of the treatment, one may exercise some judgement in performing the stratification and in choosing the K in (5.35), and one should be expected to report more than just one estimate and one interval and to examine the estimates of $\mu_{\mathbf{x},t}$ rather than concentrating only on those of $\epsilon_{t,t'}$. It must be remembered that all these estimates and intervals are somewhat biased, certainly when the $\mathbf{X}_n$ are not discrete; nevertheless, they ought to be much less biased than estimates and intervals which do not account for $\mathbf{X}_n$.

Analogous estimates and regions can be obtained for $\mathbf{E}(f(R_n)|\mathbf{X}_n = \mathbf{x}, T_n = t)$, where f is some function (e.g. $f = \mathbf{1}_{]-\infty,r]}$, $r \in \mathbb{R}$), by substituting $f(R_i)$ for R_i in all the above, assuming the appropriate moments exist.

Finally, we note that data from case-control studies do not satisfy **A2**; instead, they satisfy an analogous condition with the roles of the T_i and the R_i exchanged, which suffices to justify the use of empirical frequencies to estimate the $p_\mathbf{x}$ and $q_\mathbf{x}$ of (4.20) and hence the odds ratios $\mathrm{e}(\mathbf{x})$ in each $I_\mathbf{x}$ (cf. (4.22)). In addition, provided the sample sizes are 'large', one may be able to obtain approximate confidence intervals for $\mathrm{e}(\mathbf{x})$ (see Prob. 5.8).

5.4 The method of propensity scores

In Sect. 5.2 we have introduced the probabilities

$$f_I^{(\mathbf{x})}((t_i)_{i\in I}) = \mathbf{P}(T_i = t_i, i \in I|\mathbf{X}_i = \mathbf{x}, i \in I);$$

when $I = \{i\}$ we write them as

$$\lambda_i(t;\mathbf{x}) := \mathbf{P}(T_i = t|\mathbf{X}_i = \mathbf{x}),$$

setting $\lambda_i(t;\mathbf{x}) = 0$ whenever $\mathbf{P}(\mathbf{X}_i = \mathbf{x}) = 0$. For fixed i and t, we call $\mathbf{x} \to \lambda_i(t;\mathbf{x})$ the *propensity score function* of unit i and treatment t: it gives us the probability of unit i receiving treatment t when $\mathbf{X}_i = \mathbf{x}$. For fixed $(i, \mathbf{x})$, the function $t \to \lambda_i(t;\mathbf{x})$ is a probability mass function on $\mathcal{T}$; we call the random variables $\lambda_i(t;\mathbf{X}_i)$, $t \in \mathcal{T}$, the *propensity scores* of unit i. If l is a probability mass function on $\mathcal{T}$, we can speak of the probabilities $\mathbf{P}(\lambda_i(\,\cdot\,;\mathbf{X}_i) = l)$, where $\lambda_i(\,\cdot\,;\mathbf{x}) = l$ means that $\lambda_i(t;\mathbf{x}) = l(t)$ for all $t \in \mathcal{T}$.

The results that follow are valid when the $\mathbf{X}_i$ are discrete, and they have obvious generalizations when the $\mathbf{X}_i$ have continuous densities. The attendant methods are valid approximately when the $\mathbf{X}_i$ are continuous, with events such as $\{\mathbf{X}_i \in \mathcal{N}\}$ for small neighbourhoods $\mathcal{N} \ni \mathbf{x}$ taking the place of $\{\mathbf{X}_i = \mathbf{x}\}$.

Let l be such that $\mathbf{P}(\lambda_i(\,\cdot\,;\mathbf{X}_i) = l) > 0$; then it can be shown (Prob. 5.12 (a)) that

$$\mathbf{P}(T_i = t|\lambda_i(\,\cdot\,;\mathbf{X}_i) = l) = l(t), \tag{5.38}$$

and that if for given $t \in \mathcal{T}$ there is at least one probability mass function l such that $\mathbf{P}(\lambda_i(\,\cdot\,;\mathbf{X}_i) = l) > 0$ and $l(t) > 0$ then, with $\sum_l$ indicating summation over all such probability functions,

$$\mathbf{P}[\rho_i(V_i, \mathbf{X}_i, t) \le r] = \sum_l \mathbf{P}[R_i \le r|\lambda_i(\,\cdot\,;\mathbf{X}_i) = l, T_i = t]\mathbf{P}[\lambda_i(\,\cdot\,;\mathbf{X}_i) = l]. \tag{5.39}$$

This second property has two important consequences: Given $t, t' \in \mathcal{T}$, suppose that $\mathbf{P}(\lambda_i(\,\cdot\,;\mathbf{X}_i) = l) > 0$ and $l(t), l(t') > 0$ for at least one l. Then, by (5.39), $\mathcal{L}(\rho_i(V_i, \mathbf{X}_i, t)) \ne \mathcal{L}(\rho_i(V_i, \mathbf{X}_i, t'))$ if and only if

$$\mathcal{L}(R_i|\lambda_i(\,\cdot\,;\mathbf{X}_i) = l, T_i = t) \ne \mathcal{L}(R_i|\lambda_i(\,\cdot\,;\mathbf{X}_i) = l, T_i = t')$$

for some l, and so, in essence, the existence of a treatment effect can be studied through the function $t \to \mathcal{L}(R_i|\lambda_i(\,\cdot\,;\mathbf{X}_i) = l, T_i = t)$.

Secondly, (5.39) implies

$$\sum_{\mathbf{x}} \mathbf{E}\left(R_i \,\middle|\, \begin{matrix}\mathbf{X}_i = \mathbf{x}, \\ T_i = t\end{matrix}\right)\mathbf{P}(\mathbf{X}_i = \mathbf{x}) = \sum_l \mathbf{E}\left(R_i \,\middle|\, \begin{matrix}\lambda_i(\,\cdot\,;\mathbf{X}_i) = l, \\ T_i = t\end{matrix}\right)\mathbf{P}[\lambda_i(\,\cdot\,;\mathbf{X}_i) = l],$$

so if the distribution of $(\mathbf{X}_i, T_i, R_i)$ is independent of i, equal to that of a generic vector $(\mathbf{X}, T, R)$, so that $\lambda_i = \lambda$ too is independent of i, then, under the same assumption about t and t', the overall effect of treatment defined by (5.34) with $w(\mathbf{x}) = \mathbf{P}(\mathbf{X} = \mathbf{x})$ can be expressed as

$$\epsilon_{t,t'} = \sum_l \left\{\mathbf{E}\left(R \,\middle|\, \begin{matrix}\lambda(\,\cdot\,;\mathbf{X}) = l, \\ T = t\end{matrix}\right) - \mathbf{E}\left(R \,\middle|\, \begin{matrix}\lambda(\,\cdot\,;\mathbf{X}) = l, \\ T = t'\end{matrix}\right)\right\}\mathbf{P}(\lambda(\,\cdot\,;\mathbf{X}) = l). \tag{5.40}$$

Another identity related to (5.38) and (5.39), valid for every probability function l such that $\mathbf{P}(\lambda_i(\,\cdot\,;\mathbf{X}_i) = l) > 0$, is

$$\mathbf{P}[R_i \le r, T_i = t\,|\lambda_i(\,\cdot\,;\mathbf{X}_i) = l\,] = l(t)\frac{\sum_{\mathbf{x}}^{(l)} \mathbf{P}\left[\rho_i(V_i, \mathbf{x}, t) \le r\,|\mathbf{X}_i = \mathbf{x}\right]\mathbf{P}(\mathbf{X}_i = \mathbf{x})}{\mathbf{P}(\lambda_i(\,\cdot\,;\mathbf{X}_i) = l)},$$

where $\sum_{\mathbf{x}}^{(l)}$ indicates summation over $\mathbf{x}$ satisfying $\lambda_i(\,\cdot\,;\mathbf{x}) = l$ (Prob. 5.12 (a)); this yields (5.38) and shows that if there is no treatment effect then R_i and T_i are independent conditionally on $\lambda_i(\,\cdot\,;\mathbf{X}_i) = l$.

These results suggest that the methods of testing and estimation of Sects. 5.2 and 5.3 may be applied with $(\lambda_i(t;\mathbf{X}_i))_{t\in\mathcal{T}}$ in place of $\mathbf{X}_i$. In fact, it is evident that this is the case when $(\mathbf{X}_1, T_1, R_1), (\mathbf{X}_2, T_2, R_2), \ldots, (\mathbf{X}_N, T_N, R_N)$ are independent and identically distributed. But in order to show that the methods of Sect. 5.2 are still applicable under assumptions akin to the more general assumptions considered there, we need to state conditions which make **A0**, **A1** and **A2** valid with $\mathbf{X}_i$ replaced by $(\lambda_i(t;\mathbf{X}_i))_{t\in\mathcal{T}}$, since, for instance, our last identity alone does not guarantee that we can test for a treatment effect *within a stratum* $I_l := \{i : \lambda_i(\cdot;\mathbf{X}_i) = l\}$. The following assumption strengthens **A0** and **A1** and yields the desired results; it will be assumed in the remainder of this section:

A3 For every $I = \{i_1, i_2, \ldots, i_n\} \subset \{1, 2, \ldots, N\}$ and every $\mathbf{x}_1, \mathbf{x}_2, \ldots, \mathbf{x}_n \in \mathcal{X}$, the vectors $(U_{i_1}, U_{i_2}, \ldots, U_{i_n})$ and $(V_{i_1}, V_{i_2}, \ldots, V_{i_n})$ are independent conditionally on $\{\mathbf{X}_{i_1} = \mathbf{x}_1, \mathbf{X}_{i_2} = \mathbf{x}_2, \ldots, \mathbf{X}_{i_n} = \mathbf{x}_n\}$. Moreover, if $1 \le K \le n$, $I_1, I_2, \ldots, I_K$ form a partition of I, and $\mathbf{x}_1, \mathbf{x}_2, \ldots, \mathbf{x}_K \in \mathcal{X}$, then for $(u_i)_{i\in I}, (v_i)_{i\in I} \in \mathbb{R}^n$

$$\mathbf{P}(U_i \le u_i, V_i \le v_i, i \in I | \mathbf{X}_i = \mathbf{x}_j, i \in I_j, j = 1, 2, \ldots, K) = \\ \prod_{j=1}^{K} \mathbf{P}(U_i \le u_i, V_i \le v_i, i \in I_j | \mathbf{X}_i = \mathbf{x}_j, i \in I_j).$$

Finally, $(\mathbf{X}_1,T_1),(\mathbf{X}_2,T_2),\ldots,(\mathbf{X}_N,T_N)$ are independent and identically distributed.

By the last part of **A3**, λ_i is independent of i, so from now on we write it as λ. Also, using $\sum^{(l)}_{\mathbf{x}_i, i\in I}$ to indicate summation over $\{\mathbf{x}_i : \lambda(\cdot;\mathbf{x}_i) = l, i \in I\}$, we see that

$$\mathbf{P}(T_i = t_i, i \in I | \lambda(\cdot;\mathbf{X}_i) = l, i \in I) = \sum\nolimits^{(l)}_{\mathbf{x}_i, i\in I} \frac{\mathbf{P}(T_i = t_i, \mathbf{X}_i = \mathbf{x}_i, i \in I)}{\mathbf{P}(\lambda(\cdot;\mathbf{X}_j) = l, j \in I)} = \prod_{i\in I} l(t_i)$$

is invariant to permutations of the coordinates of $(t_i)_{i\in I}$, and hence that **A1** holds with $\{\lambda(t;\mathbf{X}_i) = l(t), t \in \mathcal{T}, i \in I\}$ in place of $\{\mathbf{X}_i = \mathbf{x}, i \in I\}$.

On the other hand, **A3** implies (Prob. 5.12 (b))

$$\mathbf{P}\left(\begin{matrix} T_i = t_i, R_i \le r_i, \\ i \in I \end{matrix} \middle| \begin{matrix} \lambda(\cdot;\mathbf{X}_i) = l, \\ i \in I \end{matrix} \right) = \psi_I^{(l)}((t_i)_{i\in I})\, \Psi_I^{(l)}((t_i)_{i\in I}, (r_i)_{i\in I}), \quad (5.41)$$

with

$$\psi_I^{(l)}((t_i)_{i\in I}) := \mathbf{P}(T_i = t_i, i \in I | \lambda(\cdot;\mathbf{X}_i) = l, i \in I),$$

$$\Psi_I^{(l)}((t_i)_{i\in I}, (r_i)_{i\in I}) := \mathbf{P}(\varrho_i(V_i, \mathbf{X}_i, t_i) \le r_i, i \in I | \lambda(\cdot;\mathbf{X}_i) = l, i \in I),$$

and from this follows that if the hypothesis of no treatment effect holds with respect to the units in $I_l := \{i : \lambda(\cdot;\mathbf{X}_i) = l\}$ then $(T_i)_{i\in I}$ and $(R_i)_{i\in I}$ are independent given $\{\lambda(\cdot;\mathbf{X}_i) = l, i \in I\}$, so the hypothesis can be tested by the methods described in Sect. 5.2 with the propensity scores in place of the background variables.

Finally, let I_1, I_2 be disjoint subsets of indices and put $I = I_1 \cup I_2$; by the second part of **A3** we have (Prob. 5.12 (c))

$$\mathbf{P}\big[U_i \le u_i, V_i \le v_i, \lambda(\cdot;\mathbf{X}_j) = l_1, \lambda(\cdot;\mathbf{X}_k) = l_2, i \in I, j \in I_1, k \in I_2\big] = \\ \mathbf{P}\big[U_i \le u_i, V_i \le v_i, \lambda(\cdot;\mathbf{X}_j) = l_1, i \in I_1\big]\mathbf{P}[U_i \le u_i, V_i \le v_i, \lambda(\cdot;\mathbf{X}_k) = l_2, i \in I_2].$$

Taking l_1 and l_2 such that $\mathbf{P}(\lambda(\cdot;\mathbf{X}_i) = l_1), \mathbf{P}(\lambda(\cdot;\mathbf{X}_i) = l_2) > 0$ and dividing both sides by $\mathbf{P}\big[\lambda(\cdot;\mathbf{X}_j) = l_1, \lambda(\cdot;\mathbf{X}_k) = l_2, j \in I_1, k \in I_2\big]$, noting that this probability equals $\mathbf{P}\big[\lambda(\cdot;\mathbf{X}_j) = l_1, i \in I_1\big]\, \mathbf{P}[\lambda(\cdot;\mathbf{X}_k) = l_2, i \in I_2]$, we conclude—the proof for K sets $I_1, I_2, \ldots, I_K$ following from this—that the second part of **A0** holds with the propensity scores in place of the background variables, and hence that inferences from different strata continue to be independent when based on propensity scores.

If λ is a known function, the method of propensity scores may be advantageous, especially if $\mathcal{T} = \{0, 1\}$, since then conditioning on $(\lambda(0; \mathbf{X}_i), \lambda(1; \mathbf{X}_i))$ amounts to conditioning on the *scalar* $\lambda(1; \mathbf{X}_i) = 1 - \lambda(0; \mathbf{X}_i)$. In this most favourable case, and with λ known, the stratification consists of sets of units

$$I_j := \left\{ i : \frac{j-1}{p} < \lambda(1; \mathbf{X}_i) \le \frac{j}{p} \right\}, \quad j = 1, 2, \ldots, p, \tag{5.42}$$

for some p. If λ is unknown then it must be estimated from the $(\mathbf{X}_i, T_i)$, the sets I_j are defined with an estimate of λ, and the correctness of the method ought to depend on the correctness of the estimate.

Suppose that $\mathcal{T} = \{0, 1\}$ and that $\mathbf{X}_i$ is discrete, so that $\lambda(1; \mathbf{X}_i)$ too is discrete. Then $\lambda(1; \mathbf{x})$ can be estimated by $\hat{\lambda}(1; \mathbf{x}) := \#J_{\mathbf{x},1} / \#I_{\mathbf{x}}$ (in the notation of Sect. 5.3), which we call the *empirical propensity score function*. If $\hat{\lambda}(1; \mathbf{X}_i)$ is used in (5.42) in place of $\lambda(1; \mathbf{X}_i)$ and attention is restricted to strata with $0 < \hat{\lambda}(1; \mathbf{x}) < 1$, one obtains what may be called a *stratification on the empirical propensity score*. The test based on the inner product statistic of Sect. 5.2.2 is, of course, not the same when carried out over such a stratification as when carried out over the finer stratification based on the $\mathbf{X}_i$, and it need not be correct. On the other hand, the estimate of the parameter $\boldsymbol{\epsilon}_{1,0}$ of (5.34) obtained by replacing the summands on the right-hand side of (5.40) by their empirical versions calculated over the strata defined by the different values of the empirical propensity score—i.e. with $\#\{i : \hat{\lambda}(1, \mathbf{X}_i) = l\} / \#\{i : 0 < \hat{\lambda}(1, \mathbf{X}_i) < 1\}$ in place of $\mathbf{P}(\lambda(1, \mathbf{X}) = l)$—is the same as our earlier estimate $\hat{\boldsymbol{\epsilon}}^{(1)}_{1,0}$ of (5.35) with $K = 1$, i.e. as the estimate of $\boldsymbol{\epsilon}_{1,0}$ obtained by stratification on $\mathbf{X}_i$. In order to see this, one observes that $\hat{\lambda}(1; \mathbf{x})$ is the propensity score function of a random vector $(\mathbf{X}^*, T^*, R^*)$ drawn from the empirical distribution of the data in the strata obtained with the empirical propensity score—that is, $\hat{\lambda}(1; \mathbf{x})$ is the probability that $T^* = 1$ given that $\mathbf{X}^* = \mathbf{x}$—and that the estimate $\hat{\boldsymbol{\epsilon}}^{(1)}_{1,0}$ is the 'theoretical' overall treatment effect pertaining to $(\mathbf{X}^*, T^*, R^*)$; but the identity (5.40) remains valid if $\mathbf{P}$ and $\mathbf{E}$ are replaced by their empirical versions and $(\lambda(1; \mathbf{X}), T, R)$ by $(\hat{\lambda}(1; \mathbf{X}^*), T^*, R^*)$, and the left-hand side is replaced by $\hat{\boldsymbol{\epsilon}}^{(1)}_{1,0}$.

Thus the method of propensity scores is advantageous when $\mathbf{X}_i$ is continuous, at least in part, and the propensity score function can be estimated by something more efficient—hence relying on stronger assumptions—than its empirical version.

In any case, propensity scores have yet another property, the so-called 'balancing property', which may be used to check whether a stratification is appropriate and is useful even when the $\mathbf{X}_i$ are continuous (Prob. 5.12 (d)):

$$\mathbf{P}[\mathbf{X}_i = \mathbf{x}, i \in I \,|\, T_i = t, \lambda(\,\cdot\,; \mathbf{X}_i) = l, i \in I] = \mathbf{P}[\mathbf{X}_i = \mathbf{x}, i \in I \,|\, \lambda(\,\cdot\,; \mathbf{X}_i) = l, i \in I]$$

for l such that $\mathbf{P}(\lambda(\,\cdot\,; \mathbf{X}_i) = l) > 0$ and $\mathbf{x}$ such that $\lambda(\,\cdot\,; \mathbf{x}) = l$, so conditionally on $\{\lambda(\,\cdot\,; \mathbf{X}_i) = l, i \in I\}$ the distribution of the background variables in stratum I is independent of the corresponding treatment assignments. If $\mathcal{T} = \{0, 1\}$, this means

$$\mathbf{P}(\mathbf{X}_i = \mathbf{x}\,|T_i = 0, \lambda(1;\mathbf{X}_i) = p) = \mathbf{P}(\mathbf{X}_i = \mathbf{x}\,|\lambda(1;\mathbf{X}_i) = p)$$
$$= \mathbf{P}(\mathbf{X}_i = \mathbf{x}\,|T_i = 1, \lambda(1;\mathbf{X}_i) = p),$$

so in order to assess the quality of a stratification one checks whether the treated and control units within a stratum have approximately the same distribution of background variables.

5.5 A plan for data analyses with the basic model

Although data analyses alone seldom serve to justify assumptions, classical methods of stratification and model checking can be useful at least to detect some violations of the basic model with unconfoundedness and, if based on large data sets and guided by more or less theoretical considerations (e.g. about which distribution a certain variable is expected to follow), may strengthen the case for that model.

In the age-old tradition of 'model-building' one assumes that a 'response' is the combined result of phenomena which are predictable to some extent through 'explanatory variables' and of other, 'purely random' phenomena which cannot be predicted at all from variables other than those they affect. Thus, in the case of (2.1) with unconfoundedness, assumed from now on, the response $R_n = \rho(V_n, \mathbf{X}_n, T_n)$ is predictable through $\mathbf{X}_n$ and T_n but only in part because of the purely random V_n which is not predictable from anything that precedes it since conditionally on $(\mathbf{X}_n, T_n)$ it has the same *fixed* law.

Stratification is the empirical analogue of conditioning, so if the $(\mathbf{X}_n, T_n, R_n)$ are a random sample and the $(\mathbf{X}_n, T_n)$ are discrete then the responses of the subsample with indices in $J_{\mathbf{x},t} := \{n : \mathbf{X}_n = \mathbf{x}, T_n = t\}$ for fixed $\mathbf{x}$ and t are independent and have the same distribution function, given by $F_{\mathbf{x},t}(r) = \mathbf{P}(R_n \leq r|\mathbf{X}_n = \mathbf{x}, T_n = t)$; that is, **A2** holds, and from Chap. 2 we know that the data in $J_{\mathbf{x},t}$ allow us to estimate $F_{\mathbf{x},t}$ for varying $(\mathbf{x}, t)$ and therefore to study the effect of the treatment on the response. Similarly, the pairs (T_n, R_n) of the units in $I_\mathbf{x} := \{n : \mathbf{X}_n = \mathbf{x}\}$ are independent and have the same distribution, which can be estimated from that subsample. In order to try to detect serious violations of the assumptions and prevent consequent biases in the conclusions one can examine the responses and the pairs of treatments and responses within and across the $J_{\mathbf{x},t}$ and $I_\mathbf{x}$ in a number of ways.

Thus, one may start by splitting $J_{\mathbf{x},t}$ into two or more subsets by stratifying the data in it on a variable not in $(\mathbf{X}_n, T_n, R_n)$ and comparing the empirical distributions of the responses in the resulting subsamples, visually or formally by means of tests and their p-values, and combining the evidence gathered from the different $(\mathbf{x}, t)$ (for example, by examining probability plots of the p-values, which are expected to look like a sample of uniforms). Typically, the index n of the datum $(\mathbf{X}_n, T_n, R_n)$ can be made to correspond to a variable not in $\mathbf{X}_n$, such as the date of its collection, the geographical coordinates associated with it, or the order by which it was inserted in a data file, so this ought to be possible. If there are two types of indexing (say year and month of collection), then various splits can be made by using one index at a time, or one index for a first split and another index for splitting each set that results from the first. Differences between the distributions of the various subsamples thus obtained indicate the existence of variables not in $\mathbf{X}_n$ which are predictive of the response. This does not necessarily bias the study of treatment effect, but it may: for example, if there is a covariate invisible to us which is correlated with the indices of the observations as well as with the responses and whose distribution within $J_{\mathbf{x},t}$ differs much from that within $J_{\mathbf{x},t'}$ for $t' \neq t$, then, even if the treatment effect is zero, its estimate will tend to say it is not.

Similar examinations can be made by splitting the subsamples $I_{\mathbf{x}}$ into subsets and comparing distributions of treatments and responses from different subsets. Since

$$\mathbf{P}(R_n \leq r, T_n = t|\mathbf{X}_n = \mathbf{x}) = F_{\mathbf{x},t}(r)\mathbf{P}(T_n = t|\mathbf{X}_n = \mathbf{x}),$$

it may be enough, following the first examinations, to compare the distributions of treatments.

Prob. 5.13 describes another, indirect way of detecting departures from the basic model; it involves a response which is correlated with R_n and is also a function of $\mathbf{X}_n$ but not of T_n.

If the data have passed these tests (which are meant to be mostly informal) one may then compare the distributions of responses from $J_{\mathbf{x},t}$ with those from $J_{\mathbf{x},t'}$. If there is no evidence of differences in treatment over the whole range of $\mathbf{x}$ and yet $F_{\mathbf{x},t}$ varies with $\mathbf{x}$ in an intelligible manner, then it does not necessarily follow that we have accounted for all the covariates and that the treatment effect is practically zero since, as we know from Simpson's paradox, that could be the result of there being a confounder missing in $\mathbf{X}_n$; but the fortuitousness of such a cancellation with a large data set can be forceful and informative regarding the existence of treatment effect. An even clearer picture will emerge if $F_{\mathbf{x},t}$ is found to be a special distribution function characterized by one or two parameters depending on $\mathbf{x}$ and not on t; in fact, we think more generally that, having gone through the verifications suggested above, the more 'felicitous' the distribution functions $F_{\mathbf{x},t}$ appear, the stronger the evidence provided by the corresponding estimates of treatment effect.

To explain this, we recall that the normal, log-normal, exponential, Weibull, Poisson and other distributions occur in a variety of probability models and each of them is characterized by more or less unique, remarkable properties, among which are properties of 'maximum entropy' under constraints on the support and moments (e.g. Galambos and Kotz [38], Rao and Shanbhag [78]). For example, it is often possible to argue that a positive variable is the product of a large number of more or less dependent variables, which according to central limit theorems entails that the distribution of its logarithm is approximately normal. In an application of the basic model, it may sometimes even be reasonable to expect the logarithm R_n of a positive response to satisfy

$$R_n = \rho_1(\mathbf{X}_n, T_n) + \rho_2(V_n, \mathbf{X}_n),$$

where conditionally on $(\mathbf{X}_n, T_n) = (\mathbf{x}, t)$ the term on the right is the sum of the constant $\rho_1(\mathbf{x}, t)$ and an approximately normal random variable with mean zero and variance that may depend on $\mathbf{x}$ but not on t. This expectation can be tested with data, as many times as there are usable subsamples; and, despite the chance of rejecting a hypothesis being always < 1, the verification of approximate normality over the whole range of $(\mathbf{x}, t)$ must count for something. It will certainly be significant if the variance of the normal distribution appears constant in $(\mathbf{x}, t)$, the conditional mean $\rho_1(\mathbf{x}, t)$ being the only thing to vary; for this will strengthen the belief that by stratifying on the covariates at our disposal we have basically succeeded in representing the response as the sum of a predictable and a purely random component (indeed the 'most random' of all variables on the real line with given mean and variance). Of course, the assumption of normality of the responses within strata with constant variance throughout can be tested very stringently: for instance, it implies that $\hat{\sigma}^2_{\mathbf{x},t}/\hat{\sigma}^2_{\mathbf{x}',t'}$ (see Sect. 5.3 for the notation) has an F distribution for $\mathbf{x} \neq \mathbf{x}'$ or $t \neq t'$, and hence that applying the appropriate distribution functions to these ratios for distinct pairs of $(\mathbf{x}, t)$ and $(\mathbf{x}', t')$ yields a random sample of uniforms.

Similar examinations apply when the responses are expected to follow other distributions, such as the Poisson in the case of 'rare events', the exponential or the Weibull distribution in the case of 'waiting times', etc. The Bernoulli distribution has too little structure and cannot be expected to provide much strength of evidence, except perhaps when the 'probability of success' depends on $\mathbf{x}$ but not on t.

When the $\mathbf{X}_n$ are not discrete, the condition defining the $I_\mathbf{x}$ and $J_{\mathbf{x},t}$ involved in these procedures must have $\mathbf{X}_n = \mathbf{x}$ replaced by $\mathbf{X}_n \in \mathcal{N}_\mathbf{x}$ for some neighbourhood $\mathcal{N}_\mathbf{x}$ of $\mathbf{x}$, say

$$\mathcal{N}_\mathbf{x} =]x_1 - \varepsilon, x_1 + \varepsilon] \times \{x_2\} \times \cdots \times]x_{d-1} - \varepsilon, x_{d-1} + \varepsilon] \times \{x_d\}$$

when $\mathcal{X} = \mathbb{R}^d$, X_2 and X_d are discrete and the other coordinates of $\mathbf{X}_n$ are continuous, and then the covariates of the units in each of them may take different values. In this case we know that even if $\mathbf{P}(T_i = t_i, i \in I | \mathbf{X}_i \in \mathcal{N}_\mathbf{x}, i \in I)$ approaches $f_I^{(\mathbf{x})}((t_i)_{i \in I})$ as $\mathcal{N}_\mathbf{x}$ shrinks to $\mathbf{x}$, the versions of **A1** and **A2** with $\mathbf{X}_i \in \mathcal{N}_\mathbf{x}$ in place of $\mathbf{X}_i = \mathbf{x}$ normally do not hold exactly, and then the tests and estimates of Sects. 5.2–5.3 are somewhat biased and depend on the particular stratification used. Finer stratifications—those obtained by splitting the range of the covariates into many short intervals or singletons—conform better to the assumptions and therefore reduce the bias in the tests and estimates, but they also reduce the numbers of units in $J_{\mathbf{x},t}$ and hence the power of the tests and the efficiency of estimates. Conversely, coarser stratifications increase the power of rejecting the null at the cost of a greater chance of rejecting it incorrectly, and they also reduce the variance of estimates. Thus, as in many statistical procedures, there is a trade-off between bias and variance—in this case the result of a more or less imperfect stratification—which necessitates ways of detecting serious violations of the assumptions.

While checking the plausibility of **A2** is done essentially as described above in the case of discrete $\mathbf{X}_n$, checking whether the probabilities of **A1** can be approximated by those with $\mathbf{X}_i \in \mathcal{N}_\mathbf{x}$ in place of $\mathbf{X}_i = \mathbf{x}$, where $\mathcal{N}_\mathbf{x}$ corresponds to a particular stratification, may be based on the observation that if

$$\mathbf{P}(T_i = t_i, i \in I | \mathbf{X}_i \in \mathcal{N}_\mathbf{x}, i \in I) \approx f_I^{(\mathbf{x})}((t_i)_{i \in I}) \tag{5.43}$$

for 'small' $\mathcal{N}_{\mathbf{x}}$, then the distribution of the covariates conditionally on their being in $\mathcal{N}_{\mathbf{x}}$ is 'balanced' across different values of treatment (Prob. 5.14), i.e. it is about the same for the units in $J_{\mathbf{x},t}$ as for those in $J_{\mathbf{x},t'}$. Consequently, at least when the $(\mathbf{X}_n, T_n)$ in $I_{\mathbf{x}}$ are approximately a random sample, one should be able to judge the suitability of (5.43) by checking the approximate equality of the distributions of the covariates of units in $J_{\mathbf{x},t}$ with different values of t, for example by examining scatter plots of pairs of averages of the covariates from units in $J_{\mathbf{x},t}$ and of those from units in $J_{\mathbf{x},t'}$, the expectation being that the pairs corresponding to the various $\mathbf{x}$ should approach the diagonal line whenever the stratification is neither too fine nor too coarse. Evidently, like much else in statistical analyses, this is a procedure that relies on intuition and offers no ascertainable results.

The tests suggested here presuppose large data sets; they are taxing, perhaps even more so to the models than to statisticians, and in performing them one must allow for some imperfections (e.g. 'outliers', formal rejection of an approximate model which nevertheless should give quite accurate estimates); but they appear to be essential in any observational study based on the methods treated in this book.

5.6 Notes

Sect. 5.2 is based on Rosenbaum [83] and Ferreira [31]. The derivation of the test of Subsect. 5.2.1 is an example of the classical method of deriving tests by conditioning on sufficient statistics (the frequencies of (5.8) in our case); see, for example, pp. 145–7 of Silvey [99]. The inner products taken over strata/matched sets considered in Subsect. 5.2.2 are called 'sum statistics' by Rosenbaum ([83], p. 35); with binary responses they are known as the Mantel-Haenszel statistic (Rosenbaum [83], p. 31).

Although using inner products taken over strata to test for the effect of treatment may not be appropriate when the direction of the effect varies with the stratum, in many real-life problems the treatment is expected to work in one direction. One can always resort to a method (e.g. Fisher's; see Rice [79], p. 315) of combining the tests from different strata into a single test, but the situations in which such methods are superior to a test based on the inner product statistic ought to be quite restricted.

For the origin of the Cochran-Mantel-Haenszel statistic, described at the end of Subsect. 5.2.2, see pp. 231–3 and 295–6 of Agresti [1].

The idea of combining statistics on various responses into a single statistic with the purpose of testing more elaborate hypotheses, described in Sect. 5.2.3, is taken from Rosenbaum ([83], Sects. 9.1-2).

Subsect. 5.2.4 and Probs. 5.5, 5.6 are based on Ferreira [32] and Rosenbaum [83].

As far as the methods of Sect. 5.2 are concerned, our strata and matched sets correspond to the 'blocks' of experimental research (e.g. Box, Hunter and Hunter [12]), i.e. groups of units which are homogeneous with respect to characteristics other than the treatments of interest. Evidently, the methods described in this chapter apply to data from experiments, and 'correcting' observational data for confounding is an attempt to treat them as experimental data.

While stratifying a random sample on a set of covariates is more or less of a definite procedure, creating matched sets from a data set (not necessarily a random sample) is something more tentative: at least when some of the covariates are continuous one may consider the degree of 'similarity' (with respect to those covariates) required of the units to be matched, whether it is best to match one treated unit to several control units, several treated units to several control units, etc., and any particular combination of criteria for matching entails discarding a greater or smaller portion of the data and some loss or gain of statistical efficiency; for methods of creating matched sets when the treatment variable is binary see Chap. 10 of Rosenbaum [83].

The confidence intervals of Sect. 5.3 can be turned into tests of the null hypothesis within strata and the results of these independent tests may also be combined into a single test; alternatively, the simultaneous confidence regions provide a method of testing multiple hypotheses over strata which controls approximately the probability of at least one incorrect rejection. The workings and bias of the alternative estimators mentioned in Remark 5.2 are illustrated in Sect. 4.1 of Ferreira [31].

The method of propensity scores (Sect. 5.4) is due to Rosenbaum and Rubin [84]; a recent account of it is given by Imbens and Rubin [46]. Our exposition is based on Ferreira [31].

The examinations proposed in Sect. 5.5 have similarities with the 'diagnostic techniques' of regression analysis (e.g. Daniel and Wood [21], Jørgensen [49]), but the latter rely partly on 'fitted values' and 'residuals', which often look normally distributed even when the model that is fitted is wrong.

5.7 Problems

5.1 Suppose that too large and too small values of a discrete test statistic S provide evidence against a null hypothesis, as is the case with any of the inner products of Sect. 5.2. Let the range of S be $\Sigma = \{s_1, s_2, \ldots, s_M\}$ and its probability function under the null $\Pi = \{p_1, p_2, \ldots, p_M\}$. Say that a test of the null based on S is 'equal-tailed' if it rejects the null at level $\alpha \in]0, 1[$ whenever S falls in the region

$$]-\infty, s'_\alpha] \cup [s''_\alpha, \infty[,$$

where $s'_\alpha = \max\{s \in \Sigma : \mathbf{P}(S_0 \leq s) \leq \alpha/2\}$, $s''_\alpha = \min\{s \in \Sigma : \mathbf{P}(S_0 \geq s) \leq \alpha/2\}$ and S_0 is a random variable independent of S and with probability function Π. Explain why it makes sense to take the p-value of a two-sided test based on S as

$$p(S) := 2 \cdot \min\{\mathbf{P}(S_0 \leq S|S), \mathbf{P}(S_0 \geq S|S)\} = 2 \cdot \min\left\{\sum_{m:s_m \leq S} p_m, \sum_{m:s_m \geq S} p_m\right\}.$$

Show that $\mathbf{P}[p(S) \leq u] \leq u$ for $0 < u < 1$. Justify the procedure mentioned in the first paragraph of Subsect. 5.2.2 of testing K null hypotheses based on K independent statistics of the same sort as S.

5.2 Let $X_1, X_2, \ldots, X_m$ and $Y_1, Y_2, \ldots, Y_n$ be two independent random samples with means μ and ν and variances σ^2 and τ^2, respectively, let $\hat{\mu}_m$, $\hat{\nu}_n$ be the sample averages computed from them, and $\hat{\sigma}^2_m$, $\hat{\tau}^2_n$ the sample variances. Use the central limit theorem and Slutsky's theorem (e.g. Ferguson [30]) to show that as $m, n \to \infty$ in such a way that $m/n \to c \neq 0, \infty$ the random variable

$$Z_{m,n}(\mu, \nu) := \frac{\hat{\mu}_m - \hat{\nu}_n - (\mu - \nu)}{\sqrt{\hat{\sigma}^2_m/m + \hat{\tau}^2_n/n}}$$

converges in distribution to a standard normal random variable, while

$$T_{m,n}(\mu, \nu) := \frac{\hat{\mu}_m - \hat{\nu}_n - (\mu - \nu)}{\sqrt{v^2_{m,n}\left(\frac{1}{m} + \frac{1}{n}\right)}},$$

where

$$v^2_{m,n} := \frac{m-1}{m+n-2}\hat{\sigma}^2_m + \frac{n-1}{m+n-2}\hat{\tau}^2_n,$$

converges in distribution to a standard normal random variable times a constant which is $\neq 1$ unless $c = 1$ or $\sigma^2 = \tau^2$.[8]

5.3 Derive the formulas for the mean and variance of inner product statistics S of (5.12) under the null, given in (5.14), Remark 5.1, in the following steps:

(a) Show that, because the permutations in different strata are independent and the null distribution is obtained by conditioning on the responses, treatments and strata, it is enough to consider a generic stratum of size n, omitting the index of stratum throughout, and calculate the mean and variance of

$$\mathcal{S} := \sum_{i=1}^{n} R_{\Pi(i)} T_i,$$

where $(\Pi(1), \Pi(2), \ldots, \Pi(n))$ is a random permutation of $(1, 2, \ldots, n)$ and the R_i and T_i are fixed numbers.

(b) Show that $\mathbf{P}[\Pi(i) = p] = 1/n$ for $p = 1, \ldots, n$,

$$\mathbf{E}[R_{\Pi(i)} T_i] = T_i\, \mathbf{E}[R_{\Pi(i)}] =: T_i \bar{R},$$

and hence that $\mathbf{E}(S|\mathcal{D}) = \mathbf{E}(\mathcal{S}) = n\bar{T}\bar{R}$.

(c) Show that

$$\mathrm{Var}[R_{\Pi(i)} T_i] = T_i^2 \frac{1}{n} \sum\nolimits_{p=1}^{n} (R_p - \bar{R})^2 =: T_i^2 v^2.$$

(d) Show that if $i \neq j$, $p \neq q$, then

---

[8] This is based on pp. 208–9 of Lehmann [55].

$$\mathbf{P}[\Pi(i) = p, \Pi(j) = q] = \frac{1}{n(n-1)},$$

$$\mathbf{E}[T_i R_{\Pi(i)} T_j R_{\Pi(j)}] = \frac{T_i T_j}{n(n-1)} \sum\nolimits_{p \neq q} R_p R_q = T_i T_j \bar{R}^2 - \frac{T_i T_j}{n-1} v^2,$$

so

$$\mathrm{Cov}[R_{\Pi(i)} T_i, R_{\Pi(j)} T_j] = -\frac{T_i T_j}{n-1} v^2.$$

(e) Show that

$$\mathrm{Var}(S|\mathcal{D}) = \mathrm{Var}(\mathcal{S}) = \sum_{i,j} \mathrm{Cov}[R_{\Pi(i)} T_i, R_{\Pi(j)} T_j]$$
$$= \frac{n^2}{n-1} v^2 \frac{1}{n} \sum\nolimits_{i=1}^{n} (T_i - \bar{T})^2 =: \frac{n^2}{n-1} v^2 V^2.$$

5.4 Prove equations (5.16)–(5.18) using the sampling schemes defined in connection with them:[9]

(a) Using $\mathbf{P}^*$, $\mathbf{E}^*$, Var*, etc., for **P**, **E**, Var, etc., conditionally on $\mathcal{D}$, note that the distribution of N_{ij} conditionally on $\mathcal{D}$ is the same as that of

$$N_{ij}^* := \sum\nolimits_{m=1}^{N_{\cdot j}} \mathbf{1}_{\{R_m^* = i\}},$$

where $R_1^*, \ldots, R_{N_{\cdot j}}^*$ are the responses that receive the $N_{\cdot j}$ treatment labels j among the responses drawn from the population consisting of $N_{i\cdot}$ responses equal to i and $N - N_{i\cdot}$ responses different from i. Then show that

$$\mathbf{P}^*(R_m^* = i) = \frac{N_{i\cdot}}{N}, \quad \mathrm{Var}^*(\mathbf{1}_{\{R_m^* = i\}}) = \frac{N_{i\cdot}(N - N_{i\cdot})}{N^2},$$

that for $m \neq n$

$$\mathbf{P}^*(R_m^* = i, R_n^* = i) = \frac{N_{i\cdot}(N_{i\cdot} - 1)}{N(N-1)}, \quad \mathrm{Cov}^*(\mathbf{1}_{\{R_m^* = i\}}, \mathbf{1}_{\{R_n^* = i\}}) = -\frac{N_{i\cdot}(N - N_{i\cdot})}{N^2(N-1)},$$

and deduce (5.16) from

$$\mathrm{Var}(N_{ij}|\mathcal{D}) \equiv \mathrm{Var}^*(N_{ij}^*) = \sum\nolimits_{m,n=1}^{N_{\cdot j}} \mathrm{Cov}^*(\mathbf{1}_{\{R_m^* = i\}}, \mathbf{1}_{\{R_n^* = i\}}).$$

(b) With the notation and sampling scheme of **(a)**, the joint distribution of N_{ij} and N_{kj} $(i \neq k)$ is the same as the joint distribution of N_{ij}^* and

$$N_{kj}^* := \sum\nolimits_{m=1}^{N_{\cdot j}} \mathbf{1}_{\{R_m^* = k\}}.$$

Show that for $m \neq n$

[9] Calculations of this sort, based on random variables rather than on probability functions, were made popular by W. Feller in his introduction to probability (cf. pp. 232–33 of [29]).

$$\mathbf{P}^*(R_m^* = i, R_n^* = k) = \frac{N_{i\cdot}N_{k\cdot}}{N(N-1)}, \quad \mathrm{Cov}^*(\mathbf{1}_{\{R_m^*=i\}}, \mathbf{1}_{\{R_n^*=k\}}) = \frac{N_{i\cdot}N_{k\cdot}}{N^2(N-1)},$$

find $\mathrm{Cov}^*(\mathbf{1}_{\{R_m^*=i\}}, \mathbf{1}_{\{R_m^*=k\}})$, and deduce (5.17) from

$$\mathrm{Cov}(N_{ij}, N_{kj}|\mathcal{D}) \equiv \mathrm{Cov}^*(N_{ij}^*, N_{kj}^*) = \sum_{m,n=1}^{N_{\cdot j}} \mathrm{Cov}^*(\mathbf{1}_{\{R_m^*=i\}}, \mathbf{1}_{\{R_n^*=k\}}).$$

(c) When $i \neq k$ and $j \neq l$, the joint distribution of N_{ij} and N_{kl} is the same as the joint distribution of

$$N_{ij}^* := \sum_{m=1}^{N_{\cdot j}} \mathbf{1}_{\{R_m^*(j)=i\}} \quad \text{and} \quad N_{kl}^* := \sum_{n=1}^{N_{\cdot l}} \mathbf{1}_{\{R_n^*(l)=k\}},$$

where $R_1^*(j), \ldots, R_{N_{\cdot j}}^*(j)$ are the responses that receive the $N_{\cdot j}$ treatment labels j and the $R_1^*(l), \ldots, R_{N_{\cdot l}}^*(l)$ those that receive the $N_{\cdot l}$ treatment labels l among the responses drawn from the population consisting of $N_{i\cdot}$ responses equal to i, $N_{k\cdot}$ equal to k, and $N - N_{i\cdot} - N_{k\cdot}$ different from i and from k.

Show that for all m, n

$$\mathbf{P}^*(R_m^*(j) = i, R_n^*(l) = k) = \frac{N_{i\cdot}N_{k\cdot}}{N(N-1)},$$

$$\mathrm{Cov}^*(\mathbf{1}_{\{R_m^*(j)=i\}}, \mathbf{1}_{\{R_n^*(l)=k\}}) = \frac{N_{i\cdot}N_{k\cdot}}{N^2(N-1)},$$

and deduce (5.18) from

$$\mathrm{Cov}(N_{ij}, N_{kl}|\mathcal{D}) \equiv \mathrm{Cov}^*(N_{ij}^*, N_{kl}^*) = \sum_{m=1}^{N_{\cdot j}} \sum_{n=1}^{N_{\cdot l}} \mathrm{Cov}^*(\mathbf{1}_{\{R_m^*(j)=i\}}, \mathbf{1}_{\{R_n^*(l)=k\}}).$$

5.5 **(a)** Show that if the (B_j, e_j) of (5.19) are allowed to have an arbitrary distribution then the model of Subsect. 5.2.4 for a single stratum amounts to a version of the basic model satisfying **A0** and **A1** and having binary treatments with the same distribution as that of n random draws without replacement from a set of m 1s and a $n - m$ 0s. *Suggestion*: Let $(U_1, \ldots, U_n)$ and $(V_1, \ldots, V_n)$ be independent vectors of independent standard uniform random variables, use the first vector to define the T_j, for instance by

$$T_1 = \mathbf{1}_{\{U_1 \leq \frac{m}{n}\}}, \quad T_{j+1} = \begin{cases} 1 & \text{if } U_{j+1} \leq \frac{m - \sum_{i=1}^{j} T_i}{n-j}, \\ 0 & \text{otherwise,} \end{cases}$$

and the second vector to define the (B_j, e_j) at will. On the other hand, if ρ and the T_j are given then

$$\rho(w, T_j) = \rho(w, 0) + [\rho(w, 1) - \rho(w, 0)]\, T_j \quad (0 \leq w \leq 1).$$

(b) Prove the identities (5.20)–(5.21).
(c) Apply Jensen's inequality to $x \to \Phi(z\sqrt{x})$ for $z > 0$ to show that

$$\mathbf{E}\,\Phi\left(z\sqrt{\frac{\hat{v}(\mathbf{B})}{v(\mathbf{B})}}\right) < \Phi\left(z\sqrt{\mathbf{E}\frac{\hat{v}(\mathbf{B})}{v(\mathbf{B})}}\right).$$

Using $v(\mathbf{B}) = p\,v(\mathbf{b}_1) + q\,v(\mathbf{b}_0) + p\,q\,(\bar{\mathbf{b}}_1 - \bar{\mathbf{b}}_0)^2$, show that if the B_j are a random sample from a normal distribution we have, taking $\hat{v}(\mathbf{B}) = v(\mathbf{b}_0)$,

$$\frac{\hat{v}(\mathbf{B})}{v(\mathbf{B})} = \frac{n}{n-m}\,\frac{X_{(n-m-1)}}{X_{(n-m-1)} + X_{(m-1)} + X_{(1)}} =: \frac{n}{n-m}\mathcal{B}_{(n-m-1)/2,m/2},$$

where $X_{(n-m-1)}$, $X_{(m-1)}$ and $X_{(1)}$ are independent and have chi-square distributions, and $\mathcal{B}_{(n-m-1)/2,m/2}$ has the beta distribution with parameters $(n-m-1)/2$ and $m/2$.[10] Conclude that

$$\mathbf{E}\frac{\hat{v}(\mathbf{B})}{v(\mathbf{B})} = \frac{n}{n-m}\mathbf{E}\,\mathcal{B}_{\frac{n-m-1}{2},\frac{m}{2}} < \frac{n}{n-m}\,\frac{n-m-1}{n-1} < 1$$

for all $n > n-m \geq 1$, and hence that for arbitrarily large n the (unconditional) coverage of the interval based on (5.22) with $v(\mathbf{B})$ replaced by $\hat{v}(\mathbf{B})$ tends to be less than the nominal level.

(d) Carry out a simulation study of the coverage and length of the conservative intervals of (5.28) and (5.31) and of the ideal but unobservable intervals of (5.24) and (5.30).[11]

5.6 Coats and Walter [16] compared the rates of death as the result of a fall or jump into the railway track in two types of underground stations, one type equipped with a 'drainage pit' under the tracks which should help the victim escape death, the other without such a pit. The pits were introduced for engineering reasons, not for reasons related to the likelihood of an accident in a station, and the general population ignored their existence, so it is plausible that the assignment of 'treatments' (provision of the station with a pit) in this study was close to being random. The data consist of the numbers of accidents and deaths resulting from 53 accidents per type of station: in stations with a pit there were 32 accidents of which 14 resulted in death, and in stations without pit there were 21 accidents of which 16 resulted in death.[12]

In order to relate these data to model (5.19), let B_j be the indicator of death in the j-th accident in the (hypothetical) situation that the corresponding station does not possess a pit (indicated by $T_j = 0$), and let $e_j = -B_j E_j$ be the corresponding effect of the (hypothetical) pit, where E_j is a binary random variable determining whether or not the victim manages to escape. Thus, if $B_j = 0$ (no death without the pit) then $R_j = e_j = 0$, it being implicitly assumed that the presence of the pit would not in that case favour the occurrence of death, while if $T_j = 1$ and $B_j = 1$ then $R_j = 1$ if $E_j = 0$ and $R_j = 0$ if $E_j = 1$. This is a situation in which the effects tend to vary

[10] This is a well-known result relating a beta random variable with two independent chi-square random variables: see for example p. 246 and Prob. 18 (c), p. 267, of Mood, Graybill and Boes [59].

[11] For possible choices of distributions of the (B_j, e_j) see Sect. 5 of Ferreira [32].

[12] I am grateful to Prof. T.J. Coats for providing me with some information about his study.

from unit to unit and the treatment may or may not have an effect on a given unit; and one that serves to show how the effects can depend on the baseline values.

(a) The average treatment effect on the treated is $\bar{\mathbf{e}}_1 = -m^{-1}\sum_{j=1}^{m} T_j B_j E_j$ with $m = 32$ and $T_j = 1$ if and only if station j is endowed with a pit; estimate this from the data. Explain why $-\bar{\mathbf{e}}_1$ is the average number of lives saved by the pit in the 32 'treated' stations.

(b) Show that the liberal and conservative confidence intervals of nominal level 95% for $-\bar{\mathbf{e}}_1$ are $[0.078, 0.570]$ and $[0.053, 0.596]$, respectively, and conclude that the number of lives saved thanks to the presence of the pit should lie somewhere between 5 and 60%. Can this statement be generalized to stations elsewhere or to those same stations in the future, or is it restricted to the underground stations in London within a certain period of observation?[13]

(c) Show that $R_j \leq B_j$ for all j, so that

$$\bar{\mathbf{R}} = \frac{1}{n}\sum_{j=1}^{n} R_j \leq \frac{1}{n}\sum_{j=1}^{n} B_j = \bar{\mathbf{B}};$$

conclude that if $\bar{\mathbf{R}} > 1/2$ then

$$v(\mathbf{B}) \leq v(\mathbf{R}) = \bar{\mathbf{R}}(1 - \bar{\mathbf{R}}).$$

Show that in the study of Coats and Walker the right-hand side of the last inequality is the most conservative value of $v(\mathbf{B})$, and that if used in (5.22) in place of $v(\mathbf{B})$ it yields $[0.049, 0.600]$ as an interval of nominal confidence 95% for $-\bar{\mathbf{e}}_1$.

5.7 Justify the use of (5.36) as an approximation to the variance of the estimator $\hat{\boldsymbol{\epsilon}}^{(K)}_{t,t'}$ of Sect. 5.3 by deriving an approximation to $\mathrm{Cov}(\hat{\delta}_{\mathbf{x}}\hat{\omega}_K(\mathbf{x}), \hat{\delta}_{\mathbf{y}}\hat{\omega}_K(\mathbf{y})|E_K)$ for $\mathbf{x}, \mathbf{y} \in \mathcal{X}_K$ when N is 'large' and $\mathcal{X}$ is infinite:[14]

(i) Let $f : \mathbb{R}^d \to \mathbb{R}$ be continuously differentiable and for fixed $h, u \in \mathbb{R}^d$ put $\varphi(\theta) = f(u + \theta h)$. Note that $\varphi'(\theta) = \nabla f(u + \theta h) \cdot h$, the inner product of the vector h and the vector gradient ∇f (the row vector of partial derivatives of f) at the point $u + \theta h$, and verify that

$$f(u+h) - f(u) = \varphi(1) - \varphi(0) = \int_0^1 \varphi'(\theta)d\theta = \int_0^1 \nabla f(u+\theta h)d\theta \cdot h.$$

Apply this identity to the coordinates $g_1, g_2 : \mathbb{R}^d \to \mathbb{R}$ of $g = (g_1, g_2)$ to get

$$g(u+h)^{\mathrm{T}} - g(u)^{\mathrm{T}} \equiv \begin{bmatrix} g_1(u+h) - g_1(u) \\ g_2(u+h) - g_2(u) \end{bmatrix} = \int_0^1 \begin{bmatrix} \nabla g_1(u+\theta h) \\ \nabla g_2(u+\theta h) \end{bmatrix} d\theta\, h^{\mathrm{T}},$$

where the column vector h^{T} is the transpose of h regarded as a row vector, etc.

---

[13] For a slightly different method of obtaining a confidence interval in this problem see Sect. 5.5.5 of Rosenbaum [83].

[14] This is a long and rather involved problem which could be of interest to those concerned with the inaccuracy of the confidence intervals for the overall effect of the treatment defined in (5.34).

(ii) Recalling the notation involved in $v_2(\hat{\boldsymbol{\epsilon}}_{t,t'}^{(K)})$, take $u = (\delta_{\mathbf{x}}, \delta_{\mathbf{y}}, w_K(\mathbf{x}), w_K(\mathbf{y}))$ in the last identity and replace the h in it by

$$\hat{h} = (\hat{\delta}_{\mathbf{x}}, \hat{\delta}_{\mathbf{y}}, \hat{w}_K(\mathbf{x}), \hat{w}_K(\mathbf{y})) - u$$

to conclude that, to the extent that $\hat{h}$ is small and the gradients are approximately constant around u,

$$g(u+\hat{h})^{\mathrm{T}} - g(u)^{\mathrm{T}} \approx \begin{bmatrix} \nabla g_1(u) \\ \nabla g_2(u) \end{bmatrix} \hat{h}^{\mathrm{T}}.$$

(iii) Explain why for 'large' N the covariance matrix of the random vector on the left of this approximate identity is about

$$\begin{bmatrix} \nabla g_1(u) \\ \nabla g_2(u) \end{bmatrix} \mathrm{Cov}(\hat{h}^{\mathrm{T}} | E_K) \left[\nabla g_1(u)^{\mathrm{T}}\, \nabla g_2(u)^{\mathrm{T}}\right].$$

(iv) Show that for $\mathbf{x} \in \mathcal{X}_K$ we have $\mathbf{E}[\hat{\delta}_{\mathbf{x}}|E_K] = \delta_{\mathbf{x}}$, $\mathbf{E}[\hat{w}_K(\mathbf{x})|E_K] = w_K(\mathbf{x})$ and

$$\begin{aligned}\mathbf{E}[\hat{\delta}_{\mathbf{x}}\hat{w}_K(\mathbf{x})|E_K] &= \sum\nolimits_{m,m' \geq K} \mathbf{E}[\hat{\delta}_{\mathbf{x}}\mathbf{1}_{\{\#J_{\mathbf{x}}=m, \#J_{\mathbf{x}}=m'\}}\hat{w}_K(\mathbf{x})|E_K] \\ &= \sum\nolimits_{m,m' \geq K} \delta_{\mathbf{x}}\mathbf{E}[\mathbf{1}_{\{\#J_{\mathbf{x},t}=m, \#J_{\mathbf{x},t'}=m'\}}\hat{w}_K(\mathbf{x})|E_K] \\ &= \delta_{\mathbf{x}}\mathbf{E}[\hat{w}_K(\mathbf{x})|E_K],\end{aligned}$$

so that $\hat{\delta}_{\mathbf{x}}$ and $\hat{w}_K(\mathbf{x})$ are conditionally uncorrelated; that conditionally on E_K and for $\mathbf{x} \neq \mathbf{y}$ in $\mathcal{X}_K$ the variables $\hat{\delta}_{\mathbf{x}}$ and $\hat{\delta}_{\mathbf{y}}$ are independent; that for $\mathbf{x} \in \mathcal{X}_K$

$$\begin{aligned}\mathbf{E}[\hat{\mu}_{\mathbf{x},t}^2|E_K] &= \sum_{m \geq K} \mathbf{E}[\hat{\mu}_{\mathbf{x},t}^2|E_K \cap \{\#J_{\mathbf{x},t} = m\}]\mathbf{P}(\#J_{\mathbf{x},t} = m|E_K) \\ &= \sum_{m \geq K} \left(\frac{\sigma_{\mathbf{x},t}^2}{m} + \mu_{\mathbf{x},t}^2\right)\mathbf{P}(\#J_{\mathbf{x},t} = m|E_K),\end{aligned}$$

and thus, using $\mathbf{E}[\hat{\mu}_{\mathbf{x},t}|E_K] = \mu_{\mathbf{x},t}$, that for $\mathbf{x} \in \mathcal{X}_K$

$$\mathrm{Var}(\hat{\mu}_{\mathbf{x},t}|E_K) = \sigma_{\mathbf{x},t}^2 \mathbf{E}\left[\frac{1}{\#J_{\mathbf{x},t}}\middle| E_K\right] =: \frac{\sigma_{\mathbf{x},t}^2}{M_{\mathbf{x},t,K}}.$$

The expectation involved here, which we have defined as the reciprocal of the number $M_{\mathbf{x},t,K}$, is well defined because the conditioning event and $\mathbf{x} \in \mathcal{X}_K$ together imply $\#J_{\mathbf{x},t} \geq K$. Explain why $M_{\mathbf{x},t,K}$ can, at least in principle, be computed at each set of values postulated for the unknown parameters $\mathbf{P}(\mathbf{X} = \mathbf{x}, T = t)$, $\mathbf{x} \in \mathcal{X}, t \in \mathcal{T}$. How would you try to estimate the $M_{\mathbf{x},t,K}$? (*Hint:* Under the assumption that the unknown parameters take particular values, the $\#J_{\mathbf{x},t}$ have a multinomial distribution, upon which one can base the computations.) Explain why

$$\mathrm{Cov}\left(\hat{w}_K(\mathbf{x}), \hat{w}_K(\mathbf{y})|\, E_K\right)$$

too can in principle be computed, and consider how one could estimate it from data. Collecting these results, write a formal expression for the covariance matrix of $\hat{h}$.

(v) Taking $g(u_1, u_2, u_3, u_4) = (u_1u_3, u_2u_4)$ in **(i)**-**(iii)** and using the matrix obtained in **(iv)** obtain an approximation to $\mathrm{Cov}(\hat{\delta}_{\mathbf{x}}\hat{w}_K(\mathbf{x}), \hat{\delta}_{\mathbf{y}}\hat{w}_K(\mathbf{y})|E_K)$ and thus justify the approximation

$$\mathrm{Var}(\hat{\epsilon}^{(K)}_{t,t'}) \approx \sum_{\mathbf{x}\in\mathcal{X}_K} \left(\frac{\sigma^2_{\mathbf{x},t}}{M_{\mathbf{x},t,K}} + \frac{\sigma^2_{\mathbf{x},t'}}{M_{\mathbf{x},t',K}} \right) w_K(\mathbf{x})^2 + \delta^2_{\mathbf{x}}\mathrm{Var}\left(\hat{w}_K(\mathbf{x}) \middle| E_K \right) + \sum_{\mathbf{x},\mathbf{y}\in\mathcal{X}_K,\mathbf{x}\neq\mathbf{y}} \delta_{\mathbf{x}}\delta_{\mathbf{y}}\,\mathrm{Cov}\left(\hat{w}_K(\mathbf{x}), \hat{w}_K(\mathbf{y}) \middle| E_K \right). \tag{5.44}$$

(vi) In practice it is almost impossible to estimate the parameters on the right-hand side of (5.44), but some simplifications and plausible arguments suggest a further approximation which usually works well and, if anything, is a bit conservative. Justify the following statements:

By Jensen's inequality the reciprocal of $M_{\mathbf{x},t}$ is underestimated by the reciprocal of $\mathbf{E}[\#J_{\mathbf{x},t}|E_K]$, but this might be compensated by the latter being overestimated by the reciprocal of $\mathbf{E}[\#J_{\mathbf{x},t}]$ ($\mathbf{x} \in \mathcal{X}_K$), which suggests that

$$\frac{\sigma^2_{\mathbf{x},t}}{M_{\mathbf{x},t,K}} \approx \frac{\sigma^2_{\mathbf{x},t}}{\mathbf{E}[\#J_{\mathbf{x},t}]} \approx \frac{\hat{\sigma}^2_{\mathbf{x},t}}{m_{\mathbf{x},t}}$$

and motivates the substitution of the last ratio here for $\sigma^2_{\mathbf{x},t}/M_{\mathbf{x},t,K}$ in (5.44). This presupposes the availability of $\hat{\sigma}^2_{\mathbf{x},t}$, which is defined at least when $K \geq 2$.

Now, unless $\mathcal{X}$ is finite, the $I_{\mathbf{x}}$ tend to stabilize rather than becoming arbitrarily large with increasing N; this is especially the case when the $\mathbf{X}_i$ are continuous, since then the stratifications are supposed to be refined as N increases. (Should most of the $I_{\mathbf{x}}$ appear 'large' in a given computation one would be justified in using $v_1(\hat{\epsilon}^{(K)}_{t,t'})$ as an approximation.) So in the situation considered here it seems well to assume that the distribution of $\{\#I_{\mathbf{x}} : \mathbf{x} \in \mathcal{X}\}$ converges as $N \to \infty$. Use the moment generating function of $\{\#I_{\mathbf{x}} : \mathbf{x} \in \mathcal{X}\}$ (each subset of which has a multinomial distribution) to argue that if the $\#I_{\mathbf{x}}$ converge in distribution then they converge to independent Poisson random variables, say $\{P_{\mathbf{x}} : \mathbf{x} \in \mathcal{X}\}$, and that, even if some of the latter variables be degenerate, the very process of stratification implies that N/n_K approaches a constant $c > 0$ (which depends on how rapidly the particular stratifications are refined with increasing N). These observations suggest that the last term in (5.44) be neglected, that the approximate distribution of $(N/n_K) \cdot \#I_{\mathbf{x}}$ be assumed as that of $c \cdot P_{\mathbf{x}}$ conditional on $P_{\mathbf{x}} \geq K$, and hence that for $\mathbf{x} \in \mathcal{X}_K$

$$\mathrm{Var}\left(N \cdot \hat{w}_K(\mathbf{x}) \middle| E_K \right) = \mathrm{Var}\left(\frac{N}{n_K}\#I_{\mathbf{x}} \middle| E_K \right) \approx c^2\mathrm{Var}\left(P_{\mathbf{x}} | P_{\mathbf{x}} \geq K \right).$$

Because the probability function of $P_{\mathbf{x}}$ conditional on $P_{\mathbf{x}} \geq K$ is magnified relative to that of $P_{\mathbf{x}}$ while its support is slightly restricted, its variance is $\leq \mathrm{Var}(P_{\mathbf{x}})$; use this to argue that

$$\mathrm{Var}\,(\,\hat{w}_K(\mathbf{x})|\,E_K) \lessapprox \frac{1}{n_K^2}\mathrm{Var}(P_{\mathbf{x}}) = \frac{\mathbf{E}(P_{\mathbf{x}})}{n_K^2}$$

and that the approximate inequality here is tighter for smaller K. Finally, consider that the best estimate of $\mathbf{E}(P_{\mathbf{x}})$ based on $P_{\mathbf{x}}$ is $P_{\mathbf{x}}$ itself, and justify taking

$$\hat{\delta}_{\mathbf{x}}^2 \frac{m_{\mathbf{x}}}{n_K^2} = \hat{\delta}_{\mathbf{x}}^2 \frac{\hat{w}_K(\mathbf{x})}{n_K}$$

in place of the remaining unknown terms in (5.44). The addition of the factor $(1 - \hat{w}_K(\mathbf{x})) \approx 1$ next to $\hat{w}_K(\mathbf{x})$ provides a formal correspondence with the variance of a Bernoulli random variable, though the expression (5.36) finally obtained is generally an upper bound rather than an estimate of the variance sought.

5.8 Explain how, under the condition indicated at the end of Sect. 5.3, the subset of data from a case-control study which fall in $I_{\mathbf{x}}$ can be used to calculate an approximate confidence interval for the odds ratio $\mathrm{e}(\mathbf{x})$ of (4.22) when the numbers of diseased and nondiseased units in $I_{\mathbf{x}}$ are ‘large’. Explain how one can estimate the overall measure of the effect of exposure on disease indicated at the end of Sect. 4.6 and discuss the difficulties of obtaining approximate confidence intervals for it. *Suggestion:* Use the central limit theorem and Cramér's theorem (e.g. Ferguson [30]) to derive an approximation to the distribution of $\log \hat{\mathrm{e}}(\mathbf{x})$, where $\hat{\mathrm{e}}(\mathbf{x})$ is the empirical version of $\mathrm{e}(\mathbf{x})$.

5.9 People with a certain professional background are exposed to an agent suspected of causing a debilitating, usually deadly but rare disease. If D is the indicator of a generic individual from some population ever having the disease, S the individual's sex (1 if male, 0 if female), B the indicator of the individual's putatively hazardous background (equal to 1 if the individual ever takes up a profession entailing exposure to the agent and equal to 0 otherwise), M the ‘magnitude of exposure’ to the potentially harmful agent, then S has a certain distribution and

$$M := U_1 \cdot B, \quad B = \begin{cases} 1 & \text{if } U_0 \le p_S, \\ 0 & \text{otherwise,} \end{cases}$$

$$D = \begin{cases} 1 & \text{if } V \le B\frac{e^{\beta_{0,S}+\beta_{1,S}M}}{1+e^{\beta_{0,S}+\beta_{1,S}M}} + (1-B)\frac{e^{\beta_{0,S}}}{1+e^{\beta_{0,S}}} \equiv \frac{e^{\beta_{0,S}+\beta_{1,S}M}}{1+e^{\beta_{0,S}+\beta_{1,S}M}}, \\ 0 & \text{otherwise,} \end{cases}$$

where conditionally on S the random variables U_0, U_1 and V are independent and standard uniform, and $0 < p_0 = 1 - p_1 < 1$, $\beta_{0,0}$, $\beta_{0,1}$, $\beta_{1,0}$ and $\beta_{1,1}$ are numbers.

Then

$$\mathbf{P}(D = 1|B = b, M = m, S = s) = b\frac{e^{\beta_{0,s}+\beta_{1,s}m}}{1 + e^{\beta_{0,s}+\beta_{1,s}m}} + (1-b)\frac{e^{\beta_{0,s}}}{1 + e^{\beta_{0,s}}}$$

and

$$\mathbf{P}(D=1|B=b,S=s)=\frac{b}{\beta_{1,s}}\log\left(\frac{1+e^{\beta_{0,s}+\beta_{1,s}}}{1+e^{\beta_{0,s}}}\right)+(1-b)\frac{e^{\beta_{0,s}}}{1+e^{\beta_{0,s}}}$$

for $b, s = 0, 1$. (The second probability reduces to the first for $b = 0$, and for $b = 1$ it is obtained by integrating the first with respect to $\mathbf{P}(M \leq m|B = b, S = s)$.)

(a) There are two ways in which this model may be identified with a special case of the basic model with unconfoundedness, depending on what one regards as the treatment. Show that this is indeed so, and in each case identify the probability law that describes the causal effect of the treatment on the disease. Can you think of another response to the treatment besides disease status which could be interesting in a real-life situation?

(b) Write a computer program to simulate the sex, year of birth, background, time in service (in the hazardous profession), magnitude of exposure, disease status, age at diagnosis and year of death of a large number of people—say of 20 or 40 million people—born between 1918 and 2023, say, and in such a way that their sex, background, magnitude of exposure and disease status follow the present model and are consistent with their values on the other variables. Regard these data as a basis for a study of the effect of exposure on disease to be carried out in 2023.

The data are not realistic because they include the year of death and disease status of individuals born more recently, for most of whom such information would not have been known in 2023 (the disease status of those diagnosed after 2023 would not have been known, and neither would have been the disease status of anyone who never got the disease and was still alive in 2023); so they may be described more properly as a set of partial *life histories* of individuals born before 2023. One may, however, turn them into *realistic data* on the population at the time of observation by deleting the year of death and age at diagnosis whenever these would have been unknown by 2023, and replace the disease status of those not diagnosed with the disease up to 2023 by their, possibly provisional, status of non-diseased.

The life histories are essential for this exercise, so it is worth investing some time in simulating them. As indications of how they might look, one may say that in both sexes the histogram of age in 2023 could increase slightly up to 50 years and then decrease linearly to about 100 years; the histograms of lifetime (age at death) could peak between 80 and 90 years and have the shape of Makeham densities, whose distribution function is $F(y; \alpha, \beta, \epsilon) = 1-\exp\{-y\alpha - (\beta/\gamma)[e^{\gamma y} - 1]\}, \alpha, \beta, \gamma > 0$; that age at death among the diseased could be a decreasing convex function of magnitude of exposure; and that age at death among the diseased increases linearly with age at diagnosis.

(c) Simulate a random sample from the population of life histories using the program of **(b)**; what population does the sample represent, and can it be used to estimate a probability like $\mathbf{P}(D = d|S = s, B = b)$? What if you simulate a sample of the *realistic data*? And what if from this last sample you only use the data on those born earlier? Note the distinction between a completed life history and a life history 'in the making'. Estimate some of the probabilities from three types of samples: those drawn directly from the set of life histories, those made to look realistic, and those from the older individuals made to look realistic. Consider the accuracy and possible bias of the estimates.

(d) Although $\mathbf{P}(D = 1|S = s, B = 1)$ and $\mathbf{P}(D = 1|S = s, B = 0)$, which enable us to quantify the causal effect of the exposure to the agent on the development of the disease, can be estimated from a random sample drawn from the population of life histories, we know that if the disease is rare then with high probability at least one of their estimates will not be defined or will not be very informative unless the sample is very large. We also know that the odds ratio provides a characterization of the effect of background on disease which corresponds to the 'reverse, pseudo-effect' of disease on background,

$$\mathbf{e}(s) := \frac{O(D = 1|B = 1, S = s)}{O(D = 1|B = 0, S = s)} = \frac{O(B = 1|D = 1, S = s)}{O(B = 1|D = 0, S = s)},$$

so that it can be estimated from the probabilities $\mathbf{P}(B = b|D = d, S = s)$, and that these probabilities can be estimated by sampling directly from cases and then from controls matched to those cases rather than by sampling from the general population.

Extend your computer program to simulate realizations of a case-control study in which N cases are drawn randomly from the set of cases in the realistic data set and matched on sex and year of birth to a set of N controls drawn randomly from the realistic data set. Which estimates of the odds ratio do you think are more reliable—those based on all the cases and controls, those based on the cases and controls born earlier, or those based on the cases and controls born more recently?

5.10 **(a)** Verify in detail the statements on Berkson's paradox made in Sect. 4.6.1.

(b) In model (4.24) let $\mathbf{X}$ be one-dimensional and

$$\mathbf{P}(S = 1|\mathbf{X} = \mathbf{x}, T = t, R = r) = \frac{\exp\{\delta_1 + \delta_2\mathbf{x} + \delta_3 t + \delta_4 r + \delta_5 tr\}}{1 + \exp\{\delta_1 + \delta_2\mathbf{x} + \delta_3 t + \delta_4 r + \delta_5 tr\}};$$

verify that

$$\mathbf{P}(T = 1|\mathbf{X} = \mathbf{x}, R = r, S = 1) = \frac{\mathbf{P}(T = 1|\mathbf{X} = \mathbf{x})}{\mathbf{P}(T = 1) + \mathbf{P}(T = 0) f(r, \mathbf{x})}$$

with

$$f(r, \mathbf{x}) = \frac{\exp\{-\delta_3 - \delta_5 r\} + \exp\{\delta_1 + \delta_2\mathbf{x} + \delta_4 r\}}{1 + \exp\{\delta_1 + \delta_2\mathbf{x} + \delta_4 r\}},$$

and conclude that the data in the source are unbiased for purposes of studying the effect of T on R by means of the odds ratio if and only if $\delta_3 = \delta_5 = 0$ or $\delta_4 = \delta_5 = 0$.

Using this specification of $\mathbf{P}(S = 1|\mathbf{X} = \mathbf{x}, T = t, R = r)$ and any appropriate specifications of $\mathbf{P}(T = 1|\mathbf{X} = \mathbf{x})$, $\mathbf{P}(T' = 1|\mathbf{X} = \mathbf{x})$ and $\mathbf{P}(R = 1|\mathbf{X} = \mathbf{x}, T' = t)$, write a computer program to simulate data illustrating Berkson's paradox.

5.11 Consider the model from Prob. 5.9 anew with the difference that U_1 and V are dependent (but still independent of U_0). Can you think of a real-life situation for which this version of the model would be more realistic than the first? Under this version, could one use a variant of the method of instrumental variables to quantify and estimate the causal effect of treatment on response? *Suggestion:* Condition on S, write the model in the notation of Sect. 4.7 and consider the *numerator* of the parameter θ defined there.

5.12 **(a)** Derive the first three identities of Sect. 5.4.

(b) Derive identity (5.41).

(c) Show that

$$\mathbf{P}\begin{pmatrix} U_i \le u_i, V_i \le v_i, \lambda(\,\cdot\,;\mathbf{X}_j) = l_j, \\ i \in I, j \in I_j, j = 1, 2, \ldots, K \end{pmatrix} = \prod_{j=1}^{K} \mathbf{P}\big[U_i \le u_i, V_i \le v_i, \lambda(\,\cdot\,;\mathbf{X}_j) = l_j, i \in I_j\big].$$

Suggestion: It is enough to consider the case $K = 2$.

(d) Derive the balancing property of propensity scores.

5.13 Let $\mathbf{X}$, $T = \tau(U, \mathbf{X})$ and $R = \rho(V, \mathbf{X}, T)$ satisfy the basic model and let R' be another response which is correlated with R conditionally on $(\mathbf{X}, T)$ and satisfies $R' = \rho'(V', \mathbf{X})$ for some response function ρ' and V' standard uniform conditionally on $\mathbf{X}$. One can use the first part of these specifications to try to study the effect of a 'treatment' T—a vaccine, or exposure to a harmful substance—on a response R—contraction of an illness, or death from a certain disease within a certain period following treatment—of an individual randomly drawn from some population by 'correcting for' a vector $\mathbf{X}$—a summary of the individual's personal characteristics, medical history, lifestyle, resources (income, neighbourhood, home and household characteristics), education level, profession, etc., thought to make unconfoundedness valid to a good approximation. The last equation, in turn, may represent the effect of $\mathbf{X}$ on a second response R'—contraction of an illness other than that recorded by R, or death by a disease other than that recorded by R—thought to be unaffected by the treatment and not of direct interest but which may be used to detect serious violations of unconfoundedness.

(a) Explain why in general V and V' are dependent conditionally on $(\mathbf{X}, T)$.

(b) Show that if U and V' are independent conditionally on $\mathbf{X}$ then $\mathcal{L}(R'|\mathbf{X} = \mathbf{x}, T = t)$ is constant in t for all $\mathbf{x}$.

(c) By **(b)**, if $\mathcal{L}(R'|\mathbf{X} = \mathbf{x}, T = t_1) \neq \mathcal{L}(R'|\mathbf{X} = \mathbf{x}, T = t_2)$ for some $t_1 \neq t_2$ and some $\mathbf{x}$ then U and V' are dependent conditionally on $\mathbf{X}$; consequently U and V are generally dependent conditionally on $\mathbf{X}$, and thus unconfoundedness generally fails in the basic model: explain why this is so, and verify it by assuming that

$$U = \varphi_1(\xi_0, \xi_1, \mathbf{X}), \quad V' = \varphi_2(\xi_0, \xi_2, \mathbf{X}), \quad V = \varphi_3(\xi_2, \xi_3, \mathbf{X}),$$

where the φ_i are some functions and conditionally on $\mathbf{X}$ the ξ_i are independent standard uniforms. (Intuitively, information is transitive: if knowing something of U is knowing something of V' and knowing something of V' is knowing something of V then knowing something of U is knowing something of V.)[15]

[15] See Prob. 5.17 (a) for an application of this idea.

5.14 Suppose that for $\mathbf{x} \in \mathcal{X}$ and $\varepsilon > 0$ we can find a neighbourhood $\mathcal{N}_{\mathbf{x}}$ of $\mathbf{x}$ such that

$$\left| \frac{\mathbf{P}(T_i = t_i, i \in I | \mathbf{X}_i \in \mathcal{N}_{\mathbf{x}}, i \in I)}{\mathbf{P}(T_i = t_i, i \in I | \mathbf{X}_i \in A, i \in I)} - 1 \right| < \varepsilon$$

for every $A \subset \mathcal{N}_{\mathbf{x}}$. Show that

$$\left| \mathbf{P}(\mathbf{X}_i \in A, i \in I | \mathbf{X}_i \in \mathcal{N}_{\mathbf{x}}, i \in I) - \mathbf{P}(\mathbf{X}_i \in A, i \in I | \mathbf{X}_i \in \mathcal{N}_{\mathbf{x}}, T_i = t_i, i \in I) \right| < \varepsilon.$$

Conclude that if $\mathbf{P}(T_i = t_i, i \in I | \mathbf{X}_i \in \mathcal{N}_{\mathbf{x}}, i \in I)$ approaches $f_I^{(\mathbf{x})}((t_i)_{i \in I}) > 0$ as $\mathcal{N}_{\mathbf{x}}$ shrinks to $\mathbf{x}$ then the covariates with values in $\mathcal{N}_{\mathbf{x}}$ and different values of the treatment are approximately 'balanced'.

5.15 Use a computer program to simulate data from models (2.32) and (2.33) of Prob. 2.9 with the specifications of **(d)** and **(e)** and various choices of $\boldsymbol{\beta}$, ϱ, σ and θ. Fit linear regression models to data generated with and without unconfoundedness, examine the quality of the fit by 'diagnostic plots', and assess the performance of the least-squares estimators. When are the models fitted to the data the correct ones? Analyse the data by the method of stratification, studying the estimator $\hat{\epsilon}_{1,0}$ of the $\epsilon_{1,0}$ of (5.34). Does stratification yield correct estimates of the treatment effect when the linear regression model is incorrect? What if $\varrho \neq 0$? How much do the estimates depend on the particular stratification used? Assess the accuracy of the three estimators of the variance of $\hat{\epsilon}_{1,0}$ and that of the three confidence intervals for $\epsilon_{1,0}$.

5.16 Use a computer program to simulate data from the model of Prob. 2.10. Study the estimator of $\epsilon_{1,0}$ obtained by the method of stratification. Do you need to stratify on all the background variables? Does stratification typically yield correct estimates of the treatment effect? See what happens if you omit a couple of confounders in the stratification procedure. How much do the estimates depend on the particular stratification used? Assess the accuracy of the three estimators of the variance of $\hat{\epsilon}_{1,0}$ and that of the three confidence intervals for $\epsilon_{1,0}$.[16]

5.17 Freedman (pp. 1–5, 15 of [36]) gives a simplified description of the HIP (Health Insurance Plan, a group of medical practices) study on the effect of mammography—i.e. screening by X-rays—in reducing death by breast cancer in women, summarizing the results in the following table:[17]

	Group	Breast cancer		All other	
	size	No.	Rate	No.	Rate
Invitation to screening					
Screened	20,200	23	1.1	428	21
Refused	10,800	16	1.5	409	38
Total	31,000	39	1.3	837	27
Control	31,000	63	2.0	879	28

[16] For an illustration of the sort of numerical results one can get see Sect. 6 of Ferreira [31].

[17] Part of this problem requires ideas from Prob. 3.1 and Sect. 4.7.

Of the population of women insured by HIP, a sample of 31,000 was invited to four rounds of annual screening and another sample of 31,000 was kept as a control group. Of those invited to be screened, 20,200 accepted and 10,800 refused, but the study was such that no one in the control group demanded to be screened. After a five-year period deaths caused by breast cancer were counted in each group; numbers of deaths are shown in the table together with death rates expressed as number of deaths per 1000, and with numbers of deaths and rates of death from other causes.

One might expect the women who accepted the invitation to be healthier in general than those who did not; this is consistent with the rates of death from other causes being much lower among the screened than among the refusers and yet very similar in the total invited for screening and in the control group. So the effect of screening may not be properly quantified by comparing the death rate in the screened group with the death rate in the control group—1.1 with 2.0, which suggests that screening reduces death from breast cancer by about 44%. The effect of *invitation* to screening, on the other hand, should be appropriately quantified by comparing the death rates in the treatment and control groups—1.3 with 2.0, which suggests that invitation to screening reduces death from breast cancer by about 38%.[18] The last figure, however, is a somewhat diluted expression of the effect of screening because it compares death rates among women who may or may not have accepted the invitation: among women who would accept screening if offered to them, the reduction should be greater. Indeed, based on the acceptance rate one would expect about 10,800 women in the control group to have refused screening if offered, and among them about 16 deaths from breast cancer; so the number of deaths among women not invited to be screened but willing to be screened should be about 63 – 16, from which follows that the effect of screening on women willing to be screened should be to reduce the death rate of breast cancer by a proportion of about $1 - 23/(63 - 16) \approx 0.5$.[19]

Formalize this analysis with the elements of Prob. 3.1, letting I, T and R stand for the data on an arbitrary woman invited for the study, $I = 1$ for invitation to screening, $I = 0$ for no invitation to screening, $T = 1$ for actual screening, $T = 0$ for refusal of screening or no invitation to screening, $R = 1$ for death by breast cancer within the five-year period, $R = 0$ for survival, assuming that $A = 1 = P$ and that $I = 0 \Rightarrow T = 0$, ignoring the woman's background characteristics, and accordingly simplifying (3.20) to

$$\begin{cases} I = \iota(W), \\ T = \tau(U, I), \\ R = \rho(V, I, T), \end{cases} \tag{5.45}$$

with U, V and W standard uniform random variables and $\mathbf{P}(I = 1) = \mathbf{P}(I = 0)$:

(a) How would you estimate the effect of screening from the data in the table under the assumption that U, V and W are independent? Are there data which suggest

---

[18] This sort of analysis based on a suggestion of treatment is useful in general; it is known as 'intention-to-treat' analysis.

[19] This 'correction for dilution' is proposed by Freedman, Petitti and Robins [34], p. 44, who provide this and other analyses as well as many important details about the HIP study, which is not as simple as it may appear from this description.

that this assumption is unrealistic? Formalize your suspicion. *Suggestion*: Consider $\mathbf{P}(R=1|T=1)/\mathbf{P}(R=1|T=0)$, $\mathbf{P}(R=1|T=1,I=1)/\mathbf{P}(R=1|T=0,I=1)$. For the second part, use Prob. 5.13 with R' as death from causes other than breast cancer to argue that the independence of U and V is doubtful.

(b) Assuming that W is independent of (U,V), perform an 'intention-to-treat analysis', i.e. estimate the effect of the *invitation* to screening on the rate of death from breast cancer.

(c) Under the same assumption as in **(b)**, one can think of quantifying the effect of screening among women who are invited and accept to be screened by

$$\frac{\mathbf{P}(\rho(V,I,1)=1|I=1,T=1)}{\mathbf{P}(\rho(V,I,0)=1|I=1,T=1)}. \tag{5.46}$$

Interpret this ratio, show that the probability in the numerator pertains to observable events, that the one in the denominator does not, but that by assuming in addition

$$\mathbf{P}(\rho(V,1,0)=1)=\mathbf{P}(\rho(V,0,0)=1), \tag{5.47}$$

and using $\mathbf{P}(I=1)=\mathbf{P}(I=0)$ and $I=0\Rightarrow T=0$, the ratio can be written as

$$\frac{\mathbf{P}(R=1,I=1,T=1)}{\mathbf{P}(R=1,I=0,T=0)-\mathbf{P}(R=1,I=1,T=0)};$$

conclude from the data in the table that, under the assumptions stated, the effect of screening in women who are invited and accept to be screened is to reduce the rate of death from breast cancer by half, approximately. Could (5.47) be unrealistic? Suppose that $\mathbf{P}(\rho(V,1,0)=1)$ were a little smaller than $\mathbf{P}(\rho(V,0,0)=1)$; would your estimate for the reduction in the death rate be somewhat optimistic?

(d) Assume that ρ is constant in the second argument, so that I affects R only through T and we can write $\rho(V,T)$ in place of $\rho(V,I,T)$, and that, as in **(c)**, W and (U,V) are independent. Show that (5.45) is a special case of the model of Sect. 4.7. Noting that $I=0\Rightarrow T=0$ is equivalent to $\tau(u,0)=0\ \forall u$, show that

$$\theta := \frac{\mathbf{E}(R|I=1)-\mathbf{E}(R|I=0)}{\mathbf{E}(T|I=1)-\mathbf{E}(T|I=0)} = \mathbf{E}\left[\rho\left(V,1\right)-\rho\left(V,0\right)|\tau(U,1)=1\right];$$

interpret the parameter θ and use the data in the table to estimate it. Interpret $-\theta$ under the assumption that $\rho\left(v,1\right)-\rho\left(v,0\right)\le 0$ for all v.

Since θ depends on the rate of mortality, one may consider instead

$$\theta' := \frac{\mathbf{E}\left[\rho\left(V,1\right)-\rho\left(V,0\right)|\tau(U,1)=1\right]}{\mathbf{E}\left[\rho\left(V,0\right)|\tau(U,1)=1\right]}.$$

Show that the denominator of θ' can be written as

$$\frac{\mathbf{P}[\rho(V,0)=1]-\mathbf{P}\left[\rho(V,0)=1,\tau(U,1)=0\right]}{\mathbf{P}[\tau(U,1)=1]},$$

that each of the probabilities involved in this expression can be estimated from the data, and hence that an estimate of θ' is

$$\frac{39-63}{47} = -\frac{24}{47} \approx -0.51.$$

Compare this with the estimate of

$$\theta'' := \frac{\mathbf{E}[\rho(V,1)|I=1,T=1] - \mathbf{E}[\rho(V,0)|I=1,T=1]}{\mathbf{E}[\rho(V,0)|I=1,T=1]},$$

directly obtainable from that of the ratio (5.46) of (**c**), and explain the connection between the two estimates.

Suggestion: In connection with the interpretation and estimation of the denominator of θ', note that $\tau(U,1) = 1$ represents the event that a woman is a *complier*—i.e. she would accept screening if offered—and that in the control group one would expect to find about 20,200 compliers and 10,800 non-compliers. Since among the 10,800 non-compliers there should be about 16 deaths, of the 63 actual deaths in the control group about $63 - 16 = 47$ should have occurred among the compliers. So the probability of death among compliers who are not screened should be about 47/20,200. To relate the two estimates note that the independence of (U,V) and W implies

$$\mathcal{L}\left(f(V)\,|I=1,\tau(U,I)=1\right) = \mathcal{L}\left(f(V)\,|\tau(U,1)=1\right)$$

for any function f, that $I = 1 \wedge \tau(U,I) = 1 \Leftrightarrow I = 1 \wedge T = 1$, and hence that θ' and θ'' are identical.

5.18 Sagarin *et al.* [90] give a simplified description of a study to determine whether giving supplements of vitamin A to children helps to reduce child mortality: 23,682 children in northern Sumatra were randomly divided into a group of 12,094 children given vitamin A supplements and a control group of 11,588. At the end of a certain period mortality was compared in the two groups. Of the children in the treatment group 2,419 did not comply, i.e. their parents refused the supplements, but mortality among them was determined nevertheless. The results of the study are summarized in the table below, which includes mortality rates per 1000.

Carry out analyses of these data analogous to those indicated in Prob. 5.17. Note that if one allows for $\mathbf{P}(I=1) \neq \mathbf{P}(I=0)$ then the estimate of (**c**) requires a slight modification; although the study involved randomization, there seem to be too many children in the treatment group.

	Group size	Death No.	Death Rate
Assigned to receive vitamin A			
Accepted	9,675	12	1.24
Refused	2,419	34	14.06
Total	12,094	46	3.80
Control	11,588	74	6.39

5.19 A study by Raghavendra *et al.* [76] investigated the immediate effect of yoga breathing exercises on the pulmonary function of asthmatic patients. A group of 60 men and women diagnosed with asthma were trained 10 minutes a day during a week in the practice of *kapalabhati*, a yoga breathing technique, as a preparation for the study. After the preparatory week, on the day appointed for the experiment, they were divided randomly into two groups of 30: one, the treatment group, was asked to practice kapalabhati again during two periods of 10 minutes, and the other, the 'control' group, to practice the more basic technique of 'deep breathing' during the same periods. Before and immediately after practice, the patients had their forced expiration volume in the first second (FEV) and forced vital capacity (FVC, the total amount of air exhaled in forced complete exhalation) measured. Scatter plots of these data and box plots of the natural logarithms of the changes in them by treatment group are shown separately for men and women in Figs. 5.2 and 5.3. The change in an individual's FEV is the ratio of the individual's FEV after practice to the individual's FEV before practice; the definition of the change in an individual's FVC is analogous.[20]

The evidence for differences between the distributions of the measurements made before and after practice is strong in all groups except in the treated men, where the changes in FVC could be explained by chance. But while the changes in the treatment group indicate an improvement in lung function, in the control group the changes in FVC among men and the changes in FEV among women indicate a worsening. These observations may be supported by tests based on the inner product statistic with two observations per matched set, corresponding to the measurements made before and after practice on each individual, essentially as in Example 5.7, or by paired t-tests comparing the measurements or their logarithms before and after practice.

(a) Is it strictly necessary to stratify the samples on sex in order to test for the existence of an immediate effect of kapalabhati relative to that of deep breathing? Is there any advantage in doing so? Can you explain the contradictory results in the control group? (No explanation needs to be certain, even if based on detailed information about the trial.)

(b) Consider the following two models for describing the FEV or FVC data on an arbitrary individual participating in the trial:

$$R_1 = \rho(V_1, X, T), \quad R_0 = \rho(V_0, X, 0), \quad T = \tau(U),$$

and

$$R_1 = \rho_1(V_1, X, T), \quad R_0 = \rho_0(V_0, X), \quad T = \tau(U), \tag{5.48}$$

where τ, ρ, ρ_0 and ρ_1 are unknown functions, X is the individual's sex, U, V_0 and V_1 are independent standard uniform random variables conditionally on X, T is the group to which the individual is assigned—say equal to 1 if it is the kapalabhati group, equal to 0 if it is the deep breathing group—and R_0 and R_1 are the responses—i.e. the FEV or FVC measurements—before and after practice. Does any of these models seem appropriate?

[20] I am grateful to Prof. Apar Avinash Saoji for sending me the data from his study.

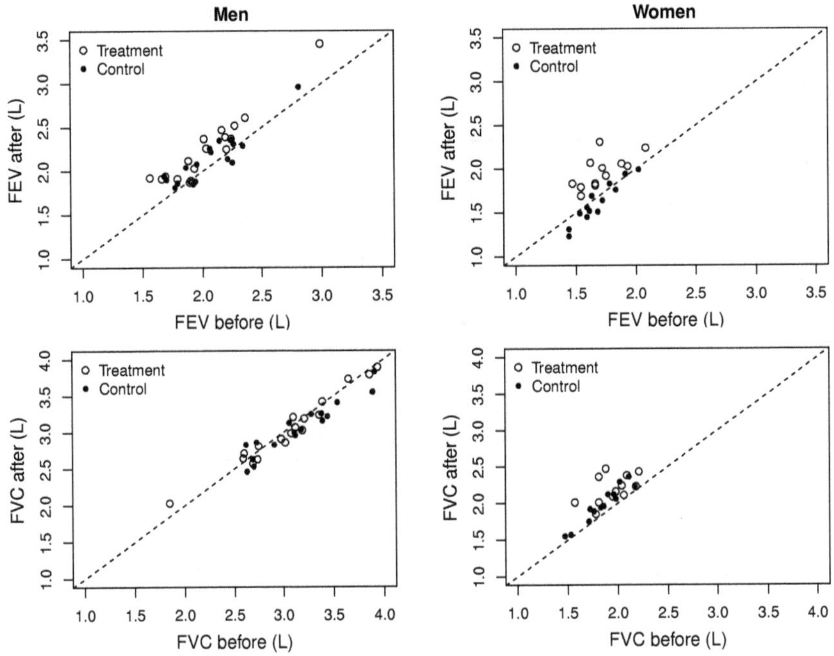

Fig. 5.2 Scatter plots of the FEV and FVC measurements before and after treatment.

As a special case of (5.48) consider

$$\rho_1(V_1, X, T) = e^{\varphi_1(V_1', X) + \psi(V_1'', X, T)}, \quad \rho_0(V_0, X) = e^{\varphi_0(V_0, X)}, \tag{5.49}$$

where the standard uniforms V_1' and V_1'' are made up of infinite disjoint sets of digits of V_1, φ_0, φ_1 and ψ are unknown functions, and φ_0 and φ_1 may or may not be equal. Interpret this specification; are the data consistent with the assumption that $\varphi_0 = \varphi_1$?

(c) Write down expressions for $\mathbf{E}[\log(R_1/R_0)|X = x, T = t]$ and

$$\mathbf{E}\big[\log(R_1/R_0)\big|X = x, T = 1\big] - \mathbf{E}\big[\log(R_1/R_0)\big|X = x, T = 0\big] \tag{5.50}$$

under the models suggested in **(b)**.

Which parameters (i.e. which theoretical quantities) would you use to quantify the immediate effects of practising kapalabhati and of practising deep breathing on FEV and FVC, and under which assumptions? Which parameters would you use to quantify the *difference* between the effects of practising kapalabhati and the effects of practising deep breathing—i.e. the effects of kapalabhati relative to those of deep breathing? Which of those parameters do you think could be *estimated* from the data? Under what theoretical conditions may the difference between the effects be considered definite rather than circumstantial, and which experimental conditions would make those conditions plausible?

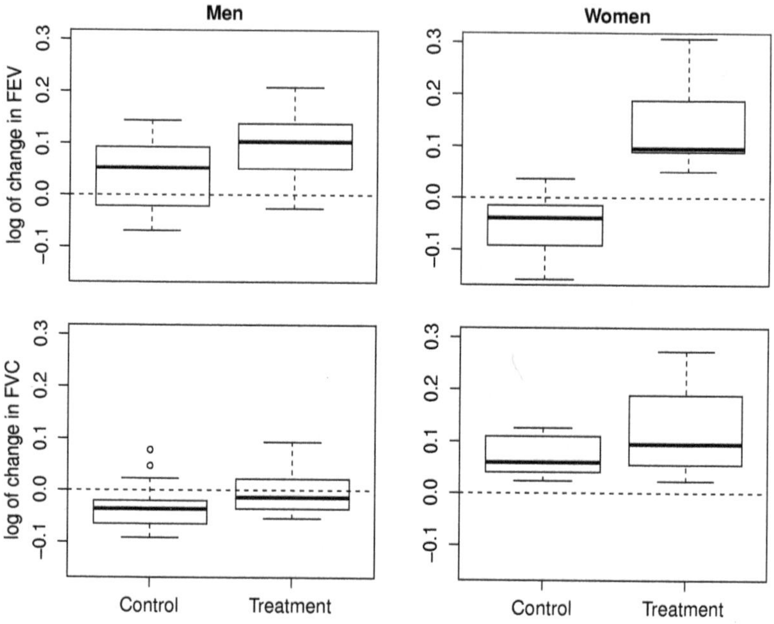

Fig. 5.3 Box plots of the logarithms of the changes in FEV and FVC measurements.

Suggestion: Notice that the experiment could be carried out a few more times with the same or with different groups, and that the participants could have been trained in deep breathing and in providing the measurements also during the preparatory week.

(d) Two-sample tests on the logarithms of the changes in FEV and FVC indicate that the changes in the treatment groups are greater than those in the control groups, the indication being very strong in the case of FEV. In order to quantify the differences between the treatment and control groups with respect to changes in FEV and FVC one may estimate the mean difference between the logarithms of changes in the two groups, given in (5.50), and compute approximate confidence intervals for it based on the first of the two random variables of Prob. 5.2. Thus, in the case of FEV the point estimates of that mean difference are 0.05 in men and 0.18 in women, with approximate 95% confidence intervals of $[0.01, 0.09]$ and $[0.13, 0.24]$; and in the case of FVC the point estimates are 0.02 in men and 0.05 in women, with approximate 95% confidence intervals of $[-0.003, 0.05]$ and $[-0.002, 0.11]$. Based on these estimates, one might say that the median percent improvement in the change in FEV in the treatment group over the control group should be about 5% in men and about 20% in women, and that the median percent improvement in the change in FVC should be about 2.5% in men and about 5% in women; under which conditions would you consider making these approximate statements? *Hint*: Consider the approximate symmetry of distributions.

(e) Two-sample tests on the logarithms of the measurements of FEV and FVC made *after practice* provide evidence of a mean difference between the groups of women but no evidence of a mean difference between the groups of men. In women, the estimate of the mean difference between the logarithms of FEV in the treatment and control groups, obtained by the approximate method indicated in **(d)**, is 0.21, with approximate 95% confidence interval of $[0.11, 0.30]$, and the mean difference between the logarithms of FVC in the treatment and control groups is 0.11, with approximate 95% confidence interval of $[0.02, 0.19]$. Explain why, under (5.48), these are estimates of the theoretical difference in (5.50). As in **(d)**, one might say that the median percent improvements in FEV and FVC of women in the treatment group over women in the control group are roughly 23% and 12%, respectively; how consistent are these with the corresponding estimates of **(d)**? Is it strictly necessary to consider the measurements made before the treatment in analyses of this sort, and is there any advantage in doing so?

(f) Can you think of improvements to the trial of Raghavendra *et al.*?

5.20 In a study about the effectiveness of a new treatment for a disease, N patients randomly drawn from an infinite population are divided randomly into a 'treatment group' of N_1 patients meant to undergo the new treatment and a 'control group' of N_0 patients meant to undergo the standard treatment (which might be no treatment at all). Sometime after the period of treatment patients are declared free of the disease, their response being labelled as 1, or diseased, their response being labelled as 0; then the generic results of the study may be summarized by the following table, where $M_{0,0}$ is the number of diseased patients in the control group, $M_{1,1}$ the number of cured patients in the treatment group, etc.:

	Response		
	0	1	Totals
Control	$M_{0,0}$	$M_{0,1}$	N_0
Treatment	$M_{1,0}$	$M_{1,1}$	N_1

(a) Show that

$$\hat{p}_0 := \frac{M_{0,1}}{N_0} \quad \text{and} \quad \hat{p}_1 := \frac{M_{1,1}}{N_1}$$

are consistent estimators (as $N_0, N_1 \to \infty$) of the probabilities p_0 and p_1 of cure by the standard and the new treatment respectively, and hence that $\hat{\varrho} := \hat{p}_1/\hat{p}_0$ is a consistent estimator of the *relative success rate of the new treatment*, $\varrho := p_1/p_0$, if $N_0/N \to c \in]0, 1[$ and $p_0 > 0$. Use Cramér's theorem (e.g. Ferguson [30]) to show that $[\hat{\varrho}_-, \hat{\varrho}_+]$ with

$$\hat{\varrho}_\pm := \hat{\varrho} \exp\left\{\pm 2\sqrt{\frac{1-\hat{p}_0}{N_0\hat{p}_0} + \frac{1-\hat{p}_1}{N_1\hat{p}_1}}\right\}$$

is a 'large sample' interval for ϱ with approximate confidence 95%.

(b) Although normally patients comply with the treatment prescribed to them, some in the treatment group may decline treatment and others in the control group may demand to join the treatment group. Suppose that $M_{1,1} = C_{1,1} + D_{0,1}$, $M_{0,1} = C_{0,1} + D_{1,1}$, $M_{1,0} = C_{1,0} + D_{0,0}$ and $M_{0,0} = C_{0,0} + D_{1,0}$, where $C_{j,k}$ denotes the number of patients with outcome k who complied with the treatment j prescribed to them, $D_{0,1}$ the number of patients cured among the defectors from the control to the treatment group, $D_{1,1}$ the number of patients cured among the defectors from the treatment to the control group, etc. Supposing that the $C_{j,k}$ and $D_{j,k}$ are observable, argue for or against the following methods of analysis:
(i) In order to estimate p_1 conservatively we replace the numerator of the $\hat{p}_1$ of **(a)** by $C_{1,1}$ and hence use

$$\hat{p}_1 := \frac{C_{1,1}}{N_1}.$$

(ii) In order to estimate p_0 in such a way as not to make it look worse than p_1 when there is a real possibility that $p_1 \le p_0$, we use

$$\hat{p}_0 := \frac{C_{0,1} + D_{0,1} + D_{0,0}}{N_0}$$

in **(a)**; but if we are certain that $p_1 \ge p_0$ then we use

$$\hat{p}_0 := \frac{C_{0,1} + D_{0,1}}{N_0}.$$

(iii) In order to estimate ϱ in such a way as to put the treatment at a disadvantage rather than at an advantage, we estimate it as indicated in **(a)** but with $\hat{p}_1$ and $\hat{p}_0$ as suggested in (i)–(ii).
(iv) As an alternative analysis to that outlined in (i)–(iii) we carry out the permutation test based on the inner product statistic (Sect. 5.2.1) and compute the interval of Sect. 5.2.4 for the average effect of the treatment on the treated, taking exactly $C_{1,1}$ of the binary responses of the treated groups as a 1 and exactly $C_{0,1} + D_{0,1} + D_{0,0}$ or exactly $C_{0,1} + D_{0,1}$ of the binary responses of the control groups as a 1.

(c) Some patients may simply abandon the study. Suppose that $M_{1,1} = C_{1,1} + L_{1,1}$, $M_{0,1} = C_{0,1} + L_{0,1}$, $M_{1,0} = C_{1,0} + L_{1,0}$ and $M_{0,0} = C_{0,0} + L_{0,0}$, where the $C_{j,k}$ are as in **(b)**, $L_{1,1}$ denotes the number of patients who left the study and were cured among those assigned to the treatment group, $L_{0,1}$ the number of patients who left the study and were cured among those assigned to the control group, etc., and the $C_{j,k}$ and the sums $L_{1,0} + L_{1,1}$ and $L_{0,0} + L_{0,1}$ are observable but the individual $L_{j,k}$ are not. Argue for or against the following methods of analysis:
(i) In order to put the treatment at a disadvantage rather than at an advantage we estimate p_1 by the $\hat{p}_1$ of (i) of **(b)**, p_0 by

$$\hat{p}_0 := \frac{C_{0,1} + L_{0,0} + L_{0,1}}{N_0},$$

and then estimate ϱ as indicated in **(a)** with these $\hat{p}_0$ and $\hat{p}_1$.
(ii) As an alternative to this, we apply (iv) of **(b)** except that we take exactly $C_{0,1} + L_{0,0} + L_{0,1}$ of the binary responses of the control groups as a 1.

(d) In a randomized controlled trial to assess the contribution of a comprehensive early intervention in psychosis to the recovery of patients, 77 patients who had experienced a first episode of psychosis were randomly divided into a 'treatment group' meant to receive the intervention and a 'control group' meant to receive standard care. The patients were followed during 1,5 years and after that they were assessed with respect to various clinical and functional outcomes. The table below gives the numbers of patients who were actually treated and of those who left the study and summarizes four of the dichotomous outcomes.[21]

	Early intervention	Standard of care	Total
No. of patients	40	37	77
No. who left the study	1	8	9
No. in remission	23	10	33
No. in employment or education	20	11	31
No. satisfied with care	19	8	27
No. with families satisfied with care	13	4	17

(i) Assuming that the 77 patients were drawn randomly from some population, use the methods outlined above to study the effect of the early intervention on each of the four outcomes. Do the permutation test and the interval for the average effect of the treatment on the treated require the assumption that the 77 patients were drawn randomly from some population?

(ii) Argue for or against the following statements:

S1: In this study, various personal and clinical characteristics of the patients were collected before the intervention, so by fitting a logistic regression model to the responses, treatments and data on some of those characteristics one could perform more powerful analyses which should reveal the effect of the intervention.

S2: Using the inner product statistic on the four outcomes simultaneously, as described in Sect. 5.2.3, should reveal the beneficial effect of the intervention (it would require the complete set of four outcomes per patient).[22]

5.21 **(a)** Let Y_0 and Y_1 be the numbers of events observed during the intervals $[0, L_0]$ and $[0, L_1]$ in two independent Poisson processes of rates μ_0 and μ_1, respectively, so that Y_0 has mean $L_0\mu_0$ and Y_1 has mean $L_1\mu_1$. Show that the distribution of Y_1 conditional on $Y_0 + Y_1 = m$ is binomial with parameters m and

$$p := \frac{L_1\mu_1}{L_0\mu_0 + L_1\mu_1},$$

that $\hat{p} := Y_1/(Y_0+Y_1)$ is an unbiased estimator of p, and suggest an estimator for the *relative rate*

[21] This is a simplified description of the study of Nishida *et al.* [63].

[22] Regarding the first statement see Freedman [35].

$$\varrho := \frac{\mu_1}{\mu_0}.$$

Show that if $[\hat{p}_-, \hat{p}_+]$ is an approximate confidence interval for p then

$$\left[\frac{L_0}{L_1}\frac{\hat{p}_-}{1-\hat{p}_-}, \frac{L_0}{L_1}\frac{\hat{p}_+}{1-\hat{p}_+}\right]$$

is an approximate confidence interval for ϱ (it becomes a point if $p_- = \hat{p}_+ = 1$ or 0).

(b) Halaby *et al.* [42] investigated the role of a 'single room policy' in reducing the transmission of multi-drug resistant bacteria in an intensive care unit. Despite continued and regularly renewed efforts to prevent and control infections, the unit had witnessed two outbreaks and a high prevalence of drug-resistant bacteria from 2001 to March 2008. After closing temporarily and being decontaminated, it reopened in April 2008, when further preventive actions were taken and the number of beds was reduced to 16. By then, however, it had already been decided that the unit would move into another building containing single rooms, each with an anteroom furnished with a washbasin and controlled ventilation, which opened in April 2009. Although the hygienic and preventive protocols remained the same in the new unit and the number of beds was kept at 16 until the end of 2011, it was thought that the improvement in facilities should help to reduce the number of bacterial transmissions. In their study, Halaby *et al.* compared the numbers of transmissions of several bacteria within the intensive care unit during the period of May to December 2008, before the move to the new building, with the homologous periods of May to December 2009 and May to December 2010, after the move. They concluded that the decrease in transmissions observed after the move could not be attributed to chance and that in the main—because of the apparent comparability of the three periods regarding protocols, staffing, number of beds and statistics on rates of admission and occupancy—it was probably the result of the improvement in facilities.

The table below shows the number of transmissions of six types of bacteria per year and month. A Poisson regression model may be fitted to a more complete data set covering all months and including the numbers of admissions per month (not used in this exercise). For instance, one may try the model specified by

$$\mathbf{E}[R|M = m, A = a, T = t] = \exp\{\alpha \log a + \beta_m + t \log \varrho\},$$

where R stands for the number of transmissions, M for the month, A for the number of admissions, T for the treatment (1 indicating the time after the move), and α, $\beta_1, \ldots, \beta_{12}$ and ϱ for unknown parameters, the latter having the same interpretation as in **(a)**. This model yields estimates of the *proportional reduction* $1 - \varrho$ in the transmission of *Citrobacter* and *Enterobacter* of 0.94 and 0.92 (with approximate 95% confidence intervals of [0.60, 0.99] and [0.77, 0.97]). Despite the rather small numbers and the many parameters, perhaps the scatter plots of Fig. 5.4 provide a rough idea about the plausibility of the fits.[23]

[23] These models and their estimation can be studied in McCullagh and Nelder [60], for example.

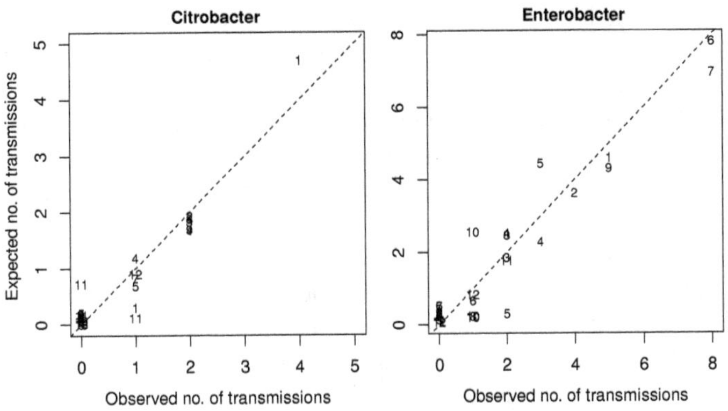

Fig. 5.4 Scatter plots comparing the observed numbers of transmissions of *Citrobacter* and *Enterobacter* with the expected numbers of transmissions estimated with the Poisson regression model; the points are labelled by the associated months.

Year	Month	Single room	*Citrobacter*	*Enterobacter*	*Morganella*	*Proteus*	*Serratia*	*Pseudomonas*
2008	5	0	1	3	0	0	0	1
2008	6	0	0	8	1	0	0	0
2008	7	0	1	8	0	0	0	0
2008	8	0	2	2	0	0	0	5
2008	9	0	2	5	0	0	0	3
2008	10	0	0	1	0	0	0	0
2008	11	0	0	2	0	0	1	0
2008	12	0	1	1	2	0	2	1
2009	5	1	0	2	0	0	0	0
2009	6	1	0	1	0	0	1	0
2009	7	1	0	0	0	1	0	1
2009	8	1	0	0	0	0	0	0
2009	9	1	0	0	0	0	0	2
2009	10	1	0	1	1	0	1	2
2009	11	1	1	0	1	0	0	2
2009	12	1	0	0	0	2	0	0
2010	5	1	0	0	0	0	0	1
2010	6	1	0	0	0	1	0	1
2010	7	1	0	0	0	0	0	0
2010	8	1	0	1	0	0	1	0
2010	9	1	0	0	0	0	0	0
2010	10	1	0	1	0	0	0	0
2010	11	1	0	0	0	0	0	0
2010	12	1	0	0	0	0	0	0

(i) Apply the method outlined in **(a)** to the total numbers of transmissions per bacterium before and after the move in order to assess the evidence of a reduction in the rates of transmission following the move. Use the inner product statistic with the data stratified by month for the same purpose. Show that even if the inner product statistic is used on the data on the *Morganella*, *Proteus*, *Serratia* and *Pseudomonas* groups simultaneously, as described in Sect. 5.2.3, it provides no evidence of changes in their rates of transmission.

(ii) Discuss the following statements, perhaps arguing for or against them:

S1: Although the estimates obtained with the Poisson regression model are more or less consistent with those obtained with the simpler Poisson model of **(a)**, the distribution of the number of transmissions could deviate considerably from the Poisson and bias the conclusions.

S2: Although the three periods are separated by months, data from neighbouring months could be correlated, and if that is the case then the test based on the inner product statistic will be biased.

S3: Even if the conditions in the unit, other than those pertaining to the single-room policy, were found to be similar before and after the move, there may have been important changes regarding aspects not considered in these analyses.

S4: The apparent lack of change in the transmission of *Morganella*, *Proteus*, *Serratia* and *Pseudomonas* bacteria supports the assumption that not much had changed besides the transfer of the unit to the new building, but some non-statistical explanation should be sought for why transmission of those bacteria may be less influenced by an improvement in conditions.

S5: Even if the reduction in the rates of transmission was caused by the single-room policy, it is impossible to disentangle the effect of that policy from the effect of concomitant causes, whose existence cannot be excluded.

5.22 (a) Although trials of new treatments typically use 'control' groups, namely groups of untreated subjects, or of subjects given the current treatment or a placebo, sometimes it is unnecessary, impossible, or even immoral not to provide the new treatment to all subjects or to make some subjects go through an invasive treatment meant to forestall a placebo effect. Discuss the situations in which the efficacy of a treatment can be established without it being necessary to compare the reactions of treated subjects with those of control subjects, considering in particular the following points:

(i) The reactions may depend on external factors, as when testing a new vaccine against a virus whose incidence and virulence may vary from season to season.[24]

(ii) The measurement of the reaction or the knowledge of the measurement may influence the reaction and thereby exaggerate the effect of the treatment or create a spurious effect, as when the measurement requires some cooperation from the subject, measurements are made before and after the treatment, and the subject's ability in providing the measurement tends to improve at every round.

---

[24] For an example of such a vaccine trial see P. Meier's paper about the trial of the poliomyelitis vaccine on pp. 3–14 of Tanur *et al.* [103], in particular the text on p. 7.

(iii) The willingness of the subjects to participate in a trial may exaggerate the effect of the treatment.

(iv) The satisfaction which subjects get from participating in a trial, the attention they get, etc., may elicit obliging reactions from them, exaggerating the effect of the treatment or producing a placebo effect.

(v) The subjects in a trial may be patients suffering from injuries or chronic illnesses (sports injuries or asthma, for example) who have resorted unsuccessfully to standard treatments in the past and whose possible recovery or improvement is ascertainable in the long run; in such cases even a moderate effect of the treatment will be evident without the need of a control group.

(b) Ventegodt *et al.* [106] reported the results of a study on the effect of a combination of 'psychodynamic therapy' with the 'Rosen method' on patients feeling physically ill. Of the 31 patients treated, 12 were regarded as 'physically well' one year after the treatment, 6 were regarded as 'physically ill', and 13 had left the study or could not be diagnosed. The condition of the patients was determined before the study, shortly after the treatment and one year after the treatment, by means of questionnaires on self-assessed physical and mental health, personal relations, etc. The 31 patients had been selected for treatment among a larger group of patients because, according to the initial questionnaire, they were 'physically ill', they could not be helped by their general practitioners, and most of them suffered from chronic pain. According to the authors of the study, psychodynamic therapy is "often very dramatic and often emotionally painful for the patient" and requires the patient "to work hard with exercises that are most often writings and readings", which "limits the number of patients who are willing to use [it]"; nevertheless, long experience shows that this type of treatment has "no severe or lasting side effects". Finally, the treatment was paid for by the patients.[25]

Use the 'intention-to-treat principle' (Probs. 6.10, 6.14) to compute a conservative estimate of the probability of success of the treatment and a confidence interval for it. Discuss the significance of the results of the study of Ventegodt *et al.*, arguing for or against the following:

(i) The study did not have a control group. If it had had one and the patients in the control group had been given about as much attention as those in the treatment group but without going through psychodynamic therapy and the Rosen method then they would probably have shown similar improvements.

(ii) Too many of the patients left the study. Those who left were probably the most difficult cases, and those who appeared to be healed would probably have been healed by a minimum of assistance not involving psychodynamic therapy or the Rosen method.

(iii) Because the patients determined their condition themselves (through the questionnaires) and they were eager to be helped and had paid for the treatment, they may have been overenthusiastic about the supposed improvements.

---

[25] I am grateful to Dr. Søren Ventegodt for providing me with some information on his study.

(iv) Even if the treatment is not harmful and the improvement in the condition of some patients after one year was genuine, only a long-term improvement would justify recommending this treatment.

(v) By most standards of conventional medicine, an improvement in health sustained during one year is worthwhile.

(vi) Self-assessment of one's health is meaningful to patients and to their relations, and to physicians as well; it has been shown to be a strong indicator of one's actual condition. Moreover, the clinical presentation associated with these patients makes objective assessment difficult.

(vii) It may be that those who did not manage to undergo the whole therapy were the most difficult cases, but because some did and improved their condition it is undeniable that the treatment is beneficial at least for some patients who until then had not been helped by conventional medicine. Moreover, since those who improved their condition went through the difficulties intrinsic to healing, and since they had had some form of assistance from conventional medicine in the past, it seems that they did require a substantial type of treatment dealing with psychological and physical aspects which the present therapy is supposed to deal with.

(viii) Although the study had no control group, the treated patients may be seen as their own 'controls'; after all, they had looked for help from conventional medicine in the past without success, and they must have seen few changes around them during the period of treatment and during the following year, apart from the treatment itself, which could explain the improvement in their health.

5.23 Consider the following method of trying to discover the cause of a disease: Patients who suffer from the disease or have died of it are identified and they, their families or their relations are asked to participate in an investigation of the cause of the disease by providing or procuring information about the patient's history (background, occupations, habits, hobbies, health problems, etc.). Assuming that enough cases and participants (say 20 patients and their families) have been secured, the investigation is carried out by a group of physicians, nurses and medical scientists, together with other persons possessing appropriate qualifications or experience, such as chemists, dieticians, psychologists, statisticians, etc., some of whom may be consulted occasionally only. The investigators conduct their enquiries during a more or less indeterminate period, individually or in group; they interview the patient or the patient's relations, write and exchange notes, meet up and discuss their observations, findings and theories with each other, as they see fit or the need arises. Assuming that the disease has at least one cause (a genetic, environmental or behavioural cause), the same cause ought to be present in most of the cases considered and could be found out. Candidates for the cause, or causes, of the disease would include personal characteristics (including genetic and physiologic characteristics), characteristics of the patient's surroundings, habits, health history of the patient's ancestors, or any set of circumstances common to many of the patients but uncommon or rare among the general population. The investigators have the possibility of checking their guesses—which will often materialize from reflections upon a small subset of cases—across all the cases. This may be difficult to do because a common cause may act in different ways in different cases; for instance, if the cause of disease is prolonged exposure to a substance then the source of exposure may be more or less hidden, and revealing it may require various and variously difficult means. Nevertheless, one may hope that a plausible cause will be ascertained in many of the cases, and the more prevalent the cause among the cases and the less prevalent among the general population the more compelling the relation of cause and effect will be, which of course will need further confirmation through the study of other cases or special populations.

Discuss the viability of this method, considering whether it would have produced results or could produce results in the following investigations:[26]

(i) Investigation of the causes of puerperal fever;

(ii) Investigation of the causes of cholera during a cholera epidemic;

(iii) Investigation of the causes of beriberi or of pellagra;

(iv) Investigation of the causes of lung cancer;

(iv) Investigation of the causes of Huntington's disease;

(v) Investigation of the causes of stomach ulcers;

(vi) Investigation of the causes of breast cancer in young mothers and in young women in general;

(vii) Investigation of the causes of Alzheimer's disease, of Parkinson's disease, of amyotrophic lateral sclerosis, or of multiple sclerosis;

(viii) Investigation of the causes of schizophrenia.

---

[26] In order to think about this question, the reader may find it useful to read Freedman's masterly essays included as Chaps. 3 and 20 in [37], the interview with Mary-Claire King in [62], the Q&A with Barry Marshall in [6], and Norbert Wiener's book [109], which emphasizes the role of the individual in scientific discovery.

Solutions to selected problems

This section gives solutions to Probs. 2.1, 2.2, 2.9, 2.10, 2.11, 2.12, 3.1, 3.6, 3.8, 3.9, 5.9, 5.10, 5.11, 5.12, 5.13, 5.14, 5.17 and 5.19.

Solution to Prob. 2.1

To alleviate notation we omit subscripts and write (X, Y, Z) for a vector satisfying the same conditions as (X_n, Z_n, Y_n) for some n, and note that the basic model we have in mind is the more general one mentioned on p. 10.

The assumptions about Y and ε mean that $Y = f(U, X)$ and $\varepsilon = g(V, X, Y)$ for some functions f and g and standard uniforms U and V conditionally on X. Then

$$Z = \varphi(Y, \varepsilon) := \alpha + \beta Y + \varepsilon, \quad \text{so} \quad Y = \varphi_\varepsilon^{-1}(Z) := -\frac{\alpha}{\beta} + \frac{Z}{\beta} - \frac{\varepsilon}{\beta} \equiv \alpha^* + \beta^* Z + \varepsilon^*.$$

(a) From the definitions of Z, Y and ε follows

$$Z = \varphi(Y, \varepsilon) = \varphi(Y, g(V, X, Y)) =: \rho(V, X, Y), \quad Y = f(U, X), \qquad (*)$$

which yields a special case of the basic model with Z playing the role of response and Y the role of treatment.

(b) Since we can 'solve for' Y in the relationship between Y and Z, we have

$$Y = \varphi_\varepsilon^{-1}(Z) = \varphi_{g(V,X,Y)}^{-1}(Z) = \varphi_{g(V,X,f(U,X))}^{-1}(Z) =: \rho'(V', X, Z)$$

in combination with

$$Z = \varphi(Y, \varepsilon) = \varphi(f(U, X), g(V, X, Y)) = \varphi(f(U, X), g(V, X, f(U, X))) =: \tau(U', X),$$

where U' and V' are obtained by alternating the digits in the decimal expansions of U and V, possibly in the same manner. These two equations satisfy the properties of the basic model except that U' and V' need not be standard uniform conditionally on X (because U and V may be dependent conditionally on X). But we may assume that their distribution functions conditional on X, denoted by F_X and G_X, are continuous and strictly increasing (this ought to be the case, but we have no proof of it), so that $U'' := F_X(U')$ and $V'' := G_X(V')$ are standard uniforms conditionally on X and $U' = F_X^{-1}(U'')$, $V' = G_X^{-1}(V'')$, and we can still write

$$Y = \rho'(V', X, Z) = \rho'\left(G_X^{-1}(V''), X, Z\right) =: \rho''(V'', X, Z)$$

and

$$Z = \tau(U', X) = \tau\left(F_X^{-1}(U''), X\right) =: \tau''(U'', X),$$

which now satisfy all the properties of the basic model.

Should F_X or G_X lack the assumed properties, we could still define triples $(\tilde{X}, \tilde{Z}, \tilde{Y})$ to replace (X, Z, Y) in the manner indicated in Sect. 2.5.

(c) For unconfoundedness to hold in the model followed by (X, Y, Z), the U and V in (*) must be independent standard uniforms independent of X. But if this is the case then the U' and V' of **(b)** cannot be independent because each of them must be a one-to-one function of (U, V), so in general unconfoundedness fails in the model followed by (X, Z, Y).

(d) We have seen that if φ is as in **(a)** then the answer is yes. In general, however, one needs to invert φ in order to write Y as a function of Z and obtain an expression for Y analogous to that derived in **(b)**; but this is possible only if for each ε the function $y \rightarrow \varphi(y, \varepsilon)$ has an inverse, which we know is not always the case.

Solution to Prob. 2.2

By assumption, $F_{\mathbf{x},t}(r) := \mathbf{P}(R \leq r | \mathbf{X} = \mathbf{x}, T = t) = \mathbf{P}(\rho(V, \mathbf{x}, t) \leq r)$ is constant in t for each r, so $\tilde{R} := \tilde{\rho}(V, \mathbf{X}, T)$ with $\tilde{\rho}(v, \mathbf{x}, t) := F_{\mathbf{x},t}^{-1}(v)$ satisfies the required properties (see Prob. 1.1). If in the basic model $(U_1, V_1), (U_2, V_2), \ldots$ are independent conditionally on $\mathbf{X}_1, \mathbf{X}_2, \ldots$ then $(\mathbf{X}_1, T_1, \tilde{R}_1), (\mathbf{X}_2, T_2, \tilde{R}_2), \ldots$ with $\tilde{R}_n := \tilde{\rho}(V_n, \mathbf{X}_n, T_n)$ follow a model equivalent to the given model.

Solution to Prob. 2.9

(b) In the first specification suggested the model reduces to the basic model with $\rho(v, x, t) := \beta_1 xt + \beta_2 t + \sigma\Phi^{-1}(v)$ and the causal effect may be defined as the function $(t, t') \rightarrow \beta_2(t - t')$: it corresponds to the difference between a 'potential response' to the forcing of a treatment t and a 'potential response' to the forcing of a treatment t'; if the treatments are binary then the causal effect may be identified with β_2. In the second specification the model reduces to the basic model with $\rho(v, x, t) := \beta_1 xt + \beta_2 t + \varphi(v, t)$ and the causal effect may be defined as the function $(t, t') \rightarrow \beta_2(t - t') + \varphi(v, t) - \varphi(v, t')$, or by $\mathcal{L}(\beta_2(t - t') + \varphi(V, t) - \varphi(V, t'))$.

(c) This requires knowledge of the linear model at the level of Jørgensen [49].

(d) It is indeed a special case of the basic model. It satisfies unconfoundedness if the U_i are independent of the V_i. For it to be a linear regression model with the ε_i as errors it must satisfy unconfoundedness and the variance of ε_i conditionally on $\{X_i = x\}$ must be independent of x, so θ must be 0.

(e) Both estimators are consistent if and only if $\varrho = 0$, i.e. under confoundedness. The model is *not* a linear regression model unless $\theta = 0$, and yet $\hat{\boldsymbol{\beta}}$ is consistent for all θ.

(f) From the expression of the limit of $\hat{\boldsymbol{\beta}}$ we see that $\hat{\beta}_1$ is consistent if and only if $\theta = -2\varrho\sigma/(3\pi - 2)$ and $\hat{\beta}_2$ is consistent if and only if $\theta = -5\varrho\sigma$, so the two are simultaneously consistent if and only if $\varrho = 0$; and under this last condition unconfoundedness holds. The model is a linear regression model with the ε_i as errors if and only if $\varrho = 0$ and $\theta = 0$. If we use equation (2.34) in (2.33) we see that when $\varrho = 0$ the (R_i, T_i, X_i) follow a linear regression model with coefficients θ, β_1 and β_2 and errors $\sigma\Phi^{-1}(V_i)$, but $\hat{\boldsymbol{\beta}}$ is derived with the understanding that we do not know how the ε_i arise (by definition, in a linear regression model the errors are unobserved); so if the parameter estimates are wrong then it is because they are based on the wrong model.

The treatment effect can be defined by $\beta_1 x + \beta_2$ as a function of x. This makes sense even without unconfoundedness because, even if T_i is 'entangled' with V_i and we cannot dictate the value of T_i, we can always consider what would happen if we could, namely how $\rho(v, x, t) := \beta_1 x(t+1) + \beta_2 t + \sigma\Phi^{-1}(v) + \theta x$ varies with t. What unconfoundedness gives us is the identity

$$\mathcal{L}(R_i | X_i = x, T_i = t) = \mathcal{L}(\rho(V_i, x, t)) = \mathcal{N}\big(\beta_1 x(t+1) + \beta_2 t + \theta x, \sigma^2\big),$$

and hence the possibility of estimating the law of the treatment effect based on the (R_i, X_i, T_i), and in particular

$$\mathbf{E}(R_i | X_i = x, T_i = 1) - \mathbf{E}(R_i | X_i = x, T_i = 0) = \beta_1 x + \beta_2.$$

Under unconfoundedness, integrating this difference with respect to the distribution of X_i ($\mathcal{N}(0, 1)$ in this case) gives us the 'overall' mean treatment effect β_2.

Solution to Prob. 2.10

(a) The model corresponds to the basic model with $\rho : [0, 1] \times \mathbb{R}^{p+q} \times \{0, 1\} \to \mathbb{R}$ defined by

$$\rho(v, \mathbf{x}, t) = \mu + \sigma\Phi^{-1}(v) + (1 - t)A_0(\mathbf{x}) + tA_1(\mathbf{x})$$

and $\tau : [0, 1] \times \mathbb{R}^{p+q} \to \{0, 1\}$ by

$$\tau(u, \mathbf{x}) \equiv \tau(u, \mathbf{u}) = \begin{cases} 1 & \text{if } u \le \Pi_1(\mathbf{u}), \\ 0 & \text{otherwise,} \end{cases}$$

where $\mathbf{u}$ denotes the vector with the first p coordinates of $\mathbf{x}$.

There is no treatment effect $\Leftrightarrow$ $\rho(v, \mathbf{x}, t)$ is constant in t $\Leftrightarrow$ $(1-t)A_0(\mathbf{x}) + tA_1(\mathbf{x})$ is independent of t $\Leftrightarrow$ $A_0(\mathbf{x}) = A_1(\mathbf{x})$ for all $\mathbf{x}$.

Unconfoundedness holds because U and V are independent conditionally on $\mathbf{X}$. In fact, it holds even if one conditions on $\mathbf{U}$ rather than on $\mathbf{X}$, as explained in the last part of this solution.

As stated in the question, conditionally on $\{T = t, \mathbf{X} = \mathbf{x}\}$ the law of R is

$$\mathcal{N}\Big(\mu + A_0(\mathbf{x})(1 - t) + A_1(\mathbf{x})t, \sigma^2\Big);$$

this is the law that characterizes the treatment effect. The mean treatment effect is

$$\mathrm{e}(\mathbf{x}) := \mathbf{E}[R | \mathbf{X} = \mathbf{x}, T = 1] - \mathbf{E}[R | \mathbf{X} = \mathbf{x}, T = 0] = A_1(\mathbf{x}) - A_0(\mathbf{x}).$$

The overall mean treatment effect, call it ϵ, is obtained by integrating $\mathrm{e}(\mathbf{x})$ with respect to the distribution of $\mathbf{X}$:

$$\epsilon = \mathbf{E}[A_1(\mathbf{X})] - \mathbf{E}[A_0(\mathbf{X})].$$

(c) Unconfoundedness holds even if we only condition on $\mathbf{U}$, since conditionally on it $\mathbf{V}$ and T are independent: i.e., conditionally on $\{\mathbf{U}=\mathbf{u}\}$ we have $\mathbf{V}=\boldsymbol{\xi}+\mathbf{B}(\mathbf{u}-\mu)$ and $T=\tau(U,\mathbf{u})$ with $\boldsymbol{\xi}$ and U independent. Moreover, conditionally on $\{T=t,\mathbf{U}=\mathbf{u}\}$ the law of R is

$$\mathcal{L}\left(\mu+\sigma\Phi^{-1}(V)+A_0(\mathbf{u},\boldsymbol{\xi}+\mathbf{B}(\mathbf{u}-\mu))\,(1-t)+A_1(\mathbf{u},\boldsymbol{\xi}+\mathbf{B}(\mathbf{u}-\mu))\,t\right),$$

so the mean treatment effect is

$$\tilde{e}(\mathbf{u}):=\mathbf{E}[A_1(\mathbf{u},\boldsymbol{\xi}+\mathbf{B}(\mathbf{u}-\mu))]-\mathbf{E}[A_0(\mathbf{u},\boldsymbol{\xi}+\mathbf{B}(\mathbf{u}-\mu))],$$

which integrated with respect to the distribution of $\mathbf{U}$ leads us back to $\boldsymbol{\epsilon}$.

Solution to Prob. 2.11

Under unconfoundedness, U_n, V_n' and V_n'' are independent given $\mathbf{X}_n$, so

$$\mathbf{P}(\min(S_n,C_n)>s|\mathbf{X}_n=\mathbf{x},T_n=t)=\mathbf{P}(S_n>s,C_n>s|\mathbf{X}_n=\mathbf{x},T_n=t)=$$

$$\mathbf{P}(\sigma(V_n'',\mathbf{x},t)>s,\chi(V_n',\mathbf{x})>s|\mathbf{X}_n=\mathbf{x},\tau(U_n,\mathbf{x})=t)=$$

$$\mathbf{P}(\sigma(V_n'',\mathbf{x},t)>s)\mathbf{P}(\chi(V_n',\mathbf{x})>s)=$$

$$\mathbf{P}(\sigma(V_n'',\mathbf{x},t)>s|\mathbf{X}_n=\mathbf{x},\tau(U_n,\mathbf{x})=t)\mathbf{P}(\chi(V_n',\mathbf{x})>s|\mathbf{X}_n=\mathbf{x})=$$

$$\mathbf{P}(S_n>s|\mathbf{X}_n=\mathbf{x},T_n=t)\mathbf{P}(C_n>s|\mathbf{X}_n=\mathbf{x}),$$

or

$$\mathbf{P}(S_n>s|\mathbf{X}_n=\mathbf{x},T_n=t)=\frac{\mathbf{P}(\min(S_n,C_n)>s|\mathbf{X}_n=\mathbf{x},T_n=t)}{\mathbf{P}(C_n>s|\mathbf{X}_n=\mathbf{x})}.$$

Both $\mathbf{P}(S_n>s|\mathbf{X}_n=\mathbf{x},T_n=t)$ and $\mathbf{P}(\min(S_n,C_n)>s|\mathbf{X}_n=\mathbf{x},T_n=t)$ reflect the effect of the treatment, and in general even $\mathbf{P}(S_n\le C_n|\mathbf{X}_n=\mathbf{x},T_n=t)$ does, but the first—the 'survival function'—may be regarded as the more interesting or direct measure of that effect; the estimation of this function from data is not obvious because what one observes in applications is $R_n=(\min(S_n,C_n),\mathbf{1}_{\{S_n\le C_n\}})$, not S_n or C_n, and it is the subject of 'survival analysis'.

Solution to Prob. 2.12

(a) Using $\mathbf{P}(U_n\le u,V_n\le v|\mathbf{X}_n=\mathbf{x})=uv$, $0\le u,v\le 1$, $\mathbf{x}:=(\mathbf{x}',\mathbf{x}'',\mathbf{x}''')$, we have

$$\mathbf{P}(U_n\le u,V_n\le v|\mathbf{X}_n'=\mathbf{x}')=\sum_{\mathbf{x}'',\mathbf{x}'''}\frac{\mathbf{P}(U_n\le u,V_n\le v,\mathbf{X}_n=\mathbf{x})}{\mathbf{P}(\mathbf{X}_n'=\mathbf{x}')}=$$

$$\sum_{\mathbf{x}'',\mathbf{x}'''}\mathbf{P}(U_n\le u,V_n\le v|\mathbf{X}_n=\mathbf{x})\mathbf{P}(\mathbf{X}_n=\mathbf{x}|\mathbf{X}_n'=\mathbf{x}')=$$

$$uv\sum_{\mathbf{x}'',\mathbf{x}'''}\mathbf{P}(\mathbf{X}_n''=\mathbf{x}'',\mathbf{X}_n'''=\mathbf{x}'''|\mathbf{X}_n'=\mathbf{x}')=uv,$$

so U_n and V_n are standard uniform and independent conditionally on $\mathbf{X}_n'$.

(b) If $\mathbf{X}_n''$ and $\mathbf{X}_n'''$ are independent conditionally on $\mathbf{X}_n'$ then, with

$$R_n = \rho(V_n, \mathbf{X}_n', \mathbf{X}_n''', T_n) =: \tilde{\rho}_{\mathbf{X}_n'''}(V_n, \mathbf{X}_n', T_n),\ T_n = \tau(U_n, \mathbf{X}_n', \mathbf{X}_n'') =: \tilde{\tau}_{\mathbf{X}_n''}(U_n, \mathbf{X}_n'),$$

we have

$$\begin{aligned}\mathcal{L}(R_n|T_n = t, \mathbf{X}_n' = \mathbf{x}) &= \mathcal{L}\left(\tilde{\rho}_{\mathbf{X}_n'''}(V_n, \mathbf{x}', t)\,\middle|\tilde{\tau}_{\mathbf{X}_n''}(U_n, \mathbf{x}') = t, \mathbf{X}_n' = \mathbf{x}'\right)\\ &= \mathcal{L}\left(\tilde{\rho}_{\mathbf{X}_n'''}(V_n, \mathbf{x}', t)\,\middle|\mathbf{X}_n' = \mathbf{x}'\right).\end{aligned}$$

The conditioning event cannot be omitted from the last expression because $\mathbf{X}_n'$ and $\mathbf{X}_n'''$ may be dependent. In the special case suggested, however, we may substitute $\xi'''(U_n''', \mathbf{X}_n')$ for $\mathbf{X}_n'''$ and write

$$\mathcal{L}\left(\tilde{\rho}_{\mathbf{X}_n'''}(V_n, \mathbf{x}', t)\,\middle|\mathbf{X}_n' = \mathbf{x}'\right) = \mathcal{L}\left(\tilde{\rho}_{\xi'''(U_n''', \mathbf{x}')}(V_n, \mathbf{x}', t)\right).$$

Finally,

$$\mathbf{P}\left(\tilde{\rho}_{\mathbf{X}_n'''}(V, \mathbf{x}', t) \le r\,\middle|\mathbf{X}_n' = \mathbf{x}'\right) = \sum_{\mathbf{x}'', \mathbf{x}'''} \frac{\mathbf{P}\big(\rho(V_n, \mathbf{X}_n', \mathbf{X}_n''', t) \le r, \mathbf{X}_n = (\mathbf{x}', \mathbf{x}'', \mathbf{x}''')\big)}{\mathbf{P}(\mathbf{X}_n' = \mathbf{x}')} =$$

$$\sum_{\mathbf{x}'', \mathbf{x}'''} \mathbf{P}\big(\rho(V_n, \mathbf{X}_n', \mathbf{X}_n''', t) \le r\,|\mathbf{X}_n = (\mathbf{x}', \mathbf{x}'', \mathbf{x}''')\big)\, \mathbf{P}\left(\mathbf{X}_n' = \mathbf{x}', \mathbf{X}_n'' = \mathbf{x}''\,\middle|\mathbf{X}_n' = \mathbf{x}'\right) =$$

$$\sum_{\mathbf{x}'', \mathbf{x}'''} \mathbf{P}\left(\rho(V_n, \mathbf{X}_n', \mathbf{X}_n''', t) \le r\,\middle|\begin{array}{l}\mathbf{X}_n = (\mathbf{x}', \mathbf{x}'', \mathbf{x}'''),\\ \tau(U_n, \mathbf{x}', \mathbf{x}'') = t\end{array}\right) \mathbf{P}\left(\mathbf{X}_n' = \mathbf{x}', \mathbf{X}_n'' = \mathbf{x}''\,\middle|\mathbf{X}_n' = \mathbf{x}'\right) =$$

$$\sum_{\mathbf{x}'', \mathbf{x}'''} \mathbf{P}(R_n \le r\,|\mathbf{X}_n = (\mathbf{x}', \mathbf{x}'', \mathbf{x}'''), T_n = t)\, \mathbf{P}(\mathbf{X}_n' = \mathbf{x}'|\mathbf{X}_n' = \mathbf{x}')\mathbf{P}(\mathbf{X}_n'' = \mathbf{x}|\mathbf{X}_n' = \mathbf{x}'),$$

using the independence of $\mathbf{X}_n''$ and $\mathbf{X}_n'''$ conditional on $\mathbf{X}_n'$.

Solution to Prob. 3.1

(c) As indicated in the question,

$$\epsilon(\mathbf{x}) = \frac{\mathbf{P}\big(R = 1, \mathbf{X} = \mathbf{x}, I = 1, T = 1, A = 1 = P\big)}{\mathbf{P}(\rho(V, \mathbf{x}, 1, 1, 1, 0) = 1)\mathbf{P}\left(\begin{smallmatrix}\mathbf{X}=\mathbf{x}, I=1,\\ A=1=P\end{smallmatrix}\right) - \mathbf{P}\left(\begin{smallmatrix}R=1, \mathbf{X}=\mathbf{x}, A=1=P,\\ I=1, T=0\end{smallmatrix}\right)}.$$

We need to show that under (3.21) the probability $\mathbf{P}(\rho(V, \mathbf{x}, 1, 1, 1, 0) = 1)$ can be written in terms of probabilities pertaining to observable events. By the second condition in (3.21) this probability is $\mathbf{P}(\rho(V, \mathbf{x}, 1, 0, 1, 0) = 1)$, which is equal to

$$\frac{\mathbf{P}(\rho(V, \mathbf{x}, 1, 0, 1, 0) = 1)}{\mathbf{P}(\mathbf{X} = \mathbf{x}, A = 1, I = 0, P = 1)}\mathbf{P}\left(\begin{array}{c}\mathbf{X} = \mathbf{x}, \alpha(W_2, \mathbf{x}) = 1,\\ \iota(W_3, \mathbf{x}, 1) = 0, \pi(W_4, \mathbf{x}, 1, 0) = 1\end{array}\right) =$$

$$\frac{\mathbf{P}(\rho(V, \mathbf{x}, A, I, P, 0) = 1, \mathbf{X} = \mathbf{x}, A = 1, I = 0, P = 1)}{\mathbf{P}(\mathbf{X} = \mathbf{x}, A = 1, I = 0, P = 1)} =$$

$$\frac{\mathbf{P}(R=1,\mathbf{X}=\mathbf{x},A=1,I=0,P=1,T=0)}{\mathbf{P}(\mathbf{X}=\mathbf{x},A=1,I=0,P=1)},$$

where in the second equality we used $I = 0 \wedge P = 1 \Rightarrow T = 0$. Using this expression in $\epsilon(\mathbf{x})$ we get

$$\begin{aligned}\epsilon(\mathbf{x}) &= \frac{\mathbf{P}(R=1,I=1,T=1|\mathbf{X}=\mathbf{x},A=1=P)}{\mathbf{P}\left(R=1,I=0,T=0\middle|\substack{\mathbf{X}=\mathbf{x},\\ A=1=P}\right)c(\mathbf{x})-\mathbf{P}\left(R=1,I=1,T=0\middle|\substack{\mathbf{X}=\mathbf{x},\\ A=1=P}\right)}\\ &= \frac{\tilde{\mathbf{P}}(R=1,I=1,T=1|\mathbf{X}=\mathbf{x})}{\tilde{\mathbf{P}}(R=1,I=0,T=0|\mathbf{X}=\mathbf{x})c(\mathbf{x})-\tilde{\mathbf{P}}(R=1,I=1,T=0|\mathbf{X}=\mathbf{x})},\end{aligned}$$

which is another way of writing the expression given in the question.

Solution to Prob. 3.6

As noted in Example 3.2,

$$\begin{aligned}\mathcal{L}\left(R\middle|\begin{matrix}T=t,X_1=x_1,\\ X_2=x_2,X_3=x_3\end{matrix}\right) &= \mathcal{L}\left(\rho(V,x_3,\varphi_5(U_5,x_2),t)\middle|\begin{matrix}\tau(U,x_3,\varphi_4(U_4,x_1))=t,\\ \varphi_3(U_3,x_1,x_2)=x_3\end{matrix}\right)\\ &= \mathcal{L}\left(\rho(V,x_3,\varphi_5(U_5,x_2),t)\right)\\ &= \mathcal{L}\left(\rho(V,x_3,\varphi_5(U_5,x_2),t)\middle|\begin{matrix}\tau(U,x_3,\varphi_4(U_4,X_1))=t,\\ \varphi_3(U_3,X_1,x_2)=x_3\end{matrix}\right)\\ &= \mathcal{L}\left(R\,|T=t,X_2=x_2,X_3=x_3\right),\end{aligned}$$

where in order to alleviate the notation we replace $\varphi_6(U_6,t)$ by t, which we may do because X_6 plays no role in these arguments.

Integrating this measure of treatment effect with respect to $\mathcal{L}(X_2|X_1=x_1,X_3=x_3)$ we get, on using the independence of (U_5,V) and (U_1,U_2,U_3),

$$\sum_{x_2}\mathbf{P}\left[\rho(V,x_3,\varphi_5(U_5,x_2),t)\le r\right]\mathbf{P}(X_2=x_2|X_1=x_1,X_3=x_3)=$$

$$\sum_{x_2}\mathbf{P}\left[\rho(V,x_3,\varphi_5(U_5,x_2),t)\le r\right]\frac{\mathbf{P}(X_1=x_1,X_2=x_2,\varphi_3(U_3,x_1,x_2)=x_3)}{\mathbf{P}(X_1=x_1,X_3=x_3)}=$$

$$\sum_{x_2}\frac{\mathbf{P}\left[\rho(V,x_3,\varphi_5(U_5,x_2),t)\le r,X_1=x_1,X_2=x_2,\varphi_3(U_3,x_1,x_2)=x_3\right]}{\mathbf{P}(X_1=x_1,X_3=x_3)}=$$

$$\frac{\mathbf{P}\left[\rho(V,x_3,\varphi_5(U_5,X_2),t)\le r,X_1=x_1,X_3=x_3\right]}{\mathbf{P}(X_1=x_1,X_3=x_3)}=$$

$$\mathbf{P}[\rho(V,x_3,X_5,t)\le r|X_1=x_1,X_3=x_3]=$$

$$\mathbf{P}[\rho(V,x_3,X_5,t)\le r|\varphi_3(U_3,x_1,X_2)=x_3],$$

which is the second measure of treatment effect of Example 3.2.

Solution to Prob. 3.8

(a) Let us ignore X_6: it plays no role in the argument and ignoring it simplifies the writing. If we condition on everything except X_7, we get

$$\mathcal{L}\left(\rho(V,X_3,X_5,X_7,T)\,\middle|\,\begin{array}{c}\tau(U,X_3,X_4,X_7)=t,X_1=x_1,X_2=x_2,\\ X_3=x_3,X_4=x_4,X_5=x_5\end{array}\right)=$$

$$\mathcal{L}\left(\rho(V,x_3,x_5,X_7,t)\,\middle|\,\begin{array}{c}\tau(U,x_3,x_4,X_7)=t,X_1=x_1,X_2=x_2,\\ \varphi_3(U_3,x_1,x_2)=x_3,\varphi_4(U_4,x_1)=x_4,\varphi_5(U_5,x_2)=x_5\end{array}\right)=$$

$$\mathcal{L}\left[\rho(V,x_3,x_5,X_7,t)\,|\tau(U,x_3,x_4,X_7)=t,\varphi_1(U_1,X_7)=x_1,\varphi_2(U_2,X_7)=x_2\right].$$

In this last expression we cannot get rid of the constraint linking X_7 and t, so t affects the response function also through its fourth argument, not just through its last; therefore, we would have to condition on X_7 as well.

Note that in the path $T \leftarrow X_7 \rightarrow R$ none of the nodes being conditioned upon points arrows in that path, so the conditions of Pearl's criterion fail.

(b) X_1 and X_2 are dependent because

$$\mathcal{L}\left(X_2\,|X_1=x_1\right)=\mathcal{L}\left(\varphi_2(U_2,X_7)\,|\varphi_1(U_1,X_7)=x_1\right)$$

depends on x_1, but according to the assumed model X_1 and X_2 are 'exogenous' and therefore independent. Thus, a test of independence on a random sample with the same distribution as (X_1,X_2) could provide evidence against the assumed model.

If X_7 points no arrows to nodes other than T and R then X_1 and X_2 are independent and cannot help us detect the anomaly in the model, nor can the other variables.

Solution to Prob. 3.9

We consider the specifications represented by Figs. 3.8 and 3.9. In the first,

$$\mathcal{L}\left(R\,|(X_1,X_3)=(x_1,x_3),T=t\right)=$$

$$\mathcal{L}\left(\rho(V,X_2,x_3,t)\,\middle|\,\begin{array}{c}\varphi_1(U_1,X_2,x_3,X_4)=x_1,\\ \varphi_3(U_3,X_2)=x_3,\tau(U,x_1,x_3,X_4)=t\end{array}\right).$$

In the conditioning event the last constraint provides X_4 with information about t and X_4, through the first constraint, passes it to X_2, which in turn acts on the second argument of ρ; so conditionally on (X_1,X_3) the treatment acts both on the second and fourth arguments of the response function.

In the second model,

$$\mathcal{L}\left(R\,|(X_1,X_3)=(x_1,x_3),T=t\right)=$$

$$\mathcal{L}\left(\rho(V,\varphi_2(U_2,x_1),x_3,t)\,\middle|\,\begin{array}{c}\varphi_1(U_1,X_4)=x_1,\varphi_3(U_3,x_1,\varphi_2(U_2,x_1))=x_3,\\ \tau(U,x_1,x_3,X_4)=t\end{array}\right)=$$

$$\mathcal{L}\left[\rho(V,\varphi_2(U_2,x_1),x_3,t)\,|\varphi_3(U_3,x_1,\varphi_2(U_2,x_1))=x_3\right]$$

is a measure of the treatment effect.

Solution to Prob. 5.9

(**a**) We may consider both B and M as treatments; B comes first, but conditionally on B the magnitude of exposure M provides the more direct cause of disease.

If M is regarded as the treatment then we can write

$$D = \rho(V, S, M) := \mathbf{1}_{[V,1]}\left(\frac{\exp\{\beta_{0,S} + \beta_{1,S}M\}}{1 + \exp\{\beta_{0,S} + \beta_{1,S}M\}}\right),$$

$$M = \tau(U, p_S) := U_1 \mathbf{1}_{[0,p_S]}(U_0),$$

where U—made up from infinite subsets of the digits in the decimal expansions of U_0 and U_1—and V are independent and standard uniform conditionally on S.

If B alone is regarded as the treatment then we can write

$$D = \tilde{\rho}(\tilde{V}, S, B) := \mathbf{1}_{[V,1]}\left(\frac{\exp\{\beta_{0,S} + \beta_{1,S}U_1 B\}}{1 + \exp\{\beta_{0,S} + \beta_{1,S}U_1 B\}}\right),$$

$$B = \tilde{\tau}(U_0, p_S) := \mathbf{1}_{[0,p_S]}(U_0),$$

where U_0 and $\tilde{V}$—made up from infinite subsets of the digits in the decimal expansions of V and U_1—are independent and standard uniform conditionally on S.

If M is regarded as the treatment then the law that describes its effect on D is $\mathcal{L}(D|S = s, M = m)$; if B is regarded as the treatment then the law that describes its effect on D is $\mathcal{L}(D|S = s, B = b)$.

Lifetime (age at death) could be another response of interest; it comes after diagnosis of the disease, so it is less direct a consequence of exposure, but it could be more meaningful in real-life since it measures the reduction in life expectancy due to the disease.

(**c**) A sample from the population of life histories is a random sample from the 'system' generating the data on (S, B, M, D), so it yields estimates of any probability concerning this random vector without bias. After being made realistic such data become partially censored (most of those born more recently have no year of death yet, and their disease status may be provisional), so they are no random sample from the system generating the data and the estimates derived from them are typically biased. If we only use the realistic data on those born very early then almost all the data represent completed histories and hence provide practically unbiased estimates.

Solution to Prob. 5.10

(**a**) We have

$$\mathbf{P}(T = t|\mathbf{X} = \mathbf{x}, R = r, S = s) = \frac{\mathbf{P}(T = t, R = r, \sigma(W, \mathbf{x}, t, R) = s|\mathbf{X} = \mathbf{x})}{\mathbf{P}(R = r, S = s|\mathbf{X} = \mathbf{x})}.$$

This probability may also be written as

$$\mathbf{P}\{\tau(U, \mathbf{x}) = t\,|\xi(\epsilon) = \mathbf{x}, \rho\,[V, \mathbf{x}, \tau'(U', \mathbf{x})] = r, \sigma(W, \mathbf{x}, T, r) = s\} =$$

$$\frac{\mathbf{P}\{\tau(U,\mathbf{x}) = t, \xi(\epsilon) = \mathbf{x}, \rho\,[V,\mathbf{x},\tau'(U',\mathbf{x})] = r, \sigma(W,\mathbf{x},t,r) = s\}}{\mathbf{P}\{\xi(\epsilon) = \mathbf{x}, \rho\,[V,\mathbf{x},\tau'(U',\mathbf{x})] = r, \sigma\,[W,\mathbf{x},\tau(U,\mathbf{x}),r] = s\}} =$$

$$\frac{\mathbf{P}\{\tau(U,\mathbf{x}) = t\}\,\mathbf{P}\{\xi(\epsilon) = \mathbf{x}\}\,\mathbf{P}\{\rho\,[V,\mathbf{x},\tau'(U',\mathbf{x})] = r\}\,\mathbf{P}\{\sigma(W,\mathbf{x},t,r) = s\}}{\mathbf{P}\{\xi(\epsilon) = \mathbf{x}\}\,\mathbf{P}\{\rho\,[V,\mathbf{x},\tau'(U',\mathbf{x})] = r\}\,\mathbf{P}\{\sigma\,[W,\mathbf{x},\tau(U,\mathbf{x}),r] = s\}} =$$

$$\frac{\mathbf{P}\{\tau(U,\mathbf{x}) = t\}\,\mathbf{P}\{\sigma(W,\mathbf{x},t,r) = s\}}{\mathbf{P}\{\sigma\,[W,\mathbf{x},\tau(U,\mathbf{x}),r] = s\}} = \frac{\mathbf{P}(T = t|\mathbf{X} = \mathbf{x})\mathbf{P}\{\sigma(W,\mathbf{x},t,r) = s\}}{\mathbf{P}\{\sigma\,[W,\mathbf{x},\tau(U,\mathbf{x}),r] = s\}},$$

which in general depends on s and therefore cannot be reduced to a function of $\mathbf{x}$, r and t, such as $\mathbf{P}(T = t|\mathbf{X} = \mathbf{x}, R = r)$ or $\mathbf{P}(T = t|\mathbf{X} = \mathbf{x})$.

If inclusion in the source is independent of T—i.e. if σ is constant in its third argument—then the expression above does simplify to $\mathbf{P}(\tau(U,\mathbf{x}) = t)$. But this last probability is

$$\begin{aligned}\mathbf{P}(T = t|\mathbf{X} = \mathbf{x}) &= \mathbf{P}\{\tau(U,\mathbf{x}) = t\,|\mathbf{X} = \mathbf{x}, \rho\,[V,\mathbf{x},\tau'(U',\mathbf{x})] = r\}) \\ &= \mathbf{P}(T = t|\mathbf{X} = \mathbf{x}, R = r).\end{aligned}$$

Now suppose that T too influences R, that is $R = \rho(V,\mathbf{X},T',T)$ in (4.24). In this case the expression of $\mathbf{P}(T = t|\mathbf{X} = \mathbf{x}, R = r, S = s)$ becomes

$$\frac{\mathbf{P}\{\tau(U,\mathbf{x}) = t\}\,\mathbf{P}\{\xi(\epsilon) = \mathbf{x}\}\,\mathbf{P}\{\rho\,[V,\mathbf{x},\tau'(U',\mathbf{x}),t] = r\}\,\mathbf{P}\{\sigma(W,\mathbf{x},t,r) = s\}}{\mathbf{P}\{\xi(\epsilon) = \mathbf{x}\}\,\mathbf{P}\{\rho\,[V,\mathbf{x},\tau'(U',\mathbf{x}),\tau(U,\mathbf{x})] = r, \sigma\,[W,\mathbf{x},\tau(U,\mathbf{x}),r] = s\}} =$$

$$\mathbf{P}(T = t|\mathbf{X} = \mathbf{x})\frac{\mathbf{P}\{\rho\,[V,\mathbf{x},\tau'(U',\mathbf{x}),t] = r\}\,\mathbf{P}\{\sigma(W,\mathbf{x},t,r) = s\}}{\mathbf{P}\{\rho\,[V,\mathbf{x},\tau'(U',\mathbf{x}),\tau(U,\mathbf{x})] = r, \sigma\,[W,\mathbf{x},\tau(U,\mathbf{x}),r] = s\}},$$

which is not just a function of $\mathbf{x}$, r and t but also of s. So the difficulty subsists, except that the dependence on r reflects the effect of T on R, but in a biased way.

Solution to Prob. 5.11

It is not implausible that in a real-life situation the factor U_1 determining the magnitude of exposure should be correlated with the factor V affecting the response (in conjunction with exposure to the agent), perhaps reflecting the fact that individuals that are careless with regard to exposure are also careless outside work and in other respects.

In order to conform to the notation used in Sect. 4.7, write δ in place of U_1, ε in place of V, I in place of B, T in place of M and R in place of D, so that

$$T = \tau_S(\delta, I) = \delta \cdot I, \qquad R = \rho_S(\varepsilon, T) = \begin{cases} 1 & \text{if } \varepsilon \le \frac{e^{\beta_{0,S}+\beta_{1,S}T}}{1+e^{\beta_{0,S}+\beta_{1,S}T}}, \\ 0 & \text{otherwise,} \end{cases}$$

with $I = \iota(U_0, S)$. The assumption is that I and (δ, ε) are independent conditionally on S but δ and ε are dependent, so that I, affecting R through T alone, serves as an instrument. Clearly, $T > 0 \Leftrightarrow I > 0$, and δ is standard uniform conditionally on $I = 1$ and $S = s$.

The numerator of the measure of treatment effect θ met with in Sect. 4.7 is

$$\begin{aligned}\tilde{\theta}(s) &:= \mathbf{E}(R|I=1,S=s) - \mathbf{E}(R|I=0,S=s)\\ &\equiv \mathbf{E}[\rho_s(\varepsilon,T)|I=1,S=s] - \mathbf{E}[\rho_s(\varepsilon,T)|I=0,S=s]\\ &= \mathbf{E}[\rho_s(\varepsilon,\delta)|I=1,S=s] - \mathbf{E}[\rho_s(\varepsilon,0)|I=0,S=s]\\ &= \mathbf{E}[\rho_s(\varepsilon,\delta)|S=s] - \mathbf{E}[\rho_s(\varepsilon,0)|S=s]\\ &= \mathbf{E}[\rho_s(\varepsilon,\delta) - \rho_s(\varepsilon,0)|S=s].\end{aligned}$$

This parameter is an overall measure of the effect of exposure on the response of an arbitrary individual relative to the response of the same individual when not exposed; and yet it quantifies the effect of I rather than that of M on the response.

To explain this, let us rewrite our model in the following version of the basic model in which the response is defined as a function ρ of I and V_0, where V_0 is a standard uniform variable which yields ε_0 and δ_0 with the same joint distribution as ε and δ and which is independent of U_0 conditionally on S:

$$I = \iota(U_0, S), \qquad R = \rho(V_0, I, S) \equiv \rho_S(\varepsilon_0, \delta_0 \cdot I) = \begin{cases} 1 & \text{if } \varepsilon_0 \le \frac{e^{\beta_{0,S}+\beta_{1,S}\delta_0 I}}{1+e^{\beta_{0,S}+\beta_{1,S}\delta_0 I}},\\ 0 & \text{otherwise.}\end{cases}$$

Regarding I as the treatment, its effect on R is characterized by $\mathcal{L}(R|I=i, S=s)$, $i = 0, 1$, and a measure of it is $\tilde{\theta}(s)$:

$$\mathbf{E}[R|I=1,S=s] - \mathbf{E}[R|I=0,S=s] = \mathbf{E}[\rho_s(\varepsilon_0,\delta_0) - \rho_s(\varepsilon_0,0)|S=s].$$

Thus, in Prob. 5.9 the independence of U_1 and V is not strictly necessary for studying the treatment effect: if U_0 is independent of (U_1, V) conditionally on S then $\tilde{\theta}(s)$ already provides us with an overall measure of the effect of exposure. It is only in order to obtain a *detailed* measure of the effect of exposure in the guise of $\mathcal{L}(R|M=m, S=s)$ for varying m that the stronger assumption is necessary.

Solution to Prob. 5.12

(a) For any probability function l for which $\mathbf{P}(\lambda_i(\,\cdot\,;\mathbf{X}_i) = l) > 0$, we have

$$\begin{aligned}\mathbf{P}(T_i = t|\lambda_i(\,\cdot\,;\mathbf{X}_i) = l) &= \sum\nolimits_{\mathbf{x}} \mathbf{P}(T_i = t, \mathbf{X}_i = \mathbf{x}|\lambda_i(\,\cdot\,;\mathbf{X}_i) = l)\\ &= \sum\nolimits_{\mathbf{x}} \frac{\mathbf{P}(T_i = t, \mathbf{X}_i = \mathbf{x}, \lambda_i(\,\cdot\,;\mathbf{x}) = l)}{\mathbf{P}(\lambda_i(\,\cdot\,;\mathbf{X}_i) = l)}\\ &= \sum_{\mathbf{x}:\lambda_i(\,\cdot\,;\mathbf{x})=l} \frac{\mathbf{P}(T_i = t, \mathbf{X}_i = \mathbf{x})}{\mathbf{P}(\lambda_i(\,\cdot\,;\mathbf{X}_i) = l)}\\ &= \sum_{\mathbf{x}:\lambda_i(\,\cdot\,;\mathbf{x})=l} \frac{\lambda_i(t;\mathbf{x})}{\mathbf{P}(\lambda_i(\,\cdot\,;\mathbf{X}_i) = l)}\mathbf{P}(\mathbf{X}_i = \mathbf{x})\\ &= l(t) \sum_{\mathbf{x}:\lambda_i(\,\cdot\,;\mathbf{x})=l} \frac{\mathbf{P}(\mathbf{X}_i = \mathbf{x})}{\mathbf{P}(\lambda_i(\,\cdot\,;\mathbf{X}_i) = l)},\end{aligned}$$

which is (5.38).

For the given $t \in \mathcal{T}$, let $\sum_l$ indicate summation over the probability functions l for which $\mathbf{P}(\lambda_i(\cdot;\mathbf{X}_i) = l) > 0$ and $l(t) > 0$; then

$$\begin{aligned}
\mathbf{P}(\rho_i(V_i, \mathbf{X}_i, t) \le r) &= \sum_{\mathbf{x}} \mathbf{P}(\rho_i(V_i, \mathbf{X}_i, t) \le r | \mathbf{X}_i = \mathbf{x}) \mathbf{P}(\mathbf{X}_i = \mathbf{x}) \\
&= \sum_{\mathbf{x}} \mathbf{P}(\rho_i(V_i, \mathbf{X}_i, t) \le r | \mathbf{X}_i = \mathbf{x}, T_i = t) \mathbf{P}(\mathbf{X}_i = \mathbf{x}) \\
&= \sum_{\mathbf{x}} \mathbf{P}(R_i \le r | \mathbf{X}_i = \mathbf{x}, T_i = t) \mathbf{P}(\mathbf{X}_i = \mathbf{x}) \\
&= \sum_{l} \sum_{\mathbf{x}:\lambda_i(\cdot;\mathbf{x})=l} \mathbf{P}(R_i \le r | \mathbf{X}_i = \mathbf{x}, T_i = t) \mathbf{P}(\mathbf{X}_i = \mathbf{x}) \\
&= \sum_{l} \sum_{\mathbf{x}:\lambda_i(\cdot;\mathbf{x})=l} \frac{\mathbf{P}(R_i \le r, \mathbf{X}_i = \mathbf{x}, T_i = t)}{\lambda_i(t;\mathbf{x})} \\
&= \sum_{l} \frac{1}{l(t)} \sum_{\mathbf{x}:\lambda_i(\cdot;\mathbf{x})=l} \mathbf{P}(R_i \le r, \mathbf{X}_i = \mathbf{x}, T_i = t) \\
&= \sum_{l} \frac{1}{l(t)} \mathbf{P}(R_i \le r, \lambda_i(\cdot;\mathbf{X}_i) = l, T_i = t) \\
&= \sum_{l} \frac{\mathbf{P}(R_i \le r, \lambda_i(\cdot;\mathbf{X}_i) = l, T_i = t)}{\mathbf{P}(T_i = t | \lambda_i(\cdot;\mathbf{X}_i) = l)} \\
&= \sum_{l} \mathbf{P}(R_i \le r | \lambda_i(\cdot;\mathbf{X}_i) = l, T_i = t) \mathbf{P}(\lambda_i(\cdot;\mathbf{X}_i) = l).
\end{aligned}$$

Using $\sum_{\mathbf{x}}^{(l)}$ to indicate summation over $\mathbf{x}$ such that $\lambda_i(\cdot;\mathbf{x}) = l$, and noting that $\lambda_i(t;\mathbf{x}) = \mathbf{P}(T_i = t|\mathbf{X}_i = \mathbf{x}) = l(t)$ for $\mathbf{x}$ under the summation sign, we have

$$\begin{aligned}
\mathbf{P}[R_i \le r, T_i = t \,|\lambda_i(\cdot;\mathbf{X}_i) = l] &= \frac{\mathbf{P}[\rho_i(V_i, \mathbf{X}_i, t) \le r, \tau_i(U_i, \mathbf{X}_i) = t, \lambda_i(\cdot;\mathbf{X}_i) = l]}{\mathbf{P}(\lambda_i(\cdot;\mathbf{X}_i) = l)} \\
&= \frac{\sum_{\mathbf{x}}^{(l)} \mathbf{P}[\rho_i(V_i, \mathbf{x}, t) \le r, \tau_i(U_i, \mathbf{x}) = t, \mathbf{X}_i = \mathbf{x}]}{\mathbf{P}(\lambda_i(\cdot;\mathbf{X}_i) = l)} \\
&= l(t) \frac{\sum_{\mathbf{x}}^{(l)} \mathbf{P}[\rho_i(V_i, \mathbf{x}, t) \le r | \mathbf{X}_i = \mathbf{x}] \mathbf{P}(\mathbf{X}_i = \mathbf{x})}{\mathbf{P}(\lambda_i(\cdot;\mathbf{X}_i) = l)}.
\end{aligned}$$

(b) Abbreviating summation over $\{\mathbf{x}_i : \lambda(\cdot, \mathbf{x}_i) = l, i \in I\}$ to $\sum_l$, using the first part of **A3** and the identity

$$\begin{aligned}
\mathbf{P}(T_i = t_i, i \in I | \mathbf{X}_i = \mathbf{x}_i, i \in I) &= \prod_{i \in I} \mathbf{P}(T_i = t_i | \mathbf{X}_i = \mathbf{x}_i) \\
&= \prod_{i \in I} \lambda(t_i; \mathbf{x}_i)
\end{aligned}$$

(which follows from the last part of **A3**), and finally noting that under the summation sign we have $\lambda(t_i;\mathbf{x}_i) = l(t_i)$, we see that

$$\mathbf{P}\left(T_i = t_i, R_i \le r_i, i \in I \middle| \lambda(\cdot, \mathbf{X}_i) = l, i \in I\right) =$$

$$\frac{\sum_l \mathbf{P}(\tau_i(U_i, \mathbf{x}_i) = t_i, \varrho_i(V_i, \mathbf{x}_i, t_i) \le r_i, \mathbf{X}_i = \mathbf{x}_i, i \in I)}{\mathbf{P}(\lambda(\cdot, \mathbf{X}_i) = l, i \in I)} =$$

$$\frac{\sum_l \mathbf{P}(\tau_i(U_i, \mathbf{x}_i) = t_i, \varrho_i(V_i, \mathbf{x}_i, t_i) \le r_i, i \in I | \mathbf{X}_i = \mathbf{x}_i, i \in I)\mathbf{P}(\mathbf{X}_i = \mathbf{x}_i, i \in I)}{\mathbf{P}(\lambda(\cdot, \mathbf{X}_i) = l, i \in I)} =$$

$$\frac{\sum_l \mathbf{P}(\varrho_i(V_i, \mathbf{x}_i, t_i) \le r_i, i \in I | \mathbf{X}_i = \mathbf{x}_i, i \in I) \prod_{j \in I} \lambda(t_j; \mathbf{x}_j)\mathbf{P}(\mathbf{X}_i = \mathbf{x}_i, i \in I)}{\mathbf{P}(\lambda(\cdot, \mathbf{X}_i) = l, i \in I)} =$$

$$\frac{\sum_l \mathbf{P}(\varrho_i(V_i, \mathbf{x}_i, t_i) \le r_i, i \in I | \mathbf{X}_i = \mathbf{x}_i, i \in I)\mathbf{P}(\mathbf{X}_i = \mathbf{x}_i, i \in I)}{\mathbf{P}(\lambda(\cdot, \mathbf{X}_i) = l, i \in I)} \prod_{j \in I} l(t_j) =$$

$$\frac{\sum_l \mathbf{P}(\varrho_i(V_i, \mathbf{X}_i, t_i) \le r_i, \mathbf{X}_i = \mathbf{x}_i, i \in I)}{\mathbf{P}(\lambda(\cdot, \mathbf{X}_i) = l, i \in I)} \prod_{j \in I} l(t_j) =$$

$$\frac{\mathbf{P}(\varrho_i(V_i, \mathbf{X}_i, t_i) \le r_i, \lambda(\cdot, \mathbf{X}_i) = l, i \in I)}{\mathbf{P}(\lambda(\cdot, \mathbf{X}_i) = l, i \in I)} \prod_{j \in I} l(t_j) =$$

$$\mathbf{P}\left(\varrho_i(V_i, \mathbf{X}_i, t_i) \le r_i, i \in I \middle| \lambda(\cdot, \mathbf{X}_i) = l, i \in I\right) \prod_{j \in I} l(t_j).$$

Thus (let $r_i \to \infty$ in the first and last terms and use the identity preceding (5.41))

$$\mathbf{P}\left(T_i = t_i, R_i \le r_i, i \in I \middle| \lambda(\cdot, \mathbf{X}_i) = l, i \in I\right) =$$

$$\mathbf{P}\left(T_i = t_i, i \in I \middle| \lambda(\cdot, \mathbf{X}_i) = l, i \in I\right) \mathbf{P}\left(\varrho_i(V_i, \mathbf{X}_i, t_i) \le r_i, i \in I \middle| \lambda(\cdot, \mathbf{X}_i) = l, i \in I\right).$$

(c) Let I_1, I_2 be disjoint subsets of indices and put $I = I_1 \cup I_2$; abbreviating summations in an obvious way, we have, on using the second and third parts of **A3**,

$$\mathbf{P}\big(U_i \le u_i, V_i \le v_i, i \in I, \lambda(\cdot, \mathbf{X}_j) = l_1, \lambda(\cdot, \mathbf{X}_k) = l_2, j \in I_1, k \in I_2\big) =$$

$$\sum_{\mathbf{x}_j : \lambda(\cdot, \mathbf{x}_j) = l_j, j = 1,2} \mathbf{P}\big(U_i \le u_i, V_i \le v_i, \mathbf{X}_j = \mathbf{x}_1, \mathbf{X}_k = \mathbf{x}_2, i \in I, j \in I_1, k \in I_2\big) =$$

$$\sum_{l_1, l_2} \mathbf{P}\big(U_i \le u_i, V_i \le v_i, i \in I \big| \mathbf{X}_i = \mathbf{x}_j, i \in I_j, j = 1,2\big)\mathbf{P}(\mathbf{X}_i = \mathbf{x}_j, i \in I_j, j = 1,2) =$$

$$\sum_{l_1, l_2} \prod_{j=1}^2 \mathbf{P}\big(U_i \le u_i, V_i \le v_i, i \in I_j \big| \mathbf{X}_i = \mathbf{x}_j, i \in I_j\big)\mathbf{P}(\mathbf{X}_i = \mathbf{x}_i, i \in I_j) =$$

$$\sum_{l_1, l_2} \prod_{j=1}^2 \mathbf{P}\left(U_i \le u_i, V_i \le v_i, \mathbf{X}_i = \mathbf{x}_j, i \in I_j\right) =$$

$$\prod_{j=1}^2 \left\{\sum_{\mathbf{x}_j : \lambda(\cdot, \mathbf{x}_j) = l_j} \mathbf{P}\left(U_i \le u_i, V_i \le v_i, \mathbf{X}_i = \mathbf{x}_j, i \in I_j\right)\right\} =$$

$$\prod_{j=1}^2 \mathbf{P}\left(U_i \le u_i, V_i \le v_i, \lambda(\cdot, \mathbf{X}_j) = l_j, i \in I_j\right).$$

(d) By the last part of **A3** it is enough to take $I = \{i\}$, which simplifies writing. If l is such that $\mathbf{P}[\lambda(\cdot, \mathbf{X}_i) = l] > 0$ and $\mathbf{x}$ such that $\lambda(\cdot; \mathbf{x}) = l$, then

$$\mathbf{P}[T_i = t | \mathbf{X}_i = \mathbf{x}, \lambda(\cdot, \mathbf{X}_i) = l] = \mathbf{P}[T_i = t | \mathbf{X}_i = \mathbf{x}] = \lambda(t; \mathbf{x}) = l(t),$$

and, by (5.38), $\mathbf{P}[T_i = t | \lambda(\cdot, \mathbf{X}_i) = l] = l(t)$. Using these identities in

$$\mathbf{P}\left(\mathbf{X}_i = \mathbf{x} \,\middle|\, \begin{matrix} T_i = t, \\ \lambda(\cdot, \mathbf{X}_i) = l \end{matrix}\right) = \mathbf{P}(\mathbf{X}_i = \mathbf{x} | \lambda(\cdot, \mathbf{X}_i) = l) \frac{\mathbf{P}(T_i = t | \mathbf{X}_i = \mathbf{x}, \lambda(\cdot, \mathbf{X}_i) = l)}{\mathbf{P}(T_i = t | \lambda(\cdot, \mathbf{X}_i) = l)}$$

gives

$$\mathbf{P}(\mathbf{X}_i = \mathbf{x} | T_i = t, \lambda(\cdot, \mathbf{X}_i) = l) = \mathbf{P}(\mathbf{X}_i = \mathbf{x} | \lambda(\cdot, \mathbf{X}_i) = l).$$

Solution to Prob. 5.13

(a) If V and V' are independent conditionally on $(\mathbf{X}, T)$ then

$$\mathcal{L}(R | \mathbf{X} = \mathbf{x}, T = t, R' = r) = \mathcal{L}(\rho(V, \mathbf{x}, t) | \mathbf{X} = \mathbf{x}, T = t, \rho'(V', \mathbf{x}) = r)$$

does not depend on r, so R and R' are independent conditionally on $(\mathbf{X}, T)$, contrary to our assumptions.

(b) If U and V' are independent conditionally on $\mathbf{X}$ then

$$\mathcal{L}(R' | \mathbf{X} = \mathbf{x}, T = t) = \mathcal{L}(\rho'(V', \mathbf{x}) | \mathbf{X} = \mathbf{x}, \tau(U, \mathbf{x}) = t)$$

does not depend on t.

(c) Suppose $\mathcal{L}(R' | \mathbf{X} = \mathbf{x}, T = t)$ varies with t. By **(b)**, U and V' are dependent conditionally on $\mathbf{X}$. If V and V' were independent conditionally on $\mathbf{X}$ then we would have, for $0 \le v, v' \le 1$,

$$vv' = \mathbf{P}(V \le v, V' \le v' | \mathbf{X} = \mathbf{x}) = \sum_t \mathbf{P}(V \le v, V' \le v' | \mathbf{X} = \mathbf{x}, T = t) \mathbf{P}(T = t | \mathbf{X} = \mathbf{x});$$

but this is usually not the case because, by **(a)**, V and V' are dependent given $(\mathbf{X}, T)$ and therefore (at least when the range of T is finite) differentiating

$$\mathbf{P}(V \le v, V' \le v' | \mathbf{X} = \mathbf{x}, T = t)$$

with respect to v and then with respect to v' does not, in general, lead to something constant in (v, v'). [If we denote the last probability by $\psi_{\mathbf{x},t}(v, v')$ and assume that the mixed partial derivatives $D_{1,2}\psi_{\mathbf{x},t}(v, v')$ and $D_{2,1}\psi_{\mathbf{x},t}(v, v')$ are constant, say equal to $k_{\mathbf{x},t}$, for $0 < v, v' < 1$, then we necessarily have $\psi_{\mathbf{x},t}(v, v') = vv'$, contradicting the conditional dependence of V and V'.] Thus, V and V' are generally dependent conditionally on $\mathbf{X}$, and U and V' are dependent conditionally on $\mathbf{X}$; therefore, U and V too should be generally dependent conditionally on $\mathbf{X}$.

According to the specifications, U and V' are generally dependent given $\mathbf{X}$ because they share ξ_0 as a potential argument, and, in general, R and R' are dependent conditionally on $(\mathbf{X}, T)$ because

$$\mathcal{L}(R|\mathbf{X}=\mathbf{x},T=t,R'=r)=\mathcal{L}\left[\rho(V,\mathbf{x},t)\,|\mathbf{X}=\mathbf{x},\tau(U,\mathbf{x})=t,\rho'(V',\mathbf{x})=r\right]=$$

$$\mathcal{L}\{\rho[\varphi_3(\xi_2,\xi_3,\mathbf{x}),\mathbf{x},t]\,|\mathbf{X}=\mathbf{x},\tau[\varphi_1(\xi_0,\xi_1,\mathbf{x}),\mathbf{x}]=t,\rho'[\varphi_2(\xi_0,\xi_2,\mathbf{x}),\mathbf{x}]=r\}$$

generally depends on r. Finally, in general

$$\mathbf{P}(V\le v,V'\le v'|\mathbf{X}=\mathbf{x})=\mathbf{P}\{\varphi_3(\xi_2,\xi_3,\mathbf{x})\le v,\varphi_2(\xi_0,\xi_2,\mathbf{x})\le v'\,|\mathbf{X}=\mathbf{x}\}\ne vv';$$

as before this implies that U and V are generally dependent conditionally on $\mathbf{X}$, which in this case is verified directly because, given $\mathbf{X}=\mathbf{x}$, U provides information about ξ_0, ξ_0 provides information about V' and ξ_2, and ξ_2 provides information about V.

Solution to Prob. 5.14

The hypothesis is that

$$-\varepsilon<\frac{\mathbf{P}(T_i=t_i,\mathbf{X}_i\in\mathcal{N}_\mathbf{x},i\in I)}{\mathbf{P}(T_i=t_i,\mathbf{X}_i\in A,i\in I)}\frac{\mathbf{P}(\mathbf{X}_i\in A,i\in I)}{\mathbf{P}(\mathbf{X}_i\in\mathcal{N}_\mathbf{x},i\in I)}-1<\varepsilon,$$

or

$$-\varepsilon\frac{\mathbf{P}(T_i=t_i,\mathbf{X}_i\in A,i\in I)}{\mathbf{P}(T_i=t_i,\mathbf{X}_i\in\mathcal{N}_\mathbf{x},i\in I)}<\frac{\mathbf{P}(\mathbf{X}_i\in A,i\in I)}{\mathbf{P}(\mathbf{X}_i\in\mathcal{N}_\mathbf{x},i\in I)}-\frac{\mathbf{P}(T_i=t_i,\mathbf{X}_i\in A,i\in I)}{\mathbf{P}(T_i=t_i,\mathbf{X}_i\in\mathcal{N}_\mathbf{x},i\in I)}$$
$$<\varepsilon\frac{\mathbf{P}(T_i=t_i,\mathbf{X}_i\in A,i\in I)}{\mathbf{P}(T_i=t_i,\mathbf{X}_i\in\mathcal{N}_\mathbf{x},i\in I)},$$

which, because $\mathbf{X}_i\in A\Rightarrow\mathbf{X}_i\in\mathcal{N}_\mathbf{x}$, implies

$$-\varepsilon<\frac{\mathbf{P}(\mathbf{X}_i\in A\cap\mathcal{N}_\mathbf{x},i\in I)}{\mathbf{P}(\mathbf{X}_i\in\mathcal{N}_\mathbf{x},i\in I)}-\frac{\mathbf{P}(T_i=t_i,\mathbf{X}_i\in A\cap\mathcal{N}_\mathbf{x},i\in I)}{\mathbf{P}(T_i=t_i,\mathbf{X}_i\in\mathcal{N}_\mathbf{x},i\in I)}<\varepsilon;$$

but this is the desired inequality.

Solution to Prob. 5.17

(a) The treatment effect may be quantified in terms of the probabilities

$$\mathbf{P}(R=1|T=1)=\mathbf{P}(R=1|T=1,I=1)=\mathbf{E}[\rho(V,1,1)],$$

$$\mathbf{P}(R=1|T=0,I=1)=\mathbf{E}[\rho(V,1,0)],\quad \mathbf{P}(R=1|T=0,I=0)=\mathbf{E}[\rho(V,0,0)]$$

and

$$\mathbf{P}(R=1|T=0)=\mathbf{P}(R=1|T=0,I=1)\mathbf{P}(I=1|T=0)+$$
$$\mathbf{P}(R=1|T=0,I=0)\mathbf{P}(I=0|T=0),$$

which according to the data in the table should be about $23/20,200$, $16/10,800$, $63/31,000$ and $(16/10,800)\times(10,800/41,800)+(63/31,000)\times(31,000/41,800)$, respectively. For example, it may be quantified by

$$\frac{\mathbf{P}(R=1|T=1)}{\mathbf{P}(R=1|T=0)}\approx 0.60\quad\text{or}\quad\frac{\mathbf{P}(R=1|T=1,I=1)}{\mathbf{P}(R=1|T=0,I=1)}\approx 0.77.$$

The last column of the table points to a statistical difference between the women who accept screening if invited for screening and those who refuse. So the assumption that conditionally on $\{T = t, I = 1\}$ the random variable representing death from causes other than breast cancer, call it $\rho'(V', 1, t)$ for some function ρ' and a standard uniform V', depends on t only through the last argument does not seem realistic; on the contrary, under the conditioning event V' ought to depend on t. But there is no *a priori* reason why such an assumption should apply to $\rho(V, 1, t)$; i.e., there is no reason why V should be taken as standard uniform conditionally on $\{T = 0, I = 1\}$. Accordingly, the second ratio does not serve to assess the effect of screening because the probabilities in it are basically incomparable, and the first ratio too does not quantify correctly the effect of screening because it involves those same probabilities.

Alternatively, using the language of Prob. 5.13 with I in place of $\mathbf{X}$, it seems reasonable to assume that death by causes other than breast cancer is not influenced by screening, so that it may be represented by $R' = \rho'(V', I)$, say. Moreover, R' should be negatively correlated with R conditionally on (I, T) because, whatever the subpopulation one considers, death caused by breast cancer excludes death by other causes; so by Prob. 5.13 (a) V and V' should be dependent conditionally on (I, T). In view of the last column in the table, (b) suggests that U and V' should be dependent conditionally on I. Finally, (c) suggests that U and V too should be dependent conditionally on I.

(b) Using the assumption and the data in the table we get

$$\frac{\mathbf{P}(\rho(V,1,T)=1)}{\mathbf{P}(\rho(V,0,T)=1)} = \frac{\mathbf{P}(R=1|I=1)}{\mathbf{P}(R=1|I=0)} \approx \frac{39}{63} \approx 0.62,$$

so inviting women for screening (irrespective of their accepting it or not) reduces their risk of dying from breast cancer within a five-year period by about 38%.

(c) We have

$$\frac{\mathbf{P}(\rho(V,I,1)=1|I=1,T=1)}{\mathbf{P}(\rho(V,I,0)=1|I=1,T=1)} = \frac{\mathbf{P}(\rho(V,I,1)=1,I=1,T=1)}{\mathbf{P}(\rho(V,I,0)=1,I=1,T=1)}$$

$$= \frac{\mathbf{P}(R=1,I=1,T=1)}{\mathbf{P}(\rho(V,I,0)=1,I=1,T=1)}.$$

While the probability in the numerator pertains to an observable event, the one in the denominator does not. However, assuming $\mathbf{P}(\rho(V,1,0)=1) = \mathbf{P}(\rho(V,0,0)=1)$ and using $\mathbf{P}(I=0) = \mathbf{P}(I=1)$, $I = 0 \Rightarrow T = 0$, and the independence of W and (U, V), we find that the latter is equal to

$$\mathbf{P}(\rho(V,I,0)=1,I=1) - \mathbf{P}(\rho(V,I,0)=1,I=1,T=0) =$$
$$\mathbf{P}(\rho(V,1,0)=1)\mathbf{P}(I=1) - \mathbf{P}(R=1,I=1,T=0) =$$
$$\mathbf{P}(\rho(V,0,0)=1)\mathbf{P}(I=0) - \mathbf{P}(R=1,I=1,T=0) =$$
$$\mathbf{P}(\rho(V,0,0)=1,I=0,T=0) - \mathbf{P}(R=1,I=1,T=0) =$$
$$\mathbf{P}(R=1,I=0,T=0) - \mathbf{P}(R=1,I=1,T=0),$$

and from the data we get

$$\frac{\mathbf{P}(\rho(V,I,1)=1|I=1,T=1)}{\mathbf{P}(\rho(V,I,0)=1|I=1,T=1)}=\frac{\mathbf{P}(R=1,I=1,T=1)}{\mathbf{P}(R=1,I=0,T=0)-\mathbf{P}(R=1,I=1,T=0)}$$

$$\approx\frac{23}{63-16}=\frac{23}{47}\approx 0.49,$$

which suggests that among women who are invited and willing to be screened the effect of screening is to reduce the death rate approximately by half.

If $\mathbf{P}(\rho(V,1,0)=1)<\mathbf{P}(\rho(V,0,0)=1)$ then the equality sign in the last identity is replaced by $>$ and the estimated reduction is somewhat optimistic.

(d) The identity is derived as in the proof leading from (4.27) to (4.28) in Sect. 4.7. The parameter θ represents the difference between the probability that a woman willing to be screened dies of breast cancer when screened and the probability that a similar woman dies of breast cancer when not screened. Under the additional assumption, $-\{\rho(V,1)-\rho(V,0)\}$ is binary, so $-\theta$ is the probability that a woman willing to be screened dies of breast cancer when not screened but survives when screened.

The estimate of θ is

$$\frac{\frac{39}{31{,}000}-\frac{63}{31{,}000}}{\frac{20{,}200}{31{,}000}}=\frac{39-63}{20,200}=-\frac{24}{20,200}\approx -0.0012$$

and the estimate of the denominator of θ' is

$$\frac{\frac{63}{31{,}000}-\frac{16}{31{,}000}}{\frac{20{,}200}{31{,}000}}=\frac{47}{20,200},$$

so that of θ' is

$$-\frac{24}{47}=\frac{23}{47}-1\approx -0.51$$

and hence corresponds to the estimate of θ'', which parameter is the ratio of question **(c)** with 1 subtracted from it. Thus, 'correcting for dilution' (cf. footnote on p. 147) in **(c)** is essentially equivalent to using the method of instrumental variables.

Solution to Prob. 5.19

(a) Because the individuals were assigned randomly to the treatments, it is not strictly necessary to analyse the data on men and women separately, but there may be differences between the effect of the treatment in the two sexes and such differences may be of interest. In this case, looking at the data on men and women separately reveals contradictory reactions in the deep breathing groups, which may have been caused by differences in the taking of measurements before and after practice or by a 'surprise effect' of deep breathing (if, as it appears, no training on this technique had been provided); so stratification on such variables as sex and age, for example, may help to detect anomalies.

(b) The first model is unrealistic because it assumes that the response function is the same before and after practice: The deep breathing group is probably no proper control group, but the model implies that practice of deep breathing has no effect on the response; and although in the kapalabhati group it allows for an effect of the treatment, the model ignores the possible effect of the measurement or time of measurement, of which the data provide evidence (see next paragraph).

The second model is realistic because it is essentially the most general model possible; it is applicable to the extent that the randomization is valid and the data can be thought of as random samples. The special case of (5.49) assumes that the effect of the practice is proportional to the response that would have been observed at the same moment without the practice, and it allows for an effect of the measurement or time of measurement as long as $\varphi_0 \neq \varphi_1$. Empirical data often support the assumption of 'proportionality', for which there are theoretical explanations as well (such as central limit theorems). Under this assumption, the data provide evidence that $\varphi_0 \neq \varphi_1$; for one would expect ψ to express consistent effects in men and women and in FEV and FVC, and yet this expectation does not reckon with the differences between the distributions of the control groups observed in Fig. 5.3.

(c) Under (5.49) with $\varphi_0 = \varphi_1$ one could quantify the effects of the treatments by

$$\mathbf{E}\big[\log(R_1/R_0)\big|X = x, T = t\big] = \mathbf{E}\left[\log\{\rho_1(V_1, x, t)/\rho_0(V_0, x)\}\right] = \mathbf{E}[\psi(V_1'', x, t)],$$

but as we have seen it seems more likely that $\varphi_0 \neq \varphi_1$. Under (5.48),

$$\mathbf{E}\big[\log(R_1/R_0)\big|X = x, T = 1\big] - \mathbf{E}\big[\log(R_1/R_0)\big|X = x, T = 0\big] =$$
$$\mathbf{E}\left[\log\{\rho_1(V_1, x, 1)/\rho_1(V_1, x, 0)\}\right] =$$
$$\mathbf{E}\big[\log R_1\big|X = x, T = 1\big] - \mathbf{E}\big[\log R_1\big|X = x, T = 0\big].$$

The first and last differences here can be estimated from the data; estimates of them are the subject of **(d)** and **(e)**. Under (5.49), even with $\varphi_0 \neq \varphi_1$, these differences are equal to

$$\mathbf{E}[\psi(V_1'', x, 1) - \psi(V_1'', x, 0)];$$

an estimate of this quantity will be definite to the extent that ψ applies to repetitions of the experiment. If carried out with independent, similar groups repetitions of the experiment would probably yield similar estimates, but with the same groups they could not be exact repetitions of the first experiment and ψ would probably reflect a sort of 'learning effect', so one would have to allow ψ to change before stabilizing.

(d), (e) If the distribution of $\log(R_1/R_0)$ is approximately symmetric and has mean μ then the median of R_1/R_0 is approximately $\exp\mu$, and that of $100 \times (R_1 - R_0)/R_0$ about $100 \times (\exp\mu - 1)$. The two sets of estimates pertain to the difference between expected values of **(c)**; they appear consistent with each other, though one could pursue the question of their identity a little further. One can estimate the difference between the effects of the treatments with the measurements taken after practice alone, but we have seen that the measurements taken before practice are useful in detecting anomalies.

References

1. Agresti, A. (2002). *Categorical Data Analysis*, 2nd ed. Wiley: Hoboken, New Jersey.
2. Angrist, J.D., Imbens, G.W. and Rubin, D.B. (1996). Identification of causal effects using instrumental variables. *Journal of the American Statistical Association*, Vol. 91, No. 434, pp. 444–455.
3. Arjas, E. and Parner, J. (2004). Causal reasoning from longitudinal data. *Scandinavian Journal of Statistics*, Vol. 31, No. 2, pp. 171–187.
4. Arjas, E. (2012). Causal inference from observational data: a Bayesian predictive approach. *Causality: Statistical Perspectives and Applications*, Ed. C. Berzuini, A.P. Dawid and L. Bernardinelli, pp. 71–84. Chichester, UK.
5. Ash, R.B. and Gardner, M.F. (1975). *Topics in Stochastic Processes*. Academic Press, New York.
6. Azad, M. (2014). Q&A with Barry Marshall. *Nature*, Vol. 514, S6–S7.
7. Basawa, I.V. and Prakasa Rao, B.L.S. (1980). *Statistical Inference for Stochastic Processes*. Academic Press, New York.
8. Beran, R. (1988). The impact of the bootstrap on statistical algorithms and theory. *Statistical Science*, Vol. 18, No. 2, pp. 175–184.
9. Berkson, J. (1946). Limitations of the application of fourfold table analysis to hospital data. *Biometrics Bulletin*, Vol. 2, No. 3, pp. 47–53.
10. Billingsley, P. (1961). *Statistical Inference for Markov Processes*. The University of Chicago Press.
11. Billingsley, P. (1995). *Probability and Measure*, 3rd ed. Wiley, New York.
12. Box, G.E.P., Hunter, S. and Hunter, W.G. (2005). *Statistics for Experimenters: Design, Innovation and Discovery*, 2nd ed. Wiley, New Jersey.
13. Breiman, L. (2001). Random forests. *Machine Learning*, Vol. 45, Issue 1, pp. 5–32.
14. Brémaud, P. (1998). *Markov Chains: Gibbs Fields, Monte Carlo Simulation, and Queues*. Springer-Verlag.
15. Brockwell, P.J. and Davis, R.A. (1991). *Time series: Theory and Methods*, 2nd ed. Springer-Verlag.
16. Coats, T.J. and Walter, D.P. (1999). Effect of station design on death in the London Underground: Observational study. *British Medical Journal*, Vol. 319, p. 957.
17. Coles, S.A. (2001). *An Introduction to Statistical Modelling of Extreme Values*. Springer-Verlag, London.
18. Cornfield, J.(1951). A method for estimating comparative rates from clinical data. Applications to cancer of the lung, breast, and cervix. *Journal of the National Cancer Institute*, Vol. 11, No. 6 (1951), pp. 1269–1275.
19. Cox, D.R. and Snell, E.J. (1980). *Applied Statistics: Principles and Examples*. Chapman & Hall, London.
20. Cox, D.R. and Wermuth, N. (2004). Causality: a statistical view. *International Statistical Review*, Vol. 72, No. 3, pp. 285–305.
21. Daniel, C. and Wood, F.S. (1971). *Fitting Equations to Data: Computer Analysis of Multifactor Data for Scientists and Engineers* (written with the assistance of J.W. Gorman). Wiley, New York.
22. Dawid, A. (2000). Causal inference without counterfactuals. *Journal of the American Statistical Association*, Vol. 95, No. 450, pp. 407–448.
23. Dawid, A. (2015). Statistical causality from a decision-theoretic perspective. *Annual Review of Statistics and Its Application*, Vol. 2, Issue 1, pp. 273–303.
24. Devroye, L.P., Györfi, L. and Lugosi, G. (1996). *A Probabilistic Theory of Pattern Recognition*. Springer: New York.

25. Diaconis, P. (1998). A place for philosophy? The rise of modeling in statistical science. *Quarterly of Applied Mathematics*, Vol. LVI, No. 4, pp. 797–805.
26. Dietz, K. and Heesterbeek, J.A.P. (2002). Daniel Bernoulli's epidemiological model revisited. *Mathematical Biosciences*, Vol. 180, pp. 1–21.
27. Doob, J.L. (1997). The Development of Rigor in Mathematical Probability (1900-1950). *The American Mathematical Monthly*, Vol. 103, No. 7, pp. 586–95.
28. Eriksson, K. and Haggstrom, O. (2014). Lord's paradox in a continuous setting and a regression artifact in numerical cognition research. *PLOS ONE*, Vol. 9, Issue 4.
29. Feller, W. (1968). *An Introduction to Probability Theory and its Applications*, Vol. I, 3rd ed. Wiley, New York.
30. Ferguson, T.S. (1996). *A Course in Large Sample Theory*. Chapman and Hall.
31. Ferreira, J.A. (2015). Some models and methods for the analysis of observational data. *Statistics Surveys*, Vol. 9, pp. 106–208.
32. Ferreira, J.A. (2017). Confidence intervals for the treatment effect on the treated. *Journal of Nonparametric Statistics*, Vol. 29, No. 3, pp. 477–490.
33. Freedman, D.A. (1997). From association to causation via regression. *Advances in Applied Mathematics*, Vol. 18, Issue 1, pp. 59–110.
34. Freedman, D.A., Petitti, D.B. and Robins, J.M. (2004). On the efficacy of screening for breast cancer. *International Journal of Epidemiology*, Vol. 33, Issue 1, pp. 43–55.
35. Freedman, D.A. (2008). Randomization does not justify logistic regression. *Statistical Science*, Vol. 23, No. 2, pp. 237–49.
36. Freedman, D.A. (2009). *Statistical Models: Theory and Practice*, revised edition. Cambridge University Press, New York.
37. Freedman, D.A. (2010). *Statistical Models and Causal Inference: A Dialogue with the Social Sciences*. Edited by D. Collier, J.S. Sekhon and P.B. Stark Cambridge University Press, New York.
38. Galambos, J. and Kotz, S. (1978). *Characterization of Probability Distributions*. Lecture Notes in Mathematics, Vol. 675, Springer-Verlag, Berlin.
39. Goldstein, H., Poole, C. and Safko, J. (2001). *Classical Mechanics*, 3rd ed. Addison-Wesley.
40. Guttorp, P. (1995). *Stochastic Modeling of Scientific Data*. Chapman & Hall, Boca Raton.
41. Györfi, L., Kholer, M., Kryzak, A. and Walk, H. (2002). *A Distribution-free Theory of Nonparametric Regression*. Springer.
42. Halaby *et al.* (2017). Impact of single room design on the spread of multi-drug resistant bacteria in an intensive care unit. *Antimicrobial Resistance and Infection Control*, Vol. 6:117.
43. Heyde, C.C. (1997). *Quasi-Likelihood and its Application: A General Approach to Optimal Parameter Estimation*. Springer-Verlag, New York.
44. Hodges, J.L. and Lehmann, E.L. (1962). Rank methods for combination of independent experiments in analysis of variance. *The Annals of Mathematical Statistics*, Vol. 33, No. 2, pp. 482–497.
45. Imbens, G.W. and Rubin, D.B. (1995). Discussion of "Causal diagrams for empirical research" by J. Pearl. *Biometrika*, Vol. 82, No. 4, pp. 694–695.
46. Imbens, G.W. and Rubin, D.B. (2015). *Causal Inference in Statistics, Social, and Biomedical Sciences: An Introduction*. Cambridge University Press, New York.
47. Isham, V. (2005). Stochastic models for epidemics. *Celebrating Statistics: Papers in honour of Sir David Cox on his 80th birthday*, Ed. A.C. Davison, Y. Dodge and N. Wermuth. Oxford University Press, New York.
48. Ito, K. and Nisio, M. (1968). On the convergence of sums of independent Banach space valued random variables. *Osaka J. Math.*, Vol. 5, No. 1, pp. 35–48.
49. Jørgensen, B. (1993). *The Theory of Linear Models*. Chapman & Hall, New York.
50. Kahane, J.-P. (1997). A century of interplay between Taylor series, Fourier series and Brownian Motion. *Bulletin of the London Mathematical Society*, Vol. 29, Issue 3, pp. 257–279.
51. Karlin, S. and Taylor, H.M. (1975). *A First Course in Stochastic Processes*, 2nd ed. Academic Press, New York.
52. Leadbetter, M.R., Lindgren, G. and Rootzen, H. (1983). *Extremes and Related Properties of Random Sequences and Processes*. Springer-Verlag, New York.

53. Lehmann, E.L. (1975). *Nonparametrics: Statistical Methods based on Ranks*. Holden-Day: San Francisco.
54. Lehmann, E.L. (1983). *Theory of Point Estimation*. Wiley, New York.
55. Lehmann, E.L. (1986). *Testing Statistical Hypotheses*, 2nd ed. Wiley, New York.
56. Lehmann, E.L. (1990). Model specification: the views of Fisher and Neyman, and later developments. *Statistical Science*, Vol. 5, No. 2, pp. 166–168.
57. Lindley, D.V. (2002). Seeing and doing: the concept of causation. *International Statistical Review*, Vol. 70, No. 2, pp. 191–197.
58. Lord, F.M. (1967). A paradox in the interpretation of group comparisons. *Psychological Bulletin*, Vol. 68, No. 5, pp. 304–305.
59. Mood, A.M., Graybill, F.A. and Boes, D.C. (1974). *Introduction to the Theory of Statistics*, 3rd ed. McGraw-Hill.
60. McCullagh, P. and Nelder, J. A. (1989). *Generalized Linear Models*, 2nd ed. Chapman & Hall, London.
61. Mosteller, F. and Tukey, J. W. (1977). *Data Analysis and Regression: A Second Course in Statistics*. Addison-Wesley.
62. Neill, U.S. (2019). A conversation with Mary-Claire King. *Journal of Clinical Investigation*, Vol. 129, No. 1, pp. 1–3.
63. Nishida, A. *et al.* (2018). A randomized controlled trial of comprehensive early intervention care in patients with first-episode psychosis in Japan: 1.5-year outcomes from the J-CAP study. *Journal of Psychiatric Research*, Vol. 102, pp. 136–141.
64. Nuzzo, R. (2014). Statistical errors. *Nature*, Vol. 506, pp. 150–152.
65. Pearl, J. (1995a). Causal diagrams for empirical research. *Biometrika*, Vol. 82, No. 4, pp. 669–710.
66. Pearl, J. (1995b). Rejoinder to discussions of "Causal diagrams for empirical research". *Biometrika*, Vol. 82, No. 4, pp. 702–710.
67. Pearl, J. (2009a). *Causality: Models, Reasoning and Inference*, 2nd ed. Cambridge University Press.
68. Pearl, J. (2009b). Causal inference in statistics: an overview. *Statistics Surveys*, Vol. 3, pp. 96–146.
69. Pearl, J. (2009c). Myth, confusion, and science in causal analysis. *Technical Report* R-348, Department of Computer Science, University of California, Los Angeles.
70. Pearl, J. (2011). Statistics and causality: separated to reunite—Commentary on Bryan Dowd's "Separated at Birth". *Health Serv. Res.*, Vol. 46, No. 2, pp. 421–429.
71. Pearl, J. (2015). Indirect confounding and causal calculus. (On three papers by Cox and Wermuth.) *Technical Report* R-457, Department of Computer Science, University of California, Los Angeles.
72. Pearl, J. (2016). Lord's Paradox Revisited—(Oh Lord! Kumbaya!). *Technical Report* R-436, Department of Computer Science, University of California, Los Angeles.
73. Pearl, J., Glymour, M. and Jewel, N.P. (2016). *Causal Inference in Statistics: a Primer*. Wiley.
74. Perković, E., Textor, J., Kalisch, M. and Maathuis, M.H. (2018). Complete graphical characterization and construction of adjustment sets in Markov equivalence classes of ancestral graphs. *Journal of Machine Learning Research*, Vol. 18, No. 220, pp. 1–62.
75. Polansky, A.M. (2000). Stabilizing bootstrap-t confidence intervals for small samples. *The Canadian Journal of Statistics*, Vol. 28, No. 3, pp. 501–506.
76. Raghavendra, P., Shetty, P., Shetty, S., Manjunath, N.K. and Avinash Saoji, A. (2016). Effect of high-frequency yoga breathing on pulmonary functions in patients with asthma: A randomized clinical trial. *Annals of Allergy, Asthma & Immunology*, Vol. 117, No. 5, pp. 550-551.
77. Rao, C.R. (1989). *Statistics and Truth: Putting Chance to Work*. International Co-operative Publishing House: Fairland, Maryland.
78. Rao, C.R. and Shanbhag, D.N. (1994). *Choquet-Deny Type Functional Equations with Applications to Stochastic Models*. Wiley: New York.
79. Rice, J.A. (1995). *Mathematical Statistics and Data Analysis*, 2nd ed. Duxbury Press.
80. Robins, J.M. (1986). A new approach to causal inference in mortality studies with a sustained exposure period—applications to control of the healthy workers survivor effect. *Mathematical Modeling*, Vol. 7, Issues 9–12, pp. 1393–1512.

81. Robins, J.M. (1995). Discussion of "Causal diagrams for empirical research" by J. Pearl. *Biometrika*, Vol. 82, No. 4, pp. 695–698.
82. Robins, J.M. and Wasserman, L. (1997). Estimation of effects of sequential treatments by reparameterizing directed acyclic graphs. *Proceedings of the Thirteenth Conference on Uncertainty in Artificial Intelligence* (UAI1997), pp. 409–420.
83. Rosenbaum, P.R. (2002). *Observational Studies*, 2nd ed. Springer.
84. Rosenbaum, P.R. and Rubin, D.B. (1983). The central role of the propensity score in observational studies for causal effects. *Biometrika*, Vol. 70, No. 1, pp. 41–55.
85. Ross, S.M. (1970). *Applied Probability Models with Optimization Applications*. Holden-Day, San Francisco.
86. Ross, S.M. (2010). *Introduction to Probability Models*, 10th ed. Academic Press.
87. Rothmann, K.J. (1986). *Modern Epidemiology*. Little, Brown & Co.: Boston/Toronto.
88. Rubin, D. (1990). Formal modes of statistical inference for causal effects. *Journal of Statistical Planning and Inference*, Vol. 25, No. 3, pp. 279–292.
89. Rubin, D. (2008). Author's reply. *Statistics in Medicine*, Vol. 28, No. 14, pp. 2741–2742.
90. Sagarin, B. J., West, S. G., Ratnikov, A., Homan, W. K., Ritchie, T. D. and Hansen, E. J. (2014). Treatment noncompliance in randomized experiments: statistical approaches and design issues. *Psychological Methods*, Vol. 19, No. 3, pp. 317–333.
91. Schilling, R.L. and Partzsch, L. (2012). *Brownian Motion: An Introduction to Stochastic Processes*, with a chapter on simulation by Björn Böttcher. De Gruyter: Berlin/Boston.
92. Scholz, F. W. and Stephens, M. A. (1987). K-sample Anderson-Darling tests. *Journal of the American Statistical Association*, Vol. 82, No. 399, pp. 918–924.
93. Senn, S. (2006). Change from baseline and analysis of covariance revisited. *Statistics in Medicine*, Vol. 25, No. 24, pp. 4334–4344.
94. Shafer, G. (1996). *The Art of Causal Conjecture*. MIT Press: Cambridge, Massachusetts.
95. Shafer, G. and Vovk, V. (2006). The sources of Kolmogorov's "Grundbegriffe". *Statistical Science*, Vol. 21, No. 1, pp. 70–98. Revised version at https://arxiv.org/pdf/1802.06071.pdf (February 2018)
96. Shehan, N., Didelez, V., Burton, P. and Tobin, M. (2008). Mendelian randomisation and causal inference in observational epidemiology. *PLOS Medicine*, Vol. 5, Issue 8, pp. 1205–1210.
97. Snoep, J.D., Morabia, A., Hernández-Diaz, S., Hernan, M.A. and Vandenbroucke, J.P. (2014). A structural approach to Berkson's fallacy and a guide to a history of opinions about it. *International Journal of Epidemiology*, Vol. 43, No. 2, pp. 515–521.
98. Shrier, I. (2008). Letter to the editor. *Statistics in Medicine*, Vol. 27, No. 14, pp. 2740–2741.
99. Silvey, S.D. (1975). *Statistical Inference*. Halsted Press.
100. Simpson, E.H. (1951). The interpretation of interaction in contingency tables. *Journal of the Royal Statistical Society*, Series B, Vol. 13, No. 2, pp. 238–241.
101. Snell, J.L. (1997). A conversation with Joe Doob. *Statistical Science*, Vol. 12, No. 4, pp. 301–311.
102. Sturtevant, A.H. (1965). *A History of Genetics*. Harper Row: New York.
103. Tanur, J.M., Mosteller, F., Kruskal, W.H., Lehmann, E.L., Link, R.F., Pieters, R.S. and Rising, G.R (1989). *Statistics: A Guide to the Unknown*, 3rd ed. Wadsworth & Brooks, California.
104. Textor, J., van der Zander B., Gilthorpe M., Liskiewicz M. and Ellison G. (2016). Robust causal inference using directed acyclic graphs: the R package 'dagitty'. *International Journal of Epidemiology*, Vol. 45, No. 6, pp. 1887–1894.
105. Trafimow, D. and Marks, M. (2015). Editorial. *Basic and Applied Social Psychology*, Vol. 37, Issue 1, pp. 1–2.
106. Ventegodt, S. *et al.* (2007). Clinical holistic medicine (mindful, short-term psychodynamic psychotherapy complemented with bodywork) in the treatment of experienced physical illness and chronic pain. *The Scientific World Journal*, Vol. 7, pp. 310–6.
107. Wasserstein, R.L. and Lazar, N.A. (2016). The ASA statement on p-values: context, process, and purpose. *The American Statistician*, Vol. 70, No. 2, pp. 129–133.
108. Wermuth, N. and Cox, D.R. (2008). Distortion of effects caused by indirect confounding. *Biometrika*, Vol. 95, No. 1, pp. 17–33.
109. Wiener, N. (1993). *Invention: The Care and Feeding of Ideas*. The MIT Press.

Subject Index

Printed in the USA
CPSIA information can be obtained
at www.ICGtesting.com
LVHW020059280124
770066LV00003B/248